T0310085

Cellular Signal Transduction in Toxicology and Pharmacology

Cellular Signal Transduction in Toxicology and Pharmacology

Data Collection, Analysis, and Interpretation

Edited by

Jonathan W. Boyd
West Virginia University
Morgantown
WV, USA

Richard R. Neubig
Michigan State University
East Lansing
MI, USA

This edition first published 2019

Registered Office
John Wiley & Sons, Inc., 111 River Street, Hoboken, NJ 07030, USA

Editorial Office
111 River Street, Hoboken, NJ 07030, USA

For details of our global editorial offices, customer services, and more information about Wiley products visit us at www.wiley.com.

Wiley also publishes its books in a variety of electronic formats and by print-on-demand. Some content that appears in standard print versions of this book may not be available in other formats.

Library of Congress Cataloging-in-Publication Data
Names: Boyd, Jonathan W., 1975– editor. | Neubig, Richard R., editor.
Title: Cellular signal transduction in toxicology and pharmacology : data collection, analysis, and interpretation / edited by Jonathan W. Boyd (West Virginia University, Morgantown, WV, US), Richard R. Neubig (Michigan State University, MI, US).
Description: First edition. | Hoboken, NJ : Wiley, 2019. | Includes bibliographical references and index. |
Identifiers: LCCN 2018059849 (print) | LCCN 2019000424 (ebook) | ISBN 9781119060253 (Adobe PDF) | ISBN 9781119060161 (ePub) | ISBN 9781119060260 (hardcover)
Subjects: LCSH: Cellular signal transduction. | Molecular toxicology. | Molecular pharmacology.
Classification: LCC QP517.C45 (ebook) | LCC QP517.C45 C4585 2019 (print) | DDC 571.7/4–dc23
LC record available at https://lccn.loc.gov/2018059849

Cover Design: Wiley
Cover Images: Courtesy of Candelaria de la Losa,
Background © Anton Todorov/Shutterstock

Set in 10/12pt Warnock by SPi Global, Pondicherry, India

Printed in the United States of America

V10014571_100919

Contents

List of Contributors

Jonathan W. Boyd
Department of Orthopaedics and Department of Physiology and Pharmacology, West Virginia University School of Medicine, Morgantown, WV, USA

Marc Birringer
Department of Oecotrophologie, Fulda University of Applied Sciences, Fulda, Hesse, Germany

Meghan Cromie
National Jewish Health, Denver, CO, USA

Patricia E. Ganey
Department of Pharmacology and Toxicology, Michigan State University, East Lansing, MI, USA

Weimin Gao
Department of Occupational and Environmental Health Sciences, West Virginia University School of Public Health, Morgantown, WV, USA

Alice Han
Chem Bio & Exposure Sci Team, Pacific Northwest National Laboratory, Richland, WA, USA

Carlo Laudanna
Department of Pathology and Diagnostics, University of Verona, Verona, Italy

Zhongwei Liu
Department of Occupational and Environmental Health Sciences, West Virginia University School of Public Health, Morgantown, WA, USA

Julie Vrana Miller
Cardno ChemRisk, Pittsburgh, PA, USA

Sean A. Misek
Department of Physiology, Michigan State University, East Lansing, MI, USA

John Morris
Resource for Biocomputing, Visualization and Informatics, University of California, San Francisco, CA, USA

Julia A. Mouch
Department of Orthopaedics, West Virginia University School of Medicine, Morgantown, WV, USA

Richard R. Neubig
Department of Pharmacology and Toxicology, Michigan State University, East Lansing, MI, USA

Robert H. Newman
Department of Biology, North Carolina A&T State University, Greensboro, NC, USA

Sakshi Pratap
Birla Institute of Technology & Science, Goa, India

Maren Prediger
Institute for Microproductions, Leibniz University, Hannover, Germany

Nicole Prince
Department of Orthopaedics, West Virginia University School of Medicine, Morgantown, WV, USA

Giovanni Scardoni
Center for Biomedical Computing, University of Verona, Verona, Italy

Song Tang
National Institute of Environmental Health, Chinese Center for Disease Control and Prevention, Beijing, China

Gabriele Tosadori
Center for Biomedical Computing, University of Verona, Verona, Italy

Qian Wang
Department of Respiratory Medicine, Jiangsu Province Hospital of Chinese Medicine, Affiliated Hospital of Nanjing University of Chinese Medicine, Nanjing, Jiangsu, China

About the Editors

Jonathan Boyd is the associate director of the Musculoskeletal Laboratory and an associate professor of orthopaedics at the West Virginia University School of Medicine (WVUSOM). He holds additional joint appointments in the Department of Physiology and Pharmacology (WVUSOM) and the Department of Occupational and Environmental Health Sciences (WVU School of Public Health). Dr. Boyd also holds appointments as a guest professor at Fulda University (Germany) and a guest researcher at the Centers for Disease Control and Prevention at the National Institute for Occupational Safety and Health (CDC/NIOSH). He received his BS Biochemistry from the University of Texas at Austin in 1998 and his PhD Environmental Toxicology from Texas Tech University in 2004. His expertise is in mechanistic toxicology, mammalian signal transduction, and musculoskeletal trauma. Dr. Boyd's research uses fundamental thermodynamic principles to investigate the pharmacodynamic responses of living systems. In general, he is interested in understanding the mammalian response to both chemical and physical stressors, and specifically, he is working toward an understanding of how humans integrate (from cellular mechanisms to physiological integration) chemical and biochemical signals in response to stimuli. His research involves multiple hierarchical levels of biological samples that include mammalian cell culture, animal models, and human subjects coupled with analytical techniques that range from single-point detectors to multidimensional imaging. The applications of his work range from toxicity screening to surgical diagnostics, and he has published over 40 papers, technical reports, and book chapters associated with his research. Finally, Jonathan is married to the love of his life, Naomi, and together they have the three best children on Earth, Jack, Lucy, and Gwen, who all fill his life with pure joy.

Richard (Rick) Neubig is professor and chair of Pharmacology & Toxicology at Michigan State University associated with the three medical colleges. Dr. Neubig received his BS in Chemistry from the University of Michigan in 1975, his MD from Harvard Medical School in the Harvard-MIT Program in Health Sciences and Technology in 1981, and his PhD in Pharmacology from Harvard

University in 1981. He trained in internal medicine at the University of Michigan Hospitals and then joined the faculty in the Departments of Pharmacology and Internal Medicine in 1984. He moved to Michigan State University in 2013 to take the position of chair of Pharmacology & Toxicology. His research has primarily focused on G-protein-coupled receptor (GPCR) signaling. His early work in the 1980s and 1990s primarily used ligand binding kinetics and biochemical studies to decipher receptor/G-protein interactions. In the mid-1990s his work took a turn to translational research and academic drug discovery. He helped establish high-throughput screening and drug discovery centers at both the University of Michigan and Michigan State University. His current work relates to therapeutic discovery in cancer, scleroderma, and other fibrotic diseases as well as mechanisms and therapeutics of genetic epilepsies. He was president of the American Society for Pharmacology and Experimental Therapeutics (ASPET) (2012–2015), and he received the Astellas and ASPET-Pharmacia awards for translational pharmacology from that organization. He served on the IUPHAR Receptor Nomenclature Committee from 2000 to 2015 and was GPCR co-chair. He was named Fellow of the American Association for the Advancement of Science (AAAS) in 2015 and is currently chair of Section S, Pharmaceutical Sciences, of the AAAS. He loves nature and the outdoors, a passion that he shares with his amazing wife, Laura Liebler. They are the lucky parents of two wonderful children, Graham and Maia, who have brought them equally wonderful spouses and a totally awesome granddaughter, Enna.

Preface

"All the world's a stage, and all the men and women are merely players; they have their exits and their entrances, and one man in his time plays many parts...." This famous phrase was penned by William Shakespeare in *As You Like It*, and he followed it with a description of what he called the seven ages, or phases, of life. While he was referring to the process of aging via a metaphor of actors playing roles in this grand production that is our existence, I often think of both the phases and roles that genes and proteins play as they enter and exit their stages of development. From an infancy represented by a group of linked nucleotides blissfully unaware of their surroundings through an awakening of transcription that is followed by the trials and tribulations of proper editing and posttranslational modifications, the next phase is a productive and functioning role in a new environment that is interacting with many other actors who are also playing their roles. However, unbeknownst to the protein, this phase of role-playing usually begins to signal the end, when all activity will decrease and virility of function will cease. Ultimately, the once-powerful protein loses its ability to interact with partners, and now without the capacity to sense and respond its surroundings, it is devolved via catabolism back into the elements of its infancy, again unaware of what the future might bring.

As pharmacologists and toxicologists today, we focus on the biological response of adding and removing actors, both heroes and villains, to the mixture of concurrent scenes that are happening on several different stages within the body (i.e. organ systems). Further, these actors are joining these ongoing scenes during different *acts* (i.e. phases), which greatly affects the role that existing genes and proteins may be playing now or in the future. Our goal is to understand how these new cast additions (or removals) might impact the overall production, and therefore we must dive into and dissect each potential role, or interaction, as it pertains to the overall structure and phases that are necessary for life. One means to achieve this is via monitoring alterations in cellular signal transduction cascades that are responsible for both functional interactions (roles) and the transitions between phases (or *acts*).

The pharmacodynamic response of cells to xenobiotics is primarily coordinated by signal transduction networks. Signal transduction proteins are embedded in early response networks that use positive and negative feedback to generate extremely diverse functions like amplification and perfect adaptation, reversible and irreversible switches, homeostasis, and oscillations. Irregular signal transduction activity has been associated with many major human diseases, including several forms of cancer, age-related diseases, and diabetes, to name a few.

From a pharmacological perspective, proteins involved in signal transduction have become one of the most intensively pursued drug targets to date for several reasons, including their direct association with a wide variety of cellular functions (both proliferation and apoptosis), and previous success for the treatment of disease (chronic myeloid leukemia). Depending on the pharmacological effect, agonists/activators or antagonists/inhibitors can be used to harness the therapeutic benefits. The steroid hormone receptors, such as the estrogen receptor, androgen receptor, and glucocorticoid receptor, are among the best examples, because both the steroid receptor agonists and antagonists have been successful clinical drugs.

From a toxicological perspective, modern evaluations have evolved to consider toxicity as a perturbation of biological pathways or networks. As such, toxicity testing approaches are shifting from common endpoint evaluations to pathway-centered approaches based on signal transduction networks, where the degree of perturbation of select networks is monitored. These new approaches are greatly increasing the information available to toxicologists, but signal transduction in toxicological research is in its infancy. Therefore, toxicologists need a resource that will guide their future study design and interpretation.

The goal of this book is to provide a comprehensive understanding of signal transduction research that is applicable to industry, academia, and government written for an audience with a background in biochemistry or cell biology. The primary aim of this book is to give the reader a solid background in signal transduction networks while supplying the tools necessary for further research in this expanding area. These tools include a discussion of experimental design, sample handling, analytical measurement techniques, data analysis, and interpretation (including novel modeling approaches). Finally, this book is organized hierarchically so that the first four chapters describe the architecture of signaling networks, from a physical basis of understanding signaling through mechanisms, physiological response, and human disease. The last half of the book is a "how-to" for incorporating signal transduction research into your portfolio with several examples, from designing signal transduction studies to measuring and monitoring signaling responses, interpreting and transforming data into usable information, an application in a toxicological study, and finally outlining critical future needs in signal transduction research.

1

Introduction to Cellular Signal Transduction

The Connection Between a Biological System and Its Surroundings

Jonathan W. Boyd[1], *Richard R. Neubig*[2], *Alice Han*[3], *and Maren Prediger*[4]

[1] *Department of Orthopaedics and Department of Physiology and Pharmacology, West Virginia University School of Medicine, Morgantown, WV, USA*
[2] *Department of Pharmacology and Toxicology, Michigan State University, East Lansing, MI, USA*
[3] *Chem Bio & Exposure Sci Team, Pacific Northwest National Laboratory, Richland, WA, USA*
[4] *Institute for Microproductions, Leibniz University, Hannover, Germany*

The rate of energy flow governs anabolic processes in chemical and biological systems. When a steady flow of external energy enters into an open system, there is a driving force to assemble mechanisms from available components in order to disperse more energy in the quest for a stationary steady state [1]. The driving force does not distinguish between mechanisms of energy transduction but favors those that disperse energy more efficiently. The mechanisms that effectively distribute energy become temporal repositories of that energy, which allows other mechanisms (i.e. energy dispersal systems) to tap into them, which further enhances energy dispersal and ultimately creates networks (Figure 1.1).

A commonplace example of this classic thermodynamic phenomenon is a hot cup of coffee reaching equilibrium with a cooler room temperature; convection waves – in addition to thermal radiation – allow the fastest distribution of energy to the local environment. In regard to ecology, times of plenty can yield massive population increases eager to consume (i.e. distribute) as much energy as possible; finally, in relevance to cellular signal transduction, many distribution networks are directly linked to energy availability via direct interaction with adenosine triphosphate (ATP). It is important to note, however, that the availability of energy on Earth is not at steady state (e.g. seasons of the year) and, in fact, is kept far from equilibrium [2]. Therefore, given an environmental change, where the rate of energy delivery changes dramatically, self-assembly processes must adapt to the new steady-state or risk infinite dissolution, via increasing entropy. Biological systems have the advantage here and are able to

Cellular Signal Transduction in Toxicology and Pharmacology: Data Collection, Analysis, and Interpretation, First Edition. Edited by Jonathan W. Boyd and Richard R. Neubig.

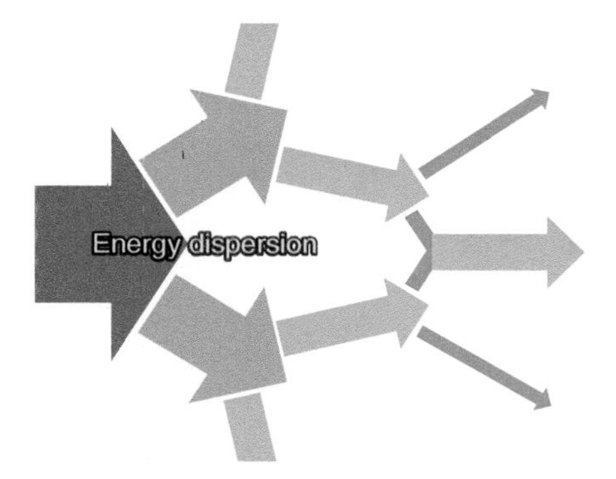

Figure 1.1 Schematic displaying energy dispersion.

re-create themselves individually through mutation or reposition themselves within a group or network for optimization in this new environment.

The means by which biological systems sense, interpret, and respond to new information from a dynamic external environment involve signal transduction. Signal transduction can be simply defined as the transformation of information from one form into an alternative form, often leading to a measurable action. While this is a very broad definition that fosters a conceptual understanding of signaling in general, it encompasses whole-organism response, including sensory perception. For example, the ability to transform a signal from a sudden change in air pressure (e.g. from a loud GROWL in the forest) into an action (e.g. "RUN!") is a process that involves multiple cell types with highly integrated signal transduction networks; in this example, cells deep within the inner ear polarize or depolarize in response to sound waves and detect the stimuli based on the magnitude and direction of hair movement. Cellular signal transduction has a much narrower scope and can be defined as a basic process that involves the conversion of a signal from outside the cell to a functional change within the cell. However, it is important to remember that the response of multicellular biological organisms necessarily involves the coordinated efforts of many cells and that cells can and do signal to each other (even though they may be spatially separated). Therefore, while this book is focused on cellular signal transduction in pharmacology and toxicology, the signaling responses discussed herein are often the foundation of both human diseases and treatment options.

Understanding the potential biological ramifications associated with altered cellular signal transduction processes is essential for research, development, testing, and evaluation of new chemical compounds for both pharmacology and toxicology. From a pharmaceutical viewpoint, the number of signal transduction proteins that are druggable targets is continuing to grow (see Santos et al. [3] for review), while toxicological sciences are beginning to appreciate the multitude of compounds that impact signaling cascades and how these impacts can/do lead to adverse health effects (see *Toxicity Testing in the 21st*

Century [4]). Given the highly integrated nature of most signal transduction networks, elucidating either the pharmacologic or toxicologic activity of new compounds can be difficult when there are no available standard operating procedures for testing their influence on signaling cascades. Challenges for scientists moving forward are as follows: (i) how to understand the relationship between seemingly disparate signaling networks (e.g. pro-survival and pro-death), (ii) determining the impacts that dose has on signaling (e.g. defining biological thresholds), and (iii) defining the temporal scale associated with response (e.g. implications of acute vs. chronic exposures), among many others. Beyond initiating discussions that we hope will facilitate research that fills exiting knowledge gaps, we hope that this book will inform the reader of currently available tools and techniques that enable them to begin to clarify the critical role that cellular signal transduction plays in the biological response to any industry-relevant compounds.

1.1 Starting Big, but Ending Small

While the definition of what constitutes life may be difficult and hotly debated [5–7], it can be defined as "the property or quality that distinguishes living organisms from dead organisms and inanimate matter, manifested in *functions such as metabolism, growth, reproduction, and response to stimuli or adaptation to the environment originating from within the organism*" [8]. From this framework, one can easily see the central role that cellular signal transduction plays in maintaining "life." Every one of the listed functions, which are all clearly associated with the existence of life, is firmly linked to the foundation of cellular signal transduction or the ability of cells (using a cell as the fundamental unit of life) to transform signals from one form into another toward functional outcomes such as growth and reproduction. Beyond simply sensing the environment, cellular signal transduction drives the processes of life by initiating, integrating, and determining biological responses to stimuli. Further, it can be argued that cellular signal transduction is the key feature that enables us to interact with both the environment and each other: it links cells together to form tissues, and tissues to organs, and finally into organisms; it enables sensory perception that provides the foundation for social relationships and even ecosystems. Without signal transduction, we would not be able to respond to our environment and would not be able to reproduce; life as we know it would not exist.

1.1.1 Key Features of Signal Transduction

All living organisms must respond to their environment, and the capabilities to do so are directly aligned with key features of cellular signal transduction. First, the capability to distinguish important information (i.e. signals) from all other information (i.e. background noise) enhances opportunities for survival;

in signal transduction, this is referred to as *specificity*. Next, this signal must be tuned and disproportionally enhanced to the point that it elicits the proper biological response; in signal transduction, this is referred to as *amplification*. This new information must be categorized and prioritized with all of the other signals that organism has previously received (or is concurrently receiving) before an action is taken; this is called signal *integration*. After a response has been mounted, there must be some way to attenuate the signal once the message has been received; in signal transduction, this is referred to as *desensitization* or *adaptation*. Examples abound, but the clearest illustration lies in this paragraph: as a reader, you were able to distinguish letters into specific words, amplify the words into short thoughts or sentences, and use previous experience in reading comprehension to integrate the sentences into useful information, and hopefully this section is succinct enough so that you do not become desensitized to the letters.

1.2 Responding to Our Environment: Sensory Perception Begins and Ends with Signal Transduction

As humans, our primary interactions with the environment are guided by five primary senses: taste, touch, smell, sight, and sound. Overall, the ability to respond to a dynamic environment enhances both short-term (e.g. safely crossing the road) and long-term survival of the species (e.g. successfully finding a mate). However, at its core, each "sense" involves our ability to discern new information from our surroundings, integrate it with existing information, and respond to changes in our local environment. This is accomplished via cellular signal transduction that is highly innervated with the central nervous system. The process begins with exogenous physical or chemical signals that come into contact with specialized cellular receptors and channels that are associated with each "sense," which can either directly or indirectly initiate the propagation of nerve impulse to the brain where the new information is integrated. Below are only a few select examples of some of the specialized receptors and channels that are directly related to how we interact with our world.

1.2.1 Taste (Gustation)

The gustatory perception in humans is categorized into five basic tastes: salty, sour, sweet, umami, and bitter. These disparate tastes are possible because of specialized gustatory receptor cells that detect and transmit signals in a concentration-dependent manner. Salty taste involves the detection of Na^+ gradient based on the ingestion of NaCl; the large concentration of Na^+ enters the cells through channels, and the result is a depolarization that is transmitted to the brain via an action potential. Sour taste is essentially a cellular response to acidic foods (those with a low pH), which increases the H^+ concentration

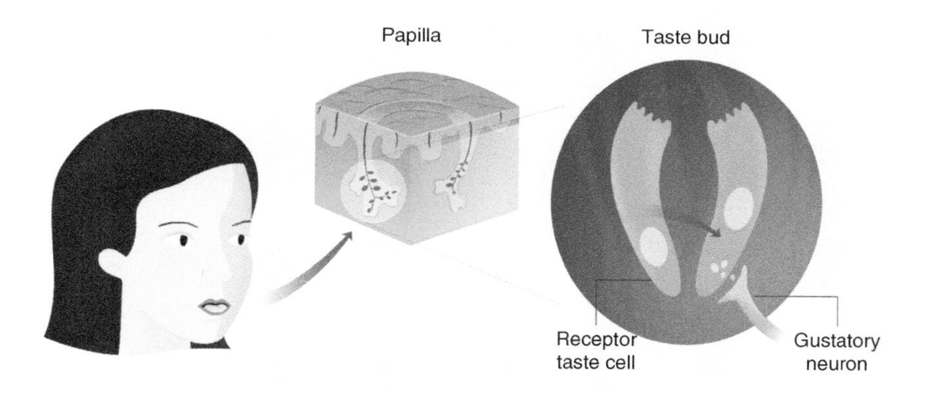

Figure 1.2 Innervation of gustatory neurons with taste buds.

that can enter the gustatory cells; just like with salty foods, the increased concentration of H^+ results in cellular depolarization and the transmission of an action potential. Sweet, umami, and bitter tastes are not the result of cation concentration-dependent depolarizations, but rather, result from functional group binding to G-protein-coupled receptors (GPCRs), which can cause depolarizations (sweet, umami, bitter) or hyperpolarizations (bitter). GPCRs will be discussed at length in future chapters, but these are highly selective receptors that play a crucial role in many disparate signaling modalities. The sweet taste is based on the gustatory cell's detection of glucose or glucose-related compounds, while the bitter taste is more complicated in that its receptors can be targeted by a number of different tastants; one such group of bitter-tasting chemical categories includes alkaloids (like those often found in coffee, beer, and tea), which can bind the GPCR and induce an action potential (Figure 1.2). For a more thorough review, please see Roper [9].

1.2.2 Smell (Olfaction)

The sense of smell is exquisitely complex, but it is similar to taste in that it is a response to chemical stimuli transmitted to the brain. As odorants are taken into the nose, they dissolve into mucus and bind to proteins that will carry them to dendrites from sensory neurons that descend from the olfactory bulb. The odorant–protein complex binds to GPCRs within the dendrites to produce a concentration-dependent membrane potential that is transmitted directly to the brain. Similar to taste, smell is primarily accomplished via specialized receptors. However, while regions of the tongue guide sensitivity based on the expression of individual localized receptors, olfactory perception appears to be based on the integration of many distinct receptors that then "imprint" different smells (Figure 1.3). For more information, please see Kaupp [10].

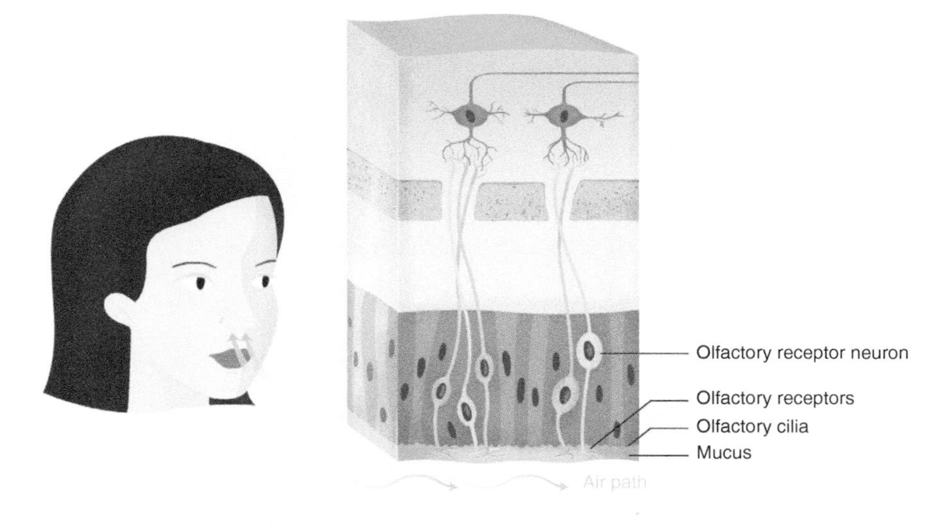

Figure 1.3 Innervation of nerves and receptors associated with odor molecules entering the nose.

1.2.3 Sight (Vision)

In humans and all vertebrate organisms, vision results from the ability to process light reflected from the surrounding environment. Incoming light enters through the pupil and is focused on the retina, which consists of specialized neurons: rods (for low-level illumination) and cones (for daylight illumination and color discrimination). To discriminate colors, there are three types of cone cells that are specialized to absorb different spectrums of light by expressing distinctive photoreceptor proteins. In response to light, a chromophore-coupled photoreceptor protein absorbs energy and undergoes a chemical rearrangement. The chromophore is 11-*cis*-retinal, and following absorption of light, energy is converted to all-*trans*-retinal; this initiates a G-protein-coupled signaling cascade, resulting in a decrease in cyclic guanosine monophosphate (cGMP) concentration that ultimately leads to the hyperpolarization of the cell via closing of ion channels. In the case of vision, cell hyperpolarization is the electrical signal that is then eventually sent to the brain via the optic nerve (Figure 1.4).

1.2.4 Sound (Audition)

The transduction of external sounds into usable information is both structurally complex (for review of audition, please see Moller [11] and Goutman et al. [12]) and operationally simple. Following exposure to alterations in air pressure,

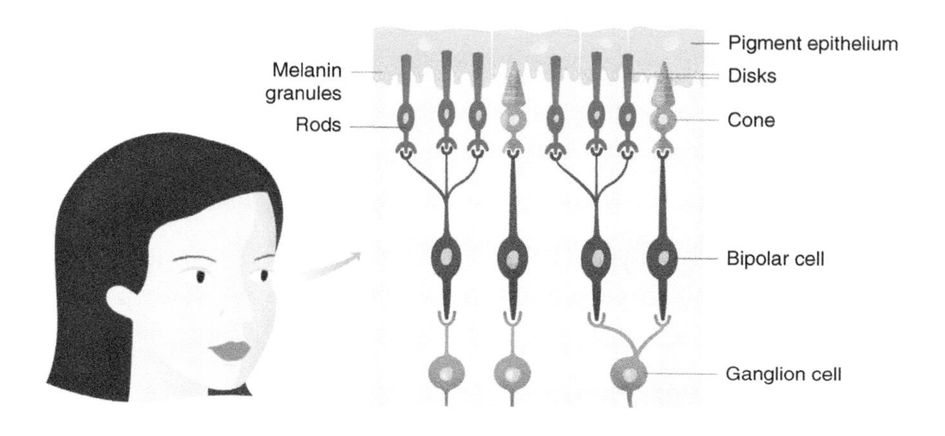

Figure 1.4 Activation of rods and cones in image formation.

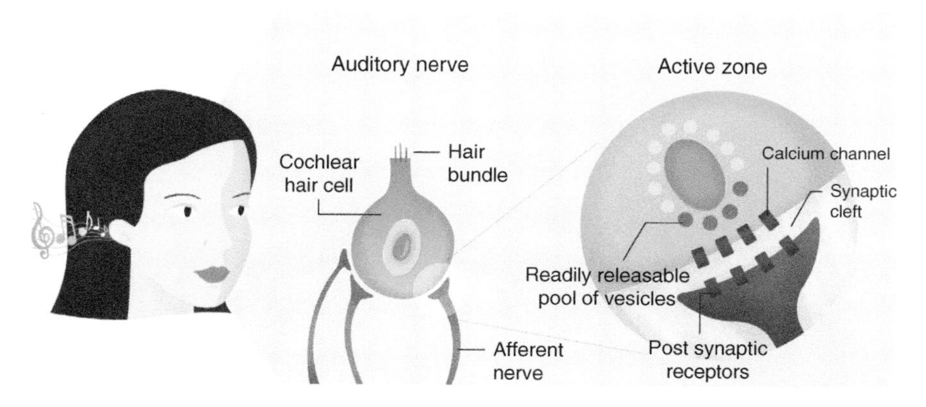

Figure 1.5 Receptors and nerves associated with the transduction of external sounds.

sound waves cause the tympanic membrane (eardrum) to vibrate at a resonant frequency that matches the external signal. This vibration is amplified across bony structures in the middle ear and transmitted as standing waves into the inner ear. In the inner ear, small hair follicles will bend as a result of the alteration in pressure, causing tension in protein tethers that open cellular ion channels, resulting in depolarization and an action potential. Groups of hairs are spaced so that different portions of the inner ear can interpret a wide range of frequencies; for humans, the typical frequency range is between 20 and 20 000 Hz (Figure 1.5).

1.2.5 Touch (Somatosensation)

The sense of "touch" actually encompasses several different categories of sensations that lead to signal transduction from the skin (for review of somatosensation, please see Lumpkin and Caterina [13]). These include response to temperature fluctuations, stretching, vibration, and movement. Signaling a change in temperature is mediated by a set of thermally sensitive transient receptor potential (TRP) cation channels, each having its own distinct temperature response region that will open a cation channel that will ultimately result in an action potential traveling along neurons. Stretch-sensitive ion channels are expressed in many types of mammalian cells, including those in the skin, muscles, and tendons. Vibration is sensed by mechanoreceptors called Merkel cells, which have elongated microvilli that can initiate a signal based on orientation (much like hair follicles of the auditory canal from above) (Figure 1.6).

1.3 Primary Transport Systems Involved in Signal Transduction

Beyond sensory perception systems that are highly innervated with direct linkages to the central nervous system, all biological interactions with the exogenous world necessitate a cellular understanding of signal transduction. This involves both extracellular and intracellular trafficking of information, which requires

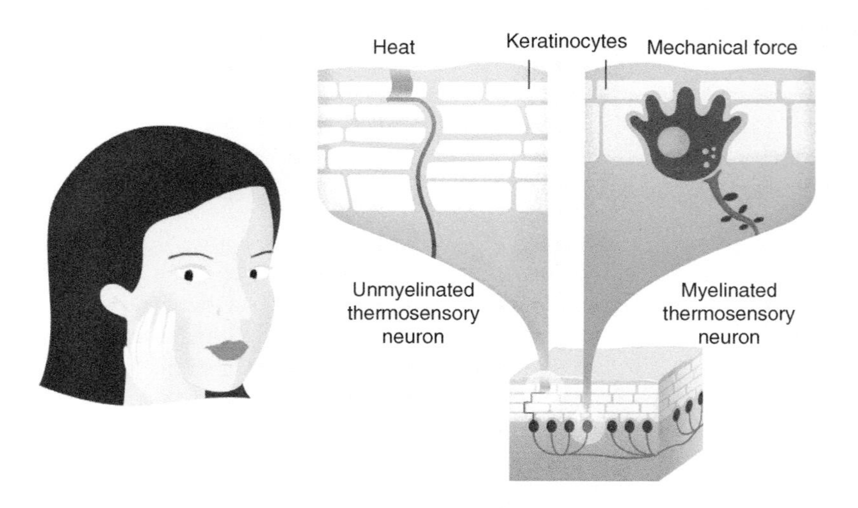

Figure 1.6 Sensory neurons associated with somatosensation.

dynamic interplay between the plasma membrane, organelles, and cellular constituents in order to optimize the response to a changing environment.

Lipophilic in nature, the cell membrane establishes a very effective barrier between the extracellular environment and the cytosol inside the cell. Only a few select molecules can freely diffuse through the semipermeable membrane. However, metabolic products from carbohydrate, protein, and lipid metabolism as well as hormones, ions, growth factors, and many more species need to transfer in and out of cells for survival. Ion channels, pumps, and transporters allow access into the cell where the molecule can initiate a series of events through interacting with other species, secondary messengers, or organelles – such as the endoplasmic reticulum (ER) or the mitochondria. Membrane receptors are found at the cell surface, where upon binding of a ligand, they activate secondary messengers in the cytoplasm, thereby propagating the signal into the cell. In return, the cell can signal to the surrounding environment through lipid vesicles and exosomes or direct interactions with the extracellular matrix, for example.

1.3.1 Ion Channels, Transporters, and Ion Pumps

Membrane transporters can be divided into three groups: ion channels, ion pumps, and coupled transporters. Ion channels passively and selectively facilitate the diffusion of ions through the membrane following their concentration gradient without energy expenditure. They consist of membrane-spanning pores, which are usually formed by four or five subunits, the most common motif in these subunits being α-helices and β-strands, and ions selectively interact with the pores to pass the bilipid barrier [14, 15]. This class of transporters is further split into voltage-gated and ligand-gated channels. Voltage-gated Na^+, K^+, or Ca^{2+} channels are smaller, with only one or two single subunits, and allow diffusion at a certain membrane potential. Ligand-gated Na^+, K^+, or Ca^{2+} channels open only upon binding of specific ligands, such as neurotransmitters. Additionally, phosphorylation of a channel can facilitate gating.

Ion pumps are so-called active transporters that use energy as driving force to move a species thermodynamically uphill against its concentration gradient and can either use the energy from ATP hydrolysis directly (primary pumps) or the energy from an already existing ion gradient (secondary pumps) that has been established by other primary pumps [16]. For example, in addition to ion flux through the respective ion channels, ATPase pumps build up gradients of Na^+, K^+, or Ca^{2+}, or H^+ across a membrane when phosphorylated by ATP, consequently reducing it to adenosine diphosphate (ADP) while transporting the ions. An example of a secondary pump is the sodium–calcium exchanger, which uses the sodium gradient to move calcium against its gradient out of the cell [17].

Species other than ions can be transported using the same principles of passive and active transport, which highlights another classification system:

uniport, symport, and antiport transport. Uniporters move one species along its gradient in a passive fashion. All ion channels are considered uniporters as well as, for example, the family of glucose transporters *GLUT*, which facilitate the influx of glucose from the extracellular fluids into the cytosol [18]. Symporters and antiporters can use both active and passive transport to move two species concurrently in coupled cotransport. When the protein translocates two species in the same direction, it is considered a symporter. The best example for symport transporters is the family of sodium-dependent glucose-coupled transporters (SGLTs), which bring glucose into the cell using the natural gradient flux of sodium into the cytosol [18]. SGLTs make use of passive sodium transport that generates enough energy to simultaneously transport glucose molecules. Antiporters, on the other hand, exchange the molecules from one side to the other. The aforementioned sodium–calcium exchange pump actively transports calcium against its gradient out of the cell while sodium comes into the cell following the electrochemical potential [17]. A more detailed description on this topic is beyond the scope of this chapter, but a remarkably in-depth classification of transporters, ion pumps, and ion channels has been published by Busch and Saier [14], and more on the mechanisms of gating ion channels can be found in Gadsby's review on ion channels and pumps [16].

1.3.2 Receptors

Receptors are proteins that react to a signal on one side of a membrane and elicit a second signal on the other side. They often span the membrane and – upon binding of the first signal – undergo a conformational change that triggers the second signal. Alternatively, they can induce an intracellular response after endocytosis. The location of receptors is not limited to the plasma membrane, as they are also found in the cytoplasm. While receptors are directly involved in the transport of information, most do not directly involve bulk transport of solutes. Due to the large number and diversity of receptors directly involved in signal transduction, much of the discussion of receptors will occur in Chapter 2, which focuses on mechanisms of key receptor families including GPCRs, receptor tyrosine kinases, and nuclear receptors, specifically their structural features and generic mechanisms of signal transduction.

1.3.3 Endocytosis

In general, the term endocytosis describes particle and macromolecular uptake into the cell from the cell's surrounding environment. This includes internalization of larger species, which is then called phagocytosis (if the larger particle is another cell or even a bacterium) and referred to as pinocytosis when fluids and molecules are engulfed in membranous vesicles. As a subform of pinocytosis, receptor-mediated endocytosis plays a crucial role in cell signaling, as it is a mechanism that internalizes ligands from specific cell surface receptors. For example, the epidermal growth factor receptor (EGFR) is internalized from clathrin-coated pits after a ligand binds, and the complex then undergoes degradation in lysosomes [19].

Clathrin-dependent endocytosis has been well characterized. Concentrated amounts of cell surface receptors are found in pitlike regions of the plasma membrane, where clathrin lines the cytosolic side of the membrane. The pits close and form vesicles inside the cytosol that can fuse with early endosomes. Fusion is facilitated by the recognition of complementary transmembrane proteins found on the vesicle membrane and the endosome membrane, such as vesicle-soluble NSF (*N*-ethylmaleimide-sensitive factor) attachment protein receptor (v-SNARE) and target-soluble NSF (*N*-ethylmaleimide-sensitive factor) attachment protein receptor (t-SNARE). Early endosomes are slightly acidic inside when compared with the cytosol, which results in ligand–receptor complexes that can dissociate. The dissociated parts can be recycled back to the membrane, degraded at the lysosome, or transported to other organelles. According to Cooper [20], a low-density lipoprotein (LDL) receptor, for example, is internalized and returns to the plasma membrane every 10 minutes.

Besides the clathrin-dependent recycling pathways, clathrin-independent routes gain interest and show an even bigger network and signaling character of endocytosis. Similar to clathrin-coated vesicles, some clathrin-independent pathways rely on caveolin-coated structures to traffic substances preferentially to cholesterol-rich membrane domains [21]; others need alternate pathways. Overall the picture emerges that endocytotic pathways vary for cell types and the species that are internalized. Importantly, with the ability to import many different molecules, endocytosis is not only responsible for the recycling of plasma membrane receptors and dissociation of their ligands but also influences processes such as cell adhesion, cell cycle progression, and cell fate determination [22, 23].

1.3.4 Exosomes

An exosome is a type of extracellular vesicle (EV) that has been recognized as an important intracellular organelle in cell-to-cell signaling. In contrast to the other two types of EVs, the exosome membrane does not come from the plasma membrane, and the vesicles are assembled inside the matrix to contain a variety of molecules such as cytokines, integrins, subunits of trimeric G proteins, elongation factors, and enzymes [24]. During exosome biogenesis, several so-called intraluminal vesicles (ILVs), which are loaded with bioactive signaling compounds, are packaged into multivesicular bodies (MVBs) in the cytosol. The MVBs then integrate into the plasma membrane and release the ILVs into the intercellular space as exosomes. It has not yet been clarified if an exosome's cargo is specific for a cell type and its proliferation state or if the molecules are engulfed spontaneously and the variances in composition are observed by chance. However, some studies suggest systematic exosome localization and tissue specificity [25–27]. Similarly, the question of how exosomes interact with the neighboring cells once released is not yet fully understood, even though phagocytotic and endocytotic pathways as well as ligand interactions seem most likely [24]. Skokos et al. [28] found that exosomes can not only signal between the same cell types, especially dendritic cells, epi- and endothelial cells, B and T cells, mast cells, mesenchymal stem cells, and cancer cells of various histotypes [29] but also between different cell types. A few distinct functions of exosomes have been studied, including immune activation and suppression as well as their potential use as transporters for proteins that could induce anticancer immune responses [30].

1.4 Key Organelles Involved in Signal Transduction

While every cellular constituent plays a role within signal transduction, some organelles are critical for both the creation and maintenance of signaling capabilities, including the mitochondria, ER, and the nucleus. Beyond their well-known functional responsibilities, these organelles also serve to integrate many disparate signals that originate from both the extracellular matrix and the cytoplasm.

1.4.1 Mitochondria

Besides being the so-called "powerhouse" of the cell, the mitochondrion has been recognized as a highly important organelle in cell signaling. While it is widely recognized for its role in energy production, its role in intracellular concentrations of signaling molecules, and acting as a space for protein–protein signaling interactions, the mitochondrion also appears to be actively involved in signaling associated with immunity, autophagy, and cell death. These three functions are intimately interwoven in regard to cell death. These responses act as the first line of defense against incoming pathogenic threats, but cell death can also be controlled through autophagy as a pro-survival mechanism [31].

Mitochondria can use pyruvate, the end product of glycolysis, and fatty acids to produce ATP. Both are broken down into acetyl-CoA by enzymes in the mitochondrial matrix and funneled into the citric acid cycle where acetyl-CoA is broken down step-by-step into CO_2 meanwhile reducing coenzymes like nicotinamide adenine dinucleotide (NAD) to a high-energy form. These coenzymes can then interact with the membrane-bound complexes of the electron transport chain to establish a proton gradient in the mitochondrial inner membrane space that ultimately drives the chemiosmotic production of ATP. Up to 15 times more ATP can be produced from a glucose molecule in the mitochondria via oxidative phosphorylation as compared to the glycolytic pathway in the cytosol [32].

Due to leakage of electrons from the electron transport chain, which react with molecular oxygen, mitochondria produce reactive oxygen species (ROS) as well. Despite the overall harmful effects of ROS on cell constituents, recent research suggests a beneficial impact for low concentrations of ROS as well as an importance in cell signaling. It has been hypothesized that these low concentrations stress the cell, causing a beneficial adaption that prolongs longevity rather than promoting cell death [33].

Additionally, the mitochondrion has been recognized as the "hub of cellular Ca^{2+} signaling" [34] and serves in taking up these ions from the extramitochondrial space, buffering the matrix, and releasing it into the cytosol of other organelles. The mitochondrial calcium uniporter is responsible for the influx of Ca^{2+} into the mitochondrial matrix, and its actions depend on the electrochemical gradient between the cytosol and the inside of the mitochondria. The release of ions, on the other hand, is handled by the earlier mentioned Na^+/Ca^{2+} exchanger. Calcium concentration in the cytosol and the mitochondrial matrix influences metabolic processes and can act as a second messenger that stimulates the production of ATP through the activation of dehydrogenases, therefore playing a part in both energy consumption and energy production. Having said that, it appears that the stimulation is finely tuned to match the energy consumption of the cell [35].

Mitochondria are known for their involvement in cellular signaling, especially in apoptosis, a form of programmed cell death. Apoptosis can be triggered in nuclear, reticular, lysosomal, and plasma membrane pathways, which all seem to converge at the point of Bcl-2-associated X protein (BAX)/Bcl-2 homologous antagonist killer (BAK) pore formation at the mitochondrial membrane. With the formation of these pores on the outer mitochondrial membrane, the permeability increases drastically. As a result, species such as cytochrome c and second mitochondria-derived activator of caspases (SMAC) are released and are able to activate caspases that ultimately lead to cell death. The production of ROS in the mitochondria has been linked to a form of nonprogrammed cell death, the receptor-interacting serine/threonine-protein kinase 3 (RIPK3)-dependent necrosis, that seems to require ROS to happen [31].

Another point of involvement in cellular signaling relates to autophagy, which is oftentimes referred to as caspase-3-independent cell death. In general, cytosolic species are engulfed by a lipid bilayer membrane to form an autophagosome that can fuse with the lysosome to degrade and recycle its contents. Studies have shown that the mitochondrial membrane can be the origin of autophagosomal membrane, in addition to the mitochondrial role in activating autophagy. Autophagy occurs constantly at low levels in the cell; however, there is increased level of activity under certain conditions, such as nutrient deprivation. If mitochondrial production of ATP is low and the cellular concentration of ATP is low as well, adenosine monophosphate-activated protein kinase (AMPK) is activated and can phosphorylate uncoordinated-51-like autophagy activating kinase 1 (ULK1) directly to upregulate autophagy or inhibit mTORC1's inhibition on ULKs by inhibitory phosphorylation of mammalian target of rapamycin complex 1 (mTORC1)'s two regulators (tuberous sclerosis complex 2 [TSC2] and regulatory-associated protein of mTOR [RAPTOR]) [31].

Lastly, mitochondria appear to stand in relation with host cells of innate immune responses. In macrophages, an immune response is triggered through pattern recognition receptors (PRRs) that recognize pathogen-associated molecular patterns (PAMPs) on microorganisms or damage-associated molecular patterns (DAMPs) from damaged cells. Several kinds of DAMPs originate in the mitochondria. PRRs bind to mitochondrial DNA, *N*-formyl peptides, ATP, and ROS. DNA fragments and ROS are able to specifically facilitate the activation of the NLR (nucleotide-binding domain, leucine-rich repeat) family, pyrin domain containing 3 (NLRP3) inflammasome. Finally, the outer mitochondrial membrane has been recognized as an important platform in the retinoic acid-inducible gene I (RIG-I)-like receptor pathway (RLR), which involves the localization, accumulation, and aggregation of the mitochondrial antiviral-signaling protein (MAVS) to the outer membrane [31].

1.4.2 Endoplasmic Reticulum

The ER is mostly known as the organelle in which protein folding occurs via chaperones in the lumen, but it is also a major signaling organelle linked to processes ranging from sterol biosynthesis to the release of arachidonic acid, apoptosis, and, lastly, the uptake and release of Ca^{2+} [36]. The latter makes the ER a platform of signal interpretation, amplification, and transmission. The morphology of the reticulum varies based on its apparent functions. When dealing with protein folding, a structure resembling flattened sacs is found, whereas the smooth ER and, in the special case of neuron and muscle cells, the sarcoplasmic reticulum are associated with calcium signaling. Besides calcium ions, secondary messengers like ROS, inositol, sterols, or 1,4,5-trisphosphate and sphingosine-1-phosphate cause signaling events in the ER. The precursor for all seven phosphoinositide species is synthesized predominantly here [37]. However, calcium ions assume a specialized role in the functions of the ER, as it is necessary for the folding chaperones calnexin and calreticulin to aid nascent proteins to find their form. While the mitochondrion is the "hub for Ca^{2+} signaling" according to Berridge [38] and Xu et al. [39], the luminal concentrations of ions in the ER are the highest in the cell, due to active transport with calcium ATPases (sarcoendoplasmic reticulum calcium transport ATPase [SERCA] pumps), making it the largest storage location for calcium. The ions can be complexed with buffer molecules to build a dynamic storage, for example, with the low-affinity and high-capacity binding proteins calsequestrin, calreticulin, and chromogranins [17, 40].

1.4.3 Nucleus

Due to its role as the central repository of genetic information, the nucleus is perhaps the key distal receiver for most signal transduction networks. The ability to initiate and halt transcription of both proliferative and degradative constructs means this organelle ultimately holds the central position within a framework of a living biological system dynamically interacting with its local environment. Transcription of genetic materials is the primary response that allows meaningful adaptations to occur within a living system, and as such, it is highly regulated at every step: from initiation through RNA elongation and termination. While a complete description of transcriptional machinery and processes is outside the scope of this book, the reader is directed to several reviews that provide additional background information [41–46].

In order to protect its crucial DNA cargo and its associated processes, signals must traverse a nuclear membrane (please see Cooper and Hausman [47] for review) that is more selective than the plasma membrane. Structurally, the inner nuclear membrane (INM) envelopes nuclear material, while the outer nuclear membrane (ONM) is continuous with the ER and the intermembrane space (between the INM and ONM) allows for unimpeded signaling between these two organelles. In contrast, direct signaling between the cytoplasm and nucleus proceeds through highly selective nuclear pores, which are made up of large transmembrane protein complexes that span the intermembrane space. These pores contain freely permeable, omnidirectional aqueous channels that allow direct passive diffusion of small, nonpolar molecules (<50 kDa) between the lumen of the nucleus and the cytoplasm. However, the polarity and size of most proteins and RNAs require selective trafficking via active transport through the pores. In addition to requiring guanosine monophosphate (GTP) hydrolysis (via the Ras-like GTPase Ran) to provide the necessary energy for transport, translocation through nuclear pores typically involves an initiating signal that triggers the opening of the pore to a size sufficient for transport. Initiating signals involve receptor proteins known as importins and exportins for importing cytoplasmic material or exporting nuclear material, respectively. An extensive review of signaling for initiation of translocation may be found in Cautain et al. [48].

Independent of biophysical barriers to signal transduction, nuclear receptors play an impactful role in the transmission of exogenous, primarily hormone-related, signals for transcription initiation. While a complete description of nuclear receptor classes is outside the scope of this book, please see Alexander et al. [49] for a review. Nuclear receptors are a specialized form of a transcription factor that bind to endogenous agonists that then initiate conformational changes in the receptor to promote transcription. There are two major subclasses of nuclear receptors: steroid and nonsteroid hormone receptors. Steroid hormone receptors are usually found in the cytoplasm in a complex with chaperone proteins that are released upon binding of endogenous ligand; this then frees the transcription factor to translocate to the nucleus, where it binds with selective regions of DNA response elements to initiate transcription. Nonsteroid hormone receptors are distributed in the nucleus and interact with other nuclear receptors and endogenous ligands that promote transcription upon conformational alterations with binding partners.

The efflux and influx of energy and information (i.e. signaling) between localized organelles allow for adaptation and growth within living systems. This is especially true for the nucleus, where transcriptional regulation plays a central part in directing many biological processes, including cell cycle progression, intracellular metabolic homeostasis, and differentiation, to name a few. However, unlocking the genetic information held within DNA involves epigenetic alterations to the suprastructure of chromosomes (for review, please see Annunziato [50] and Alberts et al. [51]). This expands the role of cell signaling into additional regulatory access points that include DNA methylation, histone modifications, promoter–enhancer interactions, and noncoding RNA (i.e. microRNA) [52]. As such, now many of the classically defined protein posttranslational modifications (e.g. methylation, phosphorylation, acetylation, etc.) are known to modify transcriptional processing elements, thereby contributing to additional layers of signaling to the process of cellular response (for review please see Arzate-Mejia et al. [53]). Critically, these complex layers of regulation have been linked to many types of diseases and potential opportunities for pharmacologic cures [54–58]. The many signal transduction factors and their associated biological implications that relate to both health and disease are covered through the next several chapters within this book.

References

1 Kaila, V. and Annila, A. (2008). Natural selection for least action. *Proc. R. Soc. A* 464 (2099): 3055–3070.
2 Kleidon, A. (2010). A basic introduction to the thermodynamics of the earth system far from equilibrium and maximum entropy production. *Philos. Trans. R. Soc. B* 365 (1545): 1303–1315.
3 Santos, R., Ursu, O., Gaulton, A. et al. (2017). A comprehensive map of molecular drug targets. *Nat. Rev. Drug Discov.* 16: 19–34.
4 National Research Council (2007). *Toxicity Testing in the 21st Century: A Vision and a Strategy*. Washington, DC: The National Academies Press.
5 Koshland, D. (2002). The seven pillars of life. *Science* 295 (5563): 2215–2216.

6 Trifonov, E. (2012). Definition of life: navigation through uncertainties. *J. Biomol. Struct. Dyn.* 29 (4): 601–602.
7 McKay, J., Williams, E., and Mathers, J. (2004). Folate and DNA methylation during *in utero* development and aging. *Biochem. Soc. Trans.* 32 (6): 1006–1007.
8 American Heritage Dictionary. 5 (2014), Boston (MA): Houghton Mifflin Harcourt; c2014. Life; [1 paragraph]. https://ahdictionary.com/word/search.html?q=life. Accessed January 15, 2018.
9 Roper, S. (2007). Signal transduction and information processing in mammalian taste buds. *Eur. J. Phys.* 454 (5): 759–776.
10 Kaupp, U. (2010). Olfactory signaling in vertebrates and insects: differences and commonalities. *Nat. Rev. Neurosci.* 11: 188–200.
11 Moller, A. (2012). *Hearing: Anatomy, Physiology, and Disorders of the Auditory System*. San Diego: Plural Publishing, Inc.
12 Goutman, J., Elgoyhen, A.B., and Gomez-Casati, M.E. (2015). Cochlear hair cells: the sound-sensing machines. *FEBS Lett.* 589 (22): 3354–3361.
13 Lumpkin, E. and Caterina, M. (2007). Mechanisms of sensory transduction in the skin. *Nature* 445: 858–865.
14 Busch, W. and Saier, M.H. (2002). The transporter classification (TC) system. *Crit. Rev. Biochem. Mol. Biol.* 37 (5): 287–337.
15 Ashcroft, F.M. (2006). From molecule to malady. *Nature* 440 (7083): 440–447.
16 Gadsby, D.C. (2009). Ion channels versus ion pumps: the principal difference, in principle. *Nat. Rev. Mol. Cell Biol.* 10 (5): 344–352.
17 Carafoli, E., Santella, L., Branca, D., and Brini, M. (2001). Generation, control, and processing of cellular calcium signals. *Crit. Rev. Biochem. Mol. Biol.* 36 (2): 107–260.
18 Wright, E.M., Hirayama, B.A., and Loo, D.F. (2007). Active sugar transport in health and disease. *J. Intern. Med.* 261 (1): 32–43.
19 Vieira, A.V., Lamaze, C., and Schmid, S.L. (1996). Control of EGF receptor signaling by Clathrin-mediated endocytosis. *Science* 274 (5295): 2086–2089.
20 Cooper, G.M. (2000). Endocytosis. In: *The Cell: A Molecular Approach*, 2e. Sunderland: Sinauer Associates.
21 Donaldson, J.G., Porat-Shliom, N., and Cohen, L.A. (2009). Clathrin-independent endocytosis: a unique platform for cell Signaling and PM Remodeling. *Cell. Signal.* 21 (1): 1–6.
22 Polo, S. and Di Fiore, P.P. (2006). Endocytosis conducts the cell signaling orchestra. *Cell* 124 (5): 897–900.
23 Grant, B. (2009). Pathways and mechanisms of endocytic recycling. *Nat. Rev. Mol. Cell Biol.* 10 (9): 597.
24 Février, B. and Raposo, G. (2004). Exosomes: endosomal-derived vesicles shipping extracellular messages. *Curr. Opin. Cell Biol.* 16 (4): 415–421.
25 Mathivanan, S., Ji, H., and Simpson, R.J. (2010). Exosomes: extracellular organelles important in intercellular communication. *J. Proteome* 73 (10): 1907–1920.

26 Shen, B., Wu, N., Yang, J., and Gould, S. (2011). Protein targeting to exosomes/microvesicles by plasma membrane anchors. *J. Biol. Chem.* 286: 14383–14395.

27 Yang, J.M. and Gould, S.J. (2013). The cis-acting signals that target proteins to exosomes and microvesicles. *Biochem. Soc. Trans.* 41 (1): 277–282.

28 Skokos, D., Botros, H.G., Demeure, C. et al. (2003). Mast cell-derived exosomes induce phenotypic and functional maturation of dendritic cells and elicit specific immune responses in vivo 1. *J. Immunol.* 170: 3037–3045.

29 Corrado, C., Raimondo, S., Chiesi, A. et al. (2013). Exosomes as intercellular Signaling organelles involved in health and disease: basic science and clinical applications. *Int. J. Mol. Sci.* 14 (3): 5338–5366.

30 Chaput, N., Schartz, N., and André, F. (2004). Exosomes as potent cell-free peptide-based vaccine. II. Exosomes in CpG adjuvants efficiently prime naive Tc1 lymphocytes leading to tumor rejection. *J. Immunol.* 172 (4): 2137–2146.

31 Tait, S.W.G. and Green, D.R. (2012). Mitochondria and cell signaling. *J. Cell Sci.* 125 (4): 807–815.

32 Alberts, B., Johnson, A., Lewis, J. et al. (2002). The mitochondrion. In: *Molecular Biology of the Cell*, 4e. New York: Garland Science.

33 Cheng, Z. and Ristow, M. (2013). Mitochondria and metabolic homeostasis. *Antioxid. Redox Signal.* 19 (3): 240–242.

34 Szabadkai, G. and Duchen, M.R. (2008). Mitochondria: the hub of cellular Ca2+ signaling. *Am. Physiol. Soc.* 23 (2): 84–94.

35 Griffiths, E.J. and Rutter, G.A. (2009). Mitochondrial calcium as a key regulator of mitochondrial ATP production in mammalian cells. *Biochim. Biophys. Acta* 1787 (11): 1324–1333.

36 Berridge, M.J. (2002). The endoplasmic reticulum: a multifunctional signaling organelle. *Cell Calcium* 32 (5–6): 235–249.

37 Di Paolo, G. and De Camilli, P. (2006). Phosphoinositides in cell regulation and membrane dynamics. *Nature* 443 (7112): 651–657.

38 Berridge, M.J. (2004). Calcium signal transduction and cellular control mechanisms. *Biochim. Biophys. Acta, Mol. Cell Res.* 1742 (1): 3–7.

39 Xu, C., Bailly-Maitre, B., and Reed, J. (2005). Endoplasmic reticulum stress: cell life and death decisions. *J. Clin. Invest.* 115 (10): 2656–2664.

40 Clapham, D.E. (1995). Calcium signaling. *Cell* 80 (2): 259–268.

41 Sainsbury, S., Bernecky, C., and Cramer, P. (2015). Structural basis of transcription initiation by RNA polymerase II. *Nat. Rev. Mol. Cell Biol.* 16 (3): 129–143.

42 Heinz, S., Romanoski, C.E., Benner, C., and Glass, C.K. (2015). The selection and function of cell type-specific enhancers. *Nat. Rev. Mol. Cell Biol.* 16 (3): 144–154.

43 Allen, B.L. and Taatjes, D.J. (2015). The mediator complex: a central integrator of transcription. *Nat. Rev. Mol. Cell Biol.* 16 (3): 155–166.

44 Jonkers, I. and Lis, J.T. (2015). Getting up to speed with transcription elongation by RNA polymerase II. *Nat. Rev. Mol. Cell Biol.* 16 (3): 167–177.

45 Venkatesh, S. and Workman, J.L. (2015). Histone exchange, chromatin structure and the regulation of transcription. *Nat. Rev. Mol. Cell Biol.* 16 (3): 178–189.

46 Porrua, O. and Libri, D. (2015). Transcription termination and the control of the transcriptome: why, where and how to stop. *Nat. Rev. Mol. Cell Biol.* 16 (3): 190–202.

47 Cooper, G. and Hausman, R.E. (eds.) (2015). *The Cell: A Molecular Approach*, 7e. Sunderland: Sinauer Associates.

48 Cautain, B., Hill, R., de Pedro, N., and Link, W. (2015). Components and regulation of nuclear transport processes. *FEBS J.* 282 (3): 445–462.

49 Alexander, S.P., Cidlowski, J.A., Kelly, E. et al. (2015). The concise guide to pharmacology 2015/16: nuclear hormone receptors. *Br. J. Pharmacol.* 172 (24): 5956–5978.

50 Annunziato, A. (2008). DNA packaging: nucleosomes and chromatin. *Nat. Educ.* 1 (1): 26.

51 Alberts, B., Johnson, A., Lewis, J. et al. (2002). Chromosomal DNA and its packaging in the chromatin Fiber. In: *Molecular Biology of the Cell*, 4e (ed. B. Alberts, A. Johnson, J. Lewis, et al.). New York: Garland Science.

52 Allis, C.D. and Jenuwein, T. (2016). The molecular hallmarks of epigenetic control. *Nat. Rev. Genet.* 17 (8): 487–500.

53 Arzate-Mejia, R.G., Valle-Garcia, D., and Recillas-Targa, F. (2011). Signaling epigenetics: novel insights on cell signaling and epigenetic regulation. *IUBMB Life* 63 (10): 881–895.

54 Worsham, M.J., Chen, K.M., Ghanem, T. et al. (2013). Epigenetic modulation of signal transduction pathways in HPV-associated HNSCC. *Otolaryngol. Head Neck Surg.* 149 (3): 409–416.

55 Ptashne, M. (2009). Binding reactions: epigenetic switches, signal transduction and cancer. *Curr. Biol.* 19 (6): R234–R241.

56 Kaminska, B., Mota, M., and Pizzi, M. (2016). Signal transduction and epigenetic mechanisms in the control of microglia activation during neuroinflammation. *Biochim. Biophys. Acta* 1862 (3): 339–351.

57 Chen, Q.W., Zhu, X.Y., Li, Y.Y., and Meng, Z.Q. (2014). Epigenetic regulation and cancer (review). *Oncol. Rep.* 31 (2): 523–532.

58 Jirtle, R.L. and Skinner, M.K. (2007). Environmental epigenomics and disease susceptibility. *Nat. Rev. Genet.* 8 (4): 253–262.

2

Mechanisms of Cellular Signal Transduction

Richard R. Neubig[1], Jonathan W. Boyd[2], Julia A. Mouch[3], and Nicole Prince[3]

[1] *Department of Pharmacology and Toxicology, Michigan State University, East Lansing, MI, USA*
[2] *Department of Orthopaedics and Department of Physiology and Pharmacology, West Virginia University School of Medicine, Morgantown, WV, USA*
[3] *Department of Orthopaedics, West Virginia University School of Medicine, Morgantown, WV, USA*

Cellular signal transduction plays a critical role in the communication of information required to establish and maintain the function of complex multicellular organisms. This usually involves some chemical mediator, either soluble or surface bound, which goes from the stimulus to the receptor. These signals can have a very short range, such as those requiring cell contact to control cell migration and tissue development as well as synaptic transmission. Longer-range signals include paracrine mechanisms that can be detected in the immediate environment of the secreting cell or those sensed throughout the body, like circulating hormones and mediators released from endocrine organs and inflammatory cells. Some signals even transfer from one organism to another (i.e. smells and pheromone signaling). In most cases, the first step in establishment of signal transduction is the recognition of the signaling molecule by a *receptor* in the target cell. Due to their key role in initiating biological communication and exquisite selectivity for one particular signal (out of the complex milieu in which the cell exists), receptors are highly sought after in the fields of pharmacology and toxicology. They are often the targets of therapeutic drugs or highly potent toxins.

Receptors are just the first step in the complex signal transduction mechanism that governs organismal communication. Receptors have an exquisite sensitivity to detect concentrations of their activating signal, often in the nanomolar to picomolar range; therefore, receptor signals need to be greatly amplified. This usually occurs by engaging a series of *transducer* or *effector* molecules to interpret and amplify their signals. Complex multicellular organisms require highly differentiated functions in specific tissues, so the variety of receptors

Cellular Signal Transduction in Toxicology and Pharmacology: Data Collection, Analysis, and Interpretation, First Edition. Edited by Jonathan W. Boyd and Richard R. Neubig.

and the range of signal transduction mechanisms and their regulatory interactions have become staggeringly complex. In this chapter, we will explore the basic building blocks of signaling mechanisms and examine the myriad of ways in which simple *signaling modules* can be utilized. The more complex level of integration and feedback control of these mechanisms in the cell will be presented in Chapter 3. Chapter 4 will extend this into the whole organism with complex interactions between multiple cell types and organs.

After introducing the broad array of receptor types and their signaling properties, we will present some organizing principles underlying diverse signaling mechanisms. Then, specific examples of signal transduction pathways will be used to illustrate those concepts. While not attempting an encyclopedic description of all signaling mechanisms, we will aim to cover many of the key ones that underlie pharmacology and toxicology. Finally, some emerging concepts in receptor mechanisms, such as allosteric modulators and functional selectivity or agonist bias, which have led to advances in therapeutic development, will also be presented. In later chapters, we will explore how the basic cellular signaling concepts relate to the pathophysiology of disease and toxicity from environmental and other exposures.

2.1 Posttranslational Modifications and Their Roles in Signal Transduction

Cellular signaling events are dictated by the interaction of signaling molecules with receptors, and it is important to note how this exquisite specificity is achieved. Cell signaling specificity is heavily reliant on the presence of posttranslational modifications (PTMs), the attachment of small chemical moieties to amino acid residues, which ultimately creates a vastly diverse proteome [1]. This complexity facilitates the range of possible cellular functionalities, and this plays a key role in both the specificity and dynamic nature of receptor interactions. While there are many different PTMs, this overview will focus on those most relevant to the field in studies of signal transduction.

2.1.1 Phosphorylation

Reversible protein phosphorylation has been well established for its role in studies of signal transduction, as it is the most common mechanism to regulate function and transmit signals in cells [2]. Phosphorylation is mediated by the activities of kinases and phosphatases, as shown in Figure 2.1. The reversible mechanism of phosphorylation involves addition of a phosphate group (PO_4) to polar R groups of amino acids via coordination with magnesium (Mg^{2+}); in eukaryotes, these residues include serine (SER), threonine (THR), and tyrosine (TYR). Following activation by stimuli, the protein of interest receives a

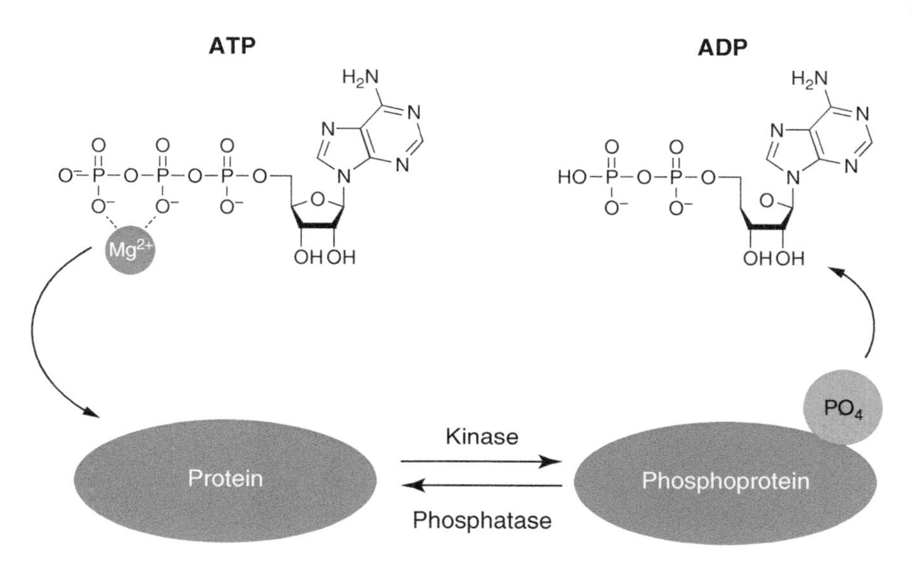

Figure 2.1 Mechanism of phosphorylation and dephosphorylation. Kinases perform a hydrolysis reaction to covalently attach a phosphate group from ATP to a protein at polar R groups of amino acids. ATP coordination with Mg^{2+} (shown in green) is essential for the transfer of a phosphate group. Phosphorylated proteins can be dephosphorylated by the action of phosphatases, which hydrolyze the phosphoric acid monoesters, which results in ADP converting back to ATP. A free hydroxyl group remains on the protein after removal of the phosphate group.

phosphate group via hydrolysis of adenosine triphosphate (ATP) due to enzymatic activity of the kinase. The anionic PO_4 closely associates with cationic Mg^{2+}, which allows facile integration within SER, THR, and TYR. Addition of PO_4 causes a shift from a hydrophobic apolar protein to a hydrophilic polar one, which causes a conformational change, allowing the protein to interact with other molecules. Different classes of kinases exist, and they are classified according to which amino acid residue the kinase phosphorylates. For example, serine/threonine kinases (STKs) phosphorylate the hydroxyl groups of serine or threonine; this group of kinases can be activated by a variety of cellular events, ranging from DNA damage to chemical signal mediation of calcium/calmodulin, cyclic adenosine monophosphate (cAMP)/cGMP, and diacylglycerol (DAG) [3]. Tyrosine kinases (TKs) act at tyrosine residues and have been found to be important mediators of a range of diverse biological processes and are involved in regulation of proliferation, differentiation, migration, metabolism, and programmed cell death, to name a few [4]. Specific examples of STKs and TKs will be discussed later in the chapter.

Phosphatases have the opposite action of kinases: they remove phosphate groups from phosphoproteins; this is achieved by hydrolyzing phosphoric acid monoesters to form a phosphate group [5]. A free hydroxyl group remains on the protein after dephosphorylation. Phosphatases are classified based on their catalytic domain, but their specific roles and targets remain widely unknown; they are typically described as playing a nonspecific role in phosphoprotein homeostasis. Recent efforts have begun to elucidate the role these enzymes play in a variety of physiological processes [6].

The interplay of phosphorylation/dephosphorylation processes can be very complex, as a single kinase or phosphatase enzyme may interact with many substrates and may function in different cell-signaling pathways. All in all, phosphorylation is one of the most influential PTMs, as it is intertwined with energy availability (via ATP) and thus reaches a broad spectrum of cell regulatory processes related to signal transduction. Due to the large breadth of phosphorylation events involved in cell signaling, details on specific roles of kinases and phosphatases will be discussed in later sections.

2.1.2 Acylation

Acylation involves the addition of an acyl group (R—O=C) to an amino acid. PTMs that fall under this category include acetylation, propionylation, butyrylation, malonylation, formylation, and succinylation; the subcategory of acylation describes the type of acyl group being added. The chemical mechanism involves acyltransferase enzymes attaching acyl groups to the protein of interest, most often at free amine or thiol groups, and this attachment can be either reversible or irreversible. While acylation events can occur at many different amino acid residues, the lysine-specific acylation events are the most well studied in the field of signal transduction and have established roles in a variety of cellular processes [7]. Lysine acetylation is one of the most commonly studied acylation events and is often regarded for its role in apoptosis, metabolism, and stress response [8]. Recently, the size of the acetylome has come close to that of phosphorylation, the most common PTM [8]; this brings to light the importance of lysine acetylation to the field of signal transduction.

Lysine acetylation is an abundant PTM in eukaryotes that involves reversible addition of an acetyl group at the primary amine of a lysine side chain. Because positively charged lysine residues play important roles in folding and function of proteins, neutralizing this charge via acetylation is of profound importance for a range of cellular processes [9]. One example of the importance of lysine acetylation in biology is histone acetylation; in this process, lysine residues of N-terminal tails protruding from histone cores are acetylated and deacetylated, essentially acting as an on/off switch for entrance into the cell growth cycle; the decision by a cell to grow and divide must intimately be connected with nutrient availability, so it is not surprising that this process is regulated by signal transduction cascades reliant on energy availability and metabolism [10].

Histone acetylation illustrates one important example of how acylation can influence the activity of a cell in a coordinated manner within signal transduction networks. In histone acetylation, acetyltransferase enzymes called histone acetyltransferases (HATs) use a carbon donor source (acetyl-CoA) to transfer an acetyl group to the free amine of a lysine residue. Once a histone is acetylated, the positive charge is removed, which allows for a more relaxed structure of chromatin, leading to greater levels of gene transcription, increased proliferation, and other altered biological effects [11]. As histone acetylation is a reversible process, histone deacetylase enzymes (HDACs) oppose the action of HATs to remove the acetyl group and restore the positive charge of lysine, which reverts histones back to a more ordered state. This type of site-specific acetylation/deacetylation of histones is central to chromatin organization and subsequently gene regulation, acting as an on/off switch for gene regulatory processes [12]. As with other types of PTMs, including other acylation events, the enzymes that regulate the acetylation/deacetylation processes can have multiple targets, and it is critical to consider that while these enzymes exhibit specificity, this specificity is intimately related to the presence of other cellular signals.

2.1.3 Alkylation

Alkyl groups are alkanes missing one hydrogen and are of the formula C_nH_{2n+1}; the incorporation of these groups into proteins increases lipophilicity to varying degrees, depending on the alkyl group being added. Reversible or irreversible alkylation can have an important impact on protein structure, as it affects backbone dihedral angles, amide bond *cis*/*trans* equilibrium, and hydrogen bonding, which are all vital components in secondary structure of proteins [13]. The most common alkyl groups transferred as a part of PTM processes are as follows: methyl groups, which are often involved in histone methylations at lysine and arginine residues, and longer lipid chains like 15-carbon farnesyl groups [14]. The larger lipid groups introduce more hydrophobicity.

Methylation is one of the most recognizable types of alkylation in the context of signal transduction. Methylation is performed by enzymes known as methyltransferases. The methyl transfer is achieved via an S_N2-like nucleophilic attack, in which a methyl group is transferred to an enzyme substrate. Numerous signal-transducing proteins, like G-proteins, contain motifs that are altered by methylation events. For example, methylation modifications of Ras proteins have been shown to regulate extracellular signal-regulated kinases (ERK) pathway activation and propagation of Ras-dependent signals [15]. Ras proteins are involved in regulation of a wide span of cellular functions but are only biologically active upon association with cell membranes. Thus, the lipophilicity of Ras has a profound impact on this interaction. Methylation and farnesylation of the C-terminus of Ras forms a lipophilic domain that interacts with the phospholipid bilayer of cell membranes, paving the way subsequent Ras-related signaling [16].

Alkylation events are not as common as the previously described phosphorylation and acylation processes but still present a type of chemical modification involved in signal transduction networks. The impacts of specific methylation events and other alkylation events will be described in later sections.

2.1.4 Glycosylation

Glycosylation is the addition of carbohydrate sugars onto substrate proteins to form sugar–protein linkages, and it is heavily involved in nutrient sensing and cell–cell communication signaling cascades. This PTM is broadly categorized into N-glycosylation (attachment at a nitrogen atom) and O-glycosylation (attachment at an oxygen atom). When glycosylation occurs, glycosyltransferase enzymes use activated sugars as cofactors to transfer the respective group to the target of interest. More than one carbohydrate moiety is often added subsequently, and the addition of these groups can result in structural changes that influence the function of the protein [17]. Glycosylation has been studied in a variety of regulatory processes ranging from protein localization, protein–protein interactions, structural stability of the cell, immune response, and modulation of cell signaling.

While glycosylation is generally not thought of in terms of signal transduction, the regulation and modification of proteins via glycosylation falls in this realm. The diversity of cell surface glycoconjugate structures (more than one sugar moiety) implies the variety of roles glycosylation can play in influencing protein function. Glycosylation events impact the correct folding of proteins as well as have physiochemical effects that alter the properties of glycoproteins. One example of the role of glycoproteins in cell signaling is shown in the activities of interleukins (ILs), glycoproteins secreted by the immune system. ILs communicate both positive and negative regulatory signals to make up parts of the innate and acquired immune responses, and they are able to do so through interactions with specific receptors (e.g. IL-1 receptor [IL-1r], epidermal growth factor receptor [EGFR], interferon-gamma receptor [IFN-gR]) [18]; in fact, many cytokines, including interferons, ILs, colony stimulation factors, and chemokines, employ glycosylation as a means to add to heterogeneity and regulate binding to their respective receptors [19]. Glycosylation of proteins is intimately involved both from the ligand and receptor perspectives, which highlights the importance of proper glycosylation; in fact, improper glycosylation is often implicated in disease [20].

Glycosylation is often regarded as complex, due to the breadth of carbohydrate groups that are added in different conformations. While many aspects of the role of glycosylation remain unknown, the identification and quantification of these structures have become of utmost importance in studying disease as well as developing therapeutics.

2.1.5 Other PTMs

This short review of PTMs aims to cover those that will be mentioned in later sections and chapters, but it is important to remember that other categories of PTMs can influence and modify signal transduction cascades. For example, hydroxylation, nitration, and sulfation involve modifying functional groups to change the reactivity of proteins [21], and palmitoylation and myristoylation describe covalent lipid modification to change the hydrophobicity of their targets [22]. Addition of peptide chains such as ubiquitin (i.e. ubiquitination) and the small ubiquitin-like modifier (e.g. SUMOylation) often serve to "tag" and communicate cellular state [23]. In many cases, several PTMs are involved in the regulation of protein function. The dynamic, regulatory nature of switching the function of a protein is essential for signal transduction cascades. While these interactions require specificity, it is important to consider the promiscuous nature of these modifying PTM enzymes; concentration dependence is often a key player in the regulation of the status of PTMs and can drastically augment or limit selectivity associated with conformation. All in all, this relationship highlights the beautifully complex nature of the cell and its ability to employ PTM regulation in signal transduction.

2.2 Receptors

The original concept of receptors as the initial step in biological signaling processes is derived from work by Langley on neurotransmitter signaling in smooth and skeletal muscle and work by Ehrlich on the specificity of toxic and immune agents [24]. The International Union of Pharmacology Receptor Nomenclature Committee [25] defines a receptor as "A cellular macromolecule, or an assembly of macromolecules, that is concerned directly and specifically in chemical signaling between and within cells." In general, receptors bind their activating ligands with high affinity and often with exquisite specificity. This definition recognizes the receptor as a functional unit that may involve multiple macromolecular (usually protein) chains and restricts the definition to receptors for which a binding event triggers or modulates a biological response. A protein that binds a ligand with high affinity to sequester it without mediating a functional consequence should be called a "binding protein" rather than a true receptor.

Our understanding of receptors has advanced tremendously from the early concept of a system to detect soluble small molecules contacting the surface of a cell. The great majority of receptors are on the cell surface and typically contain transmembrane (TM) helices that pass through the plasma membrane to transfer information into the cell. However, an increasing number of receptors are recognized that reside inside the cell (Figure 2.2). While cell

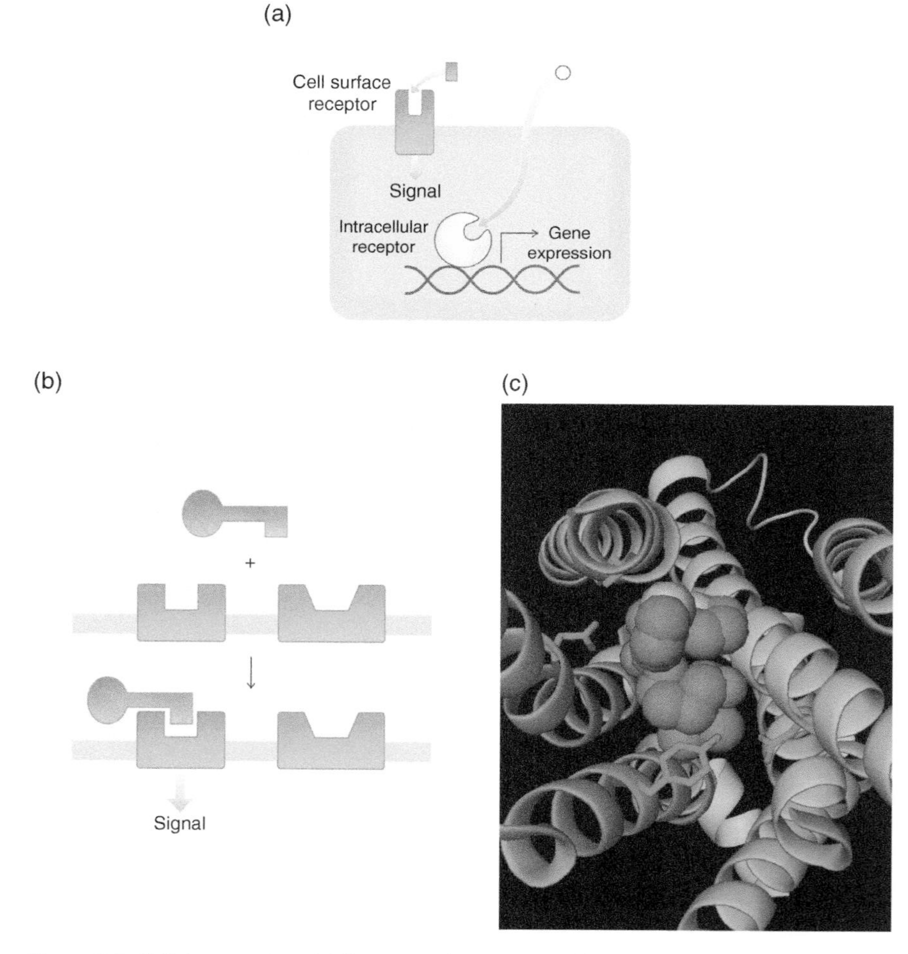

Figure 2.2 Cellular receptors. (a) Receptors detect external and internal signals from a wide range of chemical and physical mediators. (b) Structural complementarity between the activating ligand and its receptor underlies receptor selectivity. (c) Actual receptor structures are now available to demonstrate this complementarity. The asthma drug tiotropium is shown bound to the M3 muscarinic receptor (PDB: 4DAJ) with some key amino acid residues highlighted.

surface receptors are optimal for receiving signals from hydrophilic ligands or large protein molecules that cannot readily cross the plasma membrane, most intracellular receptors recognize lipophilic ligands like steroid hormones (e.g. estrogen), lipid mediators (retinoids and 15-deoxy-Δ-12,14-prostaglandin J2), bacterial products (muramyl dipeptide), and even gases (nitric oxide). The range of ligands recognized by TM receptors is also remarkable. Sensory receptors detect photons of light, odorants, and taste molecules. Receptors involved in detecting cell–cell contact and controlling organismal development (e.g. integrin and ephrin receptors) frequently recognize tethered ligands on adjacent cells, and, in some cases, the signal is transmitted in both directions [26, 27]. Some receptors respond to the proteolytic cleavage of their extracellular domain (e.g. by thrombin), and others require the coincident detection of two ligands (glutamate and glycine). This diversity in ligands is accompanied by tremendous diversity in the structure of receptors, which can be grouped by their molecular mechanisms (Figure 2.3). Thus, evolution has led to a tremendously diverse set of ligand recognition domains as well as highly diverse output signals to provide the requisite regulation of cellular functions.

2.3 Receptor Signaling Mechanisms

The range of signal outputs from receptors is, perhaps, even more remarkable. Before diving into the molecular details of complex signaling mechanisms, it is worth considering some simple patterns of signaling that are used commonly across different receptor systems.

2.3.1 Basic Principles of Signal Transduction Mechanisms

Let's consider some of the fundamental principles that underlie signal transduction mechanisms as we understand them today. These principles often apply throughout the range of steps in signal transduction (i.e. receptors, transducers, and effectors). First, signal transduction involves a transmitter (of information) and a receiver; no biological activity can occur without these two key units. Due to the diverse environmental, hormonal, and other transmitters that biological receivers respond to, most cellular signal transduction knowledge has come from research that is focused on receptors. However, this early research has opened the door to a greater understanding and has revealed that along a signal transduction cascade, cellular receptors actually become transmitters as they are merely another gateway for the overall transfer of information. For example, *selectivity and recognition* are critical not only for receptors but also for intracellular transmission of information from one step in a signaling cascade to another. *Molecular switches* are found in both heterotrimeric and low molecular weight guanosine triphosphate (GTP) binding proteins (classically involved in G-protein-coupled receptor [GPCR]-driven transduction) as well as in many kinase substrates where the activity state is determined by the addition or removal of a posttranslational phosphorylation. *Amplification* not only occurs at the level of receptors but also occurs with ion channels, effector

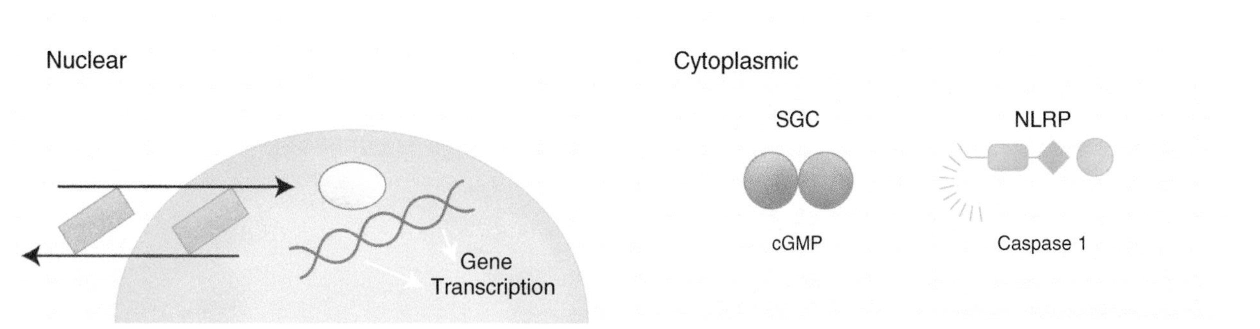

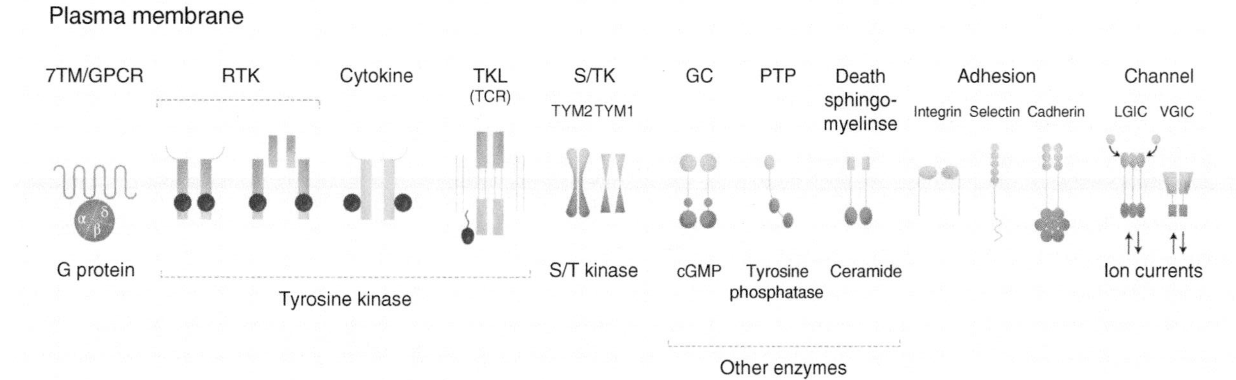

Figure 2.3 Diversity of cellular receptors. There are a tremendous number of receptor mechanisms in mammals. These include some large groups like the 7-transmembrane/G-protein-coupled receptor (7TM/GPCR) family and the receptor tyrosine kinases (RTKs). In addition, many other mechanisms are represented including tyrosine kinase-like (TKL) and T-cell receptor (TCR). Multiple other receptor systems use an extracellular ligand binding domain and an intracellular enzyme including the membrane serine/threonine kinase (S/T kinase), guanylate cyclase (GC), and protein tyrosine phosphatase (PTP) receptors. Other receptors have multiple subunits including ligand-gated ion channels (LGIC) and voltage-gated ion channels (VGIC). Intracellular receptors are often involved in controlling gene transcription in the nucleus. There is also soluble guanylate cyclase (SGC), which produces cGMP and the

enzymes, multistep kinase cascades, and gene transcription mechanisms. Furthermore, *turn-off mechanisms* are often neglected, but they are frequently regulated and often play a critical role in controlling cellular signals at all steps from transmitter to receiver. Finally, the early concept of second messenger signaling post-receptor interaction (in which the cell cytoplasm is flooded with a uniform concentration of the second messenger) is now known to be a gross oversimplification. Second messenger mechanisms, like cAMP and Ca^{2+} signaling, and kinase mechanisms, like mitogen-activated protein kinase (MAPK) signaling, are frequently governed by scaffold or transport proteins that permit *localization* of active signaling molecules to small sub-compartments of the membrane or cell. The development of powerful imaging probes and compartment-specific reporters has clearly shown that signaling mechanisms are often highly targeted inside the cell. Finally, for complex signaling mechanisms that can turn on a variety of different signal outputs, it has been recognized that unique agonist ligands can differentially activate specific output pathways from the same receptor. This idea has been termed "biased signaling" or "functional selectivity" and is of interest in the pharmaceutical sector due to the possibility for enhanced selectivity in biological responses beyond just which receptor the ligand binds to.

2.3.1.1 Selectivity and Recognition

The ability of a receptor to recognize its specific ligand in the context of other molecules present at much higher concentrations is fundamental to the process of cellular communication. Similar to the specificity of enzymes for substrates, receptors generally have binding sites for the activating ligands that provide molecular complementarity to optimize the energy and/or kinetics of binding interactions. The famous "lock and key" model of receptor recognition provides a simple analogy of this mechanism (Figure 2.2b). There are now a tremendous number of high-resolution crystal structures of membrane-bound receptors that fully validate this complementarity model (e.g. Figure 2.2c).

2.3.1.2 Flexible Modularity

The tremendous diversity in signaling mechanisms, with many different receptors, transducers, effectors, and kinase cascades, has developed through evolution by the reuse of common signaling modules and mechanisms in many different contexts. For example, the model organism, *Saccharomyces cerevisiae* or baker's yeast, has only 3 GPCR genes, but humans have about 800 when olfactory GPCRs are counted.[1] Further, humans have at least 20 different G-protein α subunits (including splice isoforms), while yeast only have 2.

---

1 There are 399 nonolfactory GPCRs according to GPCRdb, 779 including olfactory receptors at http://gpcr.usc.edu, and 1108 in the GLIDA database at Kyoto University, which appears to include receptor variants.

This tremendous expansion of the GPCR and G-protein gene families supports the much greater diversity of cell types and signaling contexts in mammals than the simpler organism, yeast.

While the number and diversity of signaling proteins is much greater in humans, the basic modular architecture underlying the mechanisms and functions is consistent throughout evolution. For example, the highly versatile 7-transmembrane (7TM)-spanning structure of GPCRs with the transducing function of the heterotrimeric G-proteins has been adapted to a wide variety of signaling contexts (Figure 2.4a). Yeast GPCRs detect pheromone peptides, but in mammals, ligands range in size from simple cations such as Ca^{2+} to amino acids and similarly sized molecules, such as the stress hormone epinephrine, to large proteins, such as follicle-stimulating hormone (FSH). Two unique receptor activation mechanisms permit detection of light and the proteolytic activity of thrombin. In both cases, the signal initiator (light or thrombin) does not act as the actual ligand that turns on the receptor. Light receptors, like rhodopsin, have a covalently bound retinal ligand that undergoes a *cis–trans* conformational change upon absorption of a photon of light. The generated *trans*-retinal is ultimately the activating ligand that causes the receptor to initiate G-protein activation and regulation of the enzyme phosphodiesterase (PDE). In the case of thrombin, the proteolytic activity of thrombin clips a short peptide from the extracellular N-terminus of the receptor to reveal a neo-N-terminus that then becomes the active ligand. This binds to the receptor to mediate G-protein activation [28–30].

In addition to the modularity afforded by the receptor, the G-proteins also have differentiated function, but their basic mechanism is shared. This was accomplished by the expansion in the number of genes and different effector enzymes and activities of G-protein effectors [29, 31]. There are four major classes of G-protein α subunits (Figure 2.4a) expressed from 16 genes and 20+ splice isoforms. The four major families are Gs, Gi/o, Gq, and G12/13. Light signaling is mediated by transducin ($G\alpha_{t1}$), which is a member of the Gi/o family. It activates a cGMP PDE, which decreases the concentration of cGMP in the visual rod outer segments and turns off an ion current as discussed below. $G\alpha_s$ was the first nonretinal G-protein subunit to be purified and characterized. Gs and its olfactory-enriched relative G_{olf} mediate activation of the adenylyl cyclase (AC) enzymes and increase levels of cAMP in the cells. A receptor for the stress hormone epinephrine (or adrenaline), the β_1-adrenergic receptor, activates Gs, increasing AC. The Gi/o family was named for its role in mediating inhibition of AC, counteracting the action of Gs; two different adrenergic receptors have opposite effects to mediate AC activity. In contrast to the β_1-adrenergic receptor, the α_2-adrenergic receptor activates Gi/o and inhibits AC. A third type of adrenergic receptor (α_1), as well as many other GPCRs, activates the Gq family of G-proteins. G_q signaling activates a lipid PDE enzyme (phospholipase C-beta [PLC-β]) that mediates Ca^{2+} signaling, as discussed

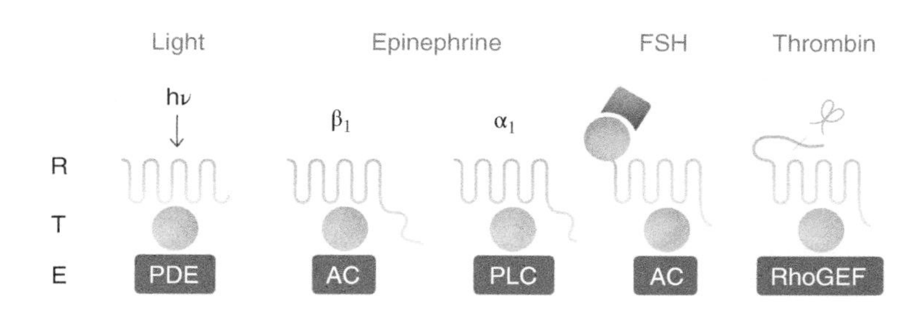

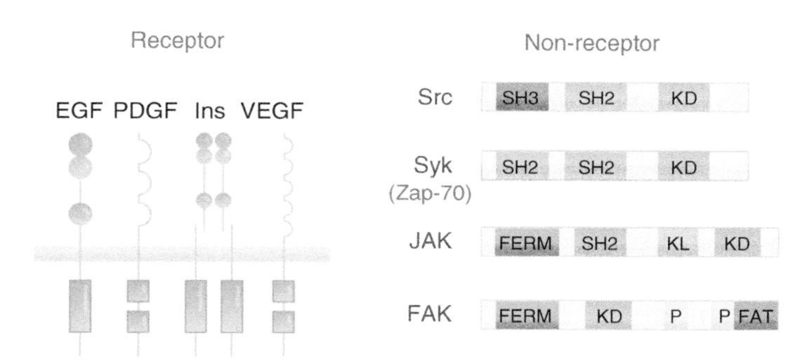

Figure 2.4 Modularity and reuse of signaling elements to generate diversity and complex regulation. (a) For GPCRs, the 7TM bundle and its coupling to the intracellular G-protein is highly conserved. The ligand binding mechanism and structure, however, have been greatly diversified through evolution. This accommodates a wide range of ligands. Also, despite structural similarities among G-proteins, they can modulate multiple downstream signals including either up- or downregulating of AC/cAMP plus activation of phosphodiesterase (PDE) in the eye, phospholipase C (PLC), and activators of the small G-protein Rho. This allows mixing and matching of different ligands and different effector outputs depending on the receptor and G-protein expressed in a cell. (b) Similarly, the tyrosine kinase domain (KD) is utilized in both membrane-bound and soluble signaling molecules. For receptor tyrosine kinases, different ligand binding domains are connected to the intracellular KD. For nonreceptor tyrosine kinases, a variety of interaction modules that facilitate the assembly of signaling complexes are employed.

later. The last G-protein family to be understood is the G12/13 family, which activates guanine nucleotide exchange factors (GEFs) for the low molecular weight GTPase signaling protein Rho.

Similarly, the role of TKs in signal transduction has also diversified to fill a tremendous number of roles in biology. Unlike GPCRs, where the basic protein structure is similar, 7TM domains with some variability in N and C termini, the TK domain structure has been incorporated in a wide variety of protein structural contexts (Figure 2.4b). It is located on the intracellular C-terminal end of *receptor tyrosine kinases* (RTK), which have a wide array of extracellular ligand binding domains coupling TK signaling to many different inputs. Also, in the soluble *non-RTKs*, the kinase domain (KD) is joined by other modular protein interaction domains. The first TK protein identified, src protein kinase, provided the name of these domains: Src homology 1 (SH1, the original name for the KD), SH2 that binds to phosphorylated tyrosines, and SH3 that binds to poly-proline stretches in other proteins. Many non-RTKs have this SH3–SH2–KD architecture. Others, such as focal adhesion kinase (FAK), include the TK domain with other interacting domains such as the FERM (4.1/ezrin/radixin/moesin) domain that is involved in cytoskeletal binding. This ability to readily mix and match different interaction domains has permitted the development of complex protein–protein interaction networks and regulatory mechanisms for the over 90 receptor and non-RTKs found in humans [32–35].

2.3.1.3 Molecular Switches

A key feature of signal transduction mechanisms is the broad utilization of molecular switches or proteins for which function is altered – usually in a binary manner – by ligand binding or a reversible posttranslational modification. While receptors were originally described in this manner, active with agonist and inactive when unoccupied or bound by antagonist, we now know that they have a richer functional repertoire than just two states. The classes of protein families acting as molecular switches range from classic receptors to kinases (via phosphorylation of tyrosine, serine, and threonine residues), but the GTP binding proteins are the best understood.

The GTP binding proteins involved in signal transduction fall into two major categories. The first group is the *heterotrimeric G-proteins* (activated by GPCRs and comprising three distinct subunits Gα, β, and γ) that play a key role in GPCR signaling as described above [29, 31]. The second group, the low-molecular-weight GTP-binding proteins, or *small GTPases* belong to the Ras superfamily of proteins that are well known for the role in cancer. They are activated by a wide variety of mechanisms at both plasma membrane and intracellular sites. Both groups of proteins undergo a critical conformational change when GTP is bound in place of guanosine diphosphate (GDP) (Figure 2.5). In general, the inactive or resting state of the protein is bound to GDP, and the mechanism of G-protein activation involves nucleotide exchange where GDP dissociates and GTP binds. Given that intracellular GTP concentrations are high

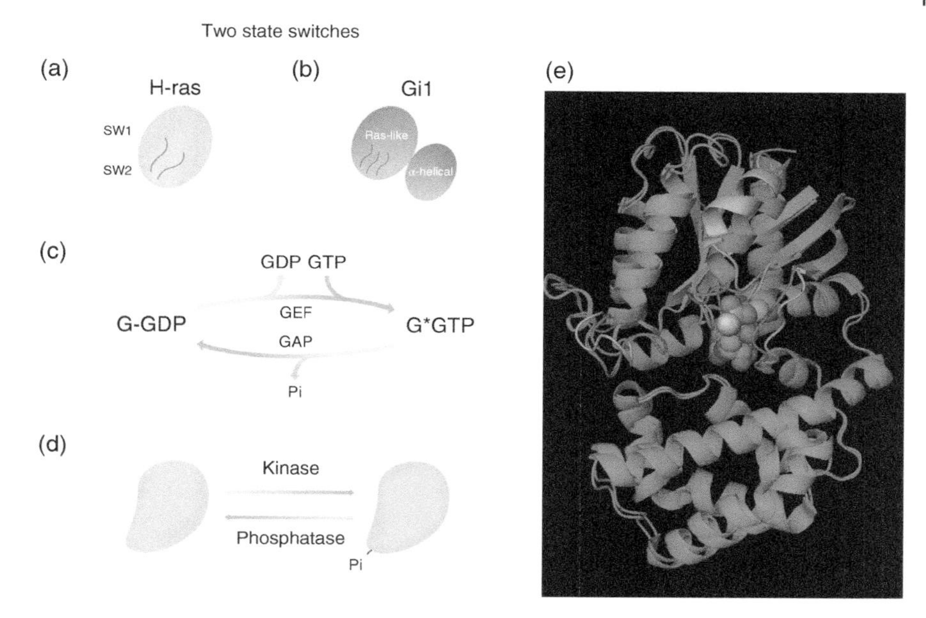

Figure 2.5 Two-state switches. (a–c) GTP binding proteins include (a) small GTPases such as H-ras and (b) heterotrimeric G-proteins. (c) The switch between active G* and inactive G state is reversible. (d) Protein phosphorylation is a widely used mechanism to switch states of proteins. Phosphorylation is mediated by protein kinases and reversed by phosphatases. (e) The crystal structure of a G-protein (Gi) in the GDP-bound inactive (cyan, PDB: 1gg2) and GTP-gammaS-bound active state (red, PDB: 1gia) illustrates the conformational change that occurs. The bulk of the protein (green) does not change much, but the three "switch" regions in cyan (inactive) and red (active) differ between the two states.

(~500 μM), the rate-limiting step for this process is usually GDP release, which is induced by an upstream activation mechanism, usually a receptor or a receptor-activated GEF.

The structural basis of activation of small GTPases and heterotrimeric G-proteins is quite similar. This is easily understood by drawing comparisons of the structures of these two protein families. The Ras superfamily GTPases have a molecular weight of about 20–25 kDa and share a common "GTPase fold" structure (Figure 2.5a). The heterotrimeric G-protein α subunits (~40 kDa) have two domains: a Ras-like domain of about 20 kDa and a "helical" domain of a similar size. When GTP is bound to the Ras oncoprotein, the conformational change that occurs is localized to two regions termed Switch I and Switch II (Figure 2.5a). G-proteins have three switch regions that change upon GTP binding, the first two of which are structurally analogous to SW1 and SW2 in Ras. This conformational change in both heterotrimeric and Ras-like G-proteins occurs because the γ-phosphate of the bound GTP pulls and extends the alpha helix of Switch II (Figure 2.5e). This relatively minor structural rearrangement forms a new binding surface on Ras and G-protein for effector molecules. Also, in the case of heterotrimeric G-proteins, it destroys a key interaction surface for the Gβγ subunit, which causes Gα-GTP dissociation from Gβγ. Finally, these proteins contain their own turn-off mechanism through their GTPase activity that hydrolyzes GTP to GDP, reversing the activating conformational change [29].

The active state of these GTP-dependent molecular switches is thus controlled by the relative rates of activation – by receptors and receptor-associated GEFs – and inactivation. While the proteins have an intrinsic GTPase activity, it is relatively slow, and the turn-off mechanism is regulated. The Ras family GTPases are nearly entirely dependent on GTPase accelerator proteins (GAPs) for turn-off since their intrinsic GTPase activity is so poor. Indeed, mutations in some Ras GAPs like the neurofibromatosis protein, NF1, can lead to the development of some cancers via simple alteration of the turn-off rate of this important molecular switch [36]. Two families of the heterotrimeric G-proteins (e.g. G_s and G_{12} family) depend largely on their *intrinsic* GTPase activity for turn-off while the deactivation of the G_i and G_q families is strongly enhanced by a family of GAPs called regulators of G-protein signaling (RGS) [37, 38]. This concept of regulated turn-off will be explored later in more detail.

2.3.1.4 GPCRs and Second Messengers

Second messengers are molecules that are initially produced or released through membrane-level receptor signals that then propagate the signal(s) throughout the cell. The classical concept, as described earlier for GPCRs, involves the binding of a first messenger, such as a hormone or neurotransmitter, to a receptor (or discriminator). This activates the transducer, a G-protein, which in turn stimulates the amplifier (adenylyl cyclase) to produce the second messenger cAMP [29, 39]. Second messengers can be either soluble (e.g. cAMP) or may be hydrophobic and constrained to the lipid membrane (e.g. DAG and other lipid second messengers, Figure 2.6).

2.3.1.4.1 Cyclic Nucleotides

Two of the best-known second messengers are the cyclic nucleotides, cAMP and cGMP. As noted above, the former is produced upon activation of AC enzymes by β-adrenergic receptors and other GPCRs in the heart and many other tissues. These GPCRs belong to the class that couples to the stimulatory G-protein (Gs), which results in the dissociation of GTP-bound $G\alpha_s$ from Gβγ. GTP-$G\alpha_s$ binds to and activates one of the membrane-bound AC enzymes to produce cAMP. The cAMP then binds to target proteins like protein kinase A (PKA) that activates downstream signaling proteins by phosphorylating them on serine or threonine residues. This simple linear cascade provides a strong amplification mechanism with catalytic steps occurring at the level of G-protein activation, cAMP production, protein phosphorylation, and subsequent steps.

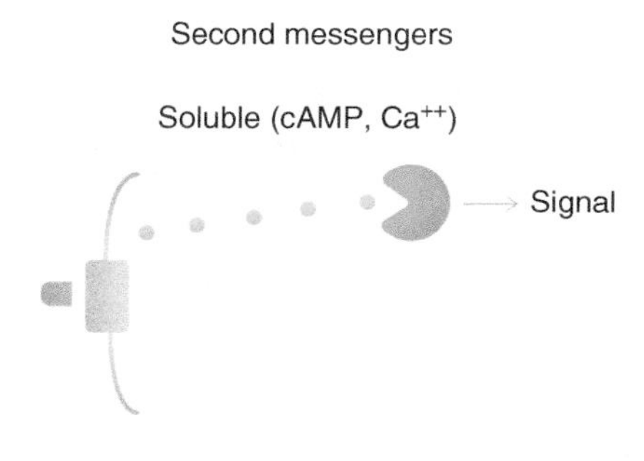

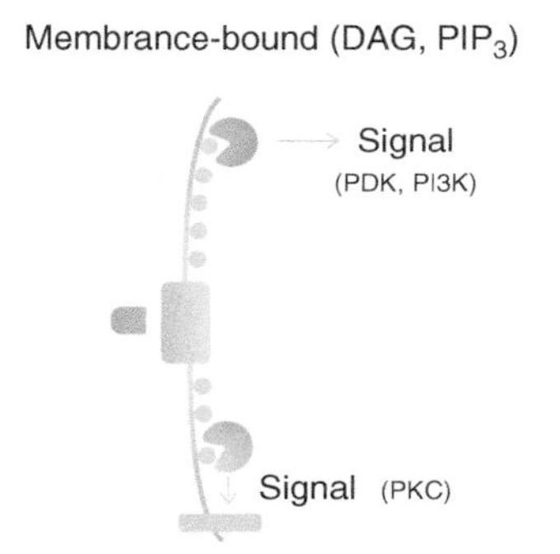

Figure 2.6 Second messengers. Receptor-activated enzymes in the cellular membrane can produce a wide array of small molecule signals that mediate their effects. The classic second messenger cAMP is soluble and can diffuse to cytoplasmic targets such as protein kinase A (PKA) and some non-kinase targets such as Epac and the cyclic nucleotide-gated channels (HCN or I_h). Other second messengers such as DAG and PIP3 are very hydrophobic and remain membrane bound. They either activate other membrane proteins or recruit cytosolic proteins like PKC or PI-3-K to the membrane. The output signals can either be retained locally on the membrane or may diffuse into the cytoplasm as well.

The intracellular concentration, duration, and distribution of cAMP also depends on the activity of surrounding PDEs, which degrade the nucleotide [40, 41]. The concentration of cAMP does not equilibrate throughout the cell, but rather pools of increased cAMP are found throughout the cell as a consequence of compartmentation [42]. Three major targets of cAMP signaling are the PKA, nucleotide-gated ion channels (HCN/I_h/I_f channels), and the exchange protein directly activated by cAMP (Epac or cAMP-GEF) [43], and the signaling is involved in pathways related to a wide variety of processes, sometimes leading to completely opposite outcomes like cell survival or death.

Another cyclic nucleotide second messenger, cGMP, shows how powerful diversification and reuse of related protein structural motifs is in providing the richness of signaling pathways. cGMP is produced by guanylyl cyclase (GC) enzymes that are evolutionarily related to the adenylyl cyclases. Also, cGMP binds to a cGMP-dependent kinase that is in the same family as PKA. Furthermore, the GC enzymes are modulated not by a G-protein but are included in the structure of a family of single-TM, dimeric, membrane-bound receptors for atrial natriuretic peptide and other ligands. Also, there is a soluble GC activated by the first messenger gas, nitric oxide. The versatile structural modules of cyclase enzyme and cyclic nucleotide binding kinase provide a Lego®-like set of mix-and-match signaling components that provides a diverse repertoire of signaling pathways. Finally, noncyclic nucleotide species such as the penta- or tetra-phosphorylated guanosine molecules, as well as to cyclic dinucleotides c-di-AMP and c-di-GMP, are also important in signaling. These were originally identified in bacteria, but the cyclic dinucleotides are now recognized to play important roles in immune modulation in mammals [44–47].

2.3.1.4.2 Lipid Second Messengers

Lipids have long been seen as building blocks for membranes [48] but are also now recognized as valuable players in signaling networks and are involved in critical aspects of cell proliferation, differentiation, and migration. Examples include inositol phospholipids, which are easily phosphorylated and dephosphorylated and are found to influence signaling mechanisms through specific interactions with their head groups [49]. These phosphoinositides are substrates for release of IP_3 that mediates calcium signaling. Depending on the number and location of phosphates on the inositol ring, they can recruit proteins to cell membranes and facilitate molecular assembly of signaling complexes (see Chapter 3, Figure 3.2). From its precursor, phosphatidylinositol, seven phosphoinositide species are synthesized through phosphorylation and dephosphorylation at positions 3, 4, and 5 on the inositol ring [49]. Phospholipases and phosphatases can hydrolyze the lipid. The former may cause signal propagation and amplification through the resulting metabolites (DAG and IP_3); the latter results in signal termination.

A number of effector enzymes for membrane-bound receptors can generate lipid second messengers. Three examples of these lipid messengers include DAG, phosphatidylinositol trisphosphate (PIP_3), and ceramide. The primary cellular source of DAG is a relatively abundant lipid, phosphatidylinositol bisphosphate (PIP_2) that is cleaved by the enzyme phospholipase C (PLC). This produces both DAG and a soluble second messenger IP_3 that links to Ca^{2+} signaling. The PLC-β isoform is activated by a G-protein ($G_{q/11}$ family), while PLC-γ is activated by TK receptors such as the PDGF receptor. Both lead to activation of the protein kinase C (PKC) that is recruited to surface or intracellular membranes upon binding to membrane-associated DAG [50].

Two lipid messengers (PIP_3 and ceramide) play opposite roles in cell survival. PIP_3 generation by phosphorylation of PIP_2 by the lipid kinase enzyme PI-3 kinase leads to recruitment and activation of the antiapoptotic protein kinase Akt. As might be expected, PI-3 kinase mutations that lead to gain of function are relatively common in human cancers (5–30%). Consequently, significant efforts are being made to target PI-3 kinase isoforms in cancer. Idelalisib has been approved for leukemia and lymphoma. Also, the lipid phosphatase and tensin homolog (PTEN), which hydrolyses PIP_3 back to PIP_2, does the opposite of PI-3 kinase. It is a well-documented tumor suppressor and is mutated in many cancers, especially cancers of the uterus and prostate. As the yin to the yang of PIP_3, another lipid messenger, ceramide, reduces cancer cell growth by a number of mechanisms. It inhibits cell proliferation and is proapoptotic. Both the synthesis and downstream signaling mechanisms of ceramide are complex and still being worked out [51], but it binds to protein kinases and proteases and can activate a family of phosphatases (ceramide-activated protein phosphatases [CAPPs]).

2.3.1.4.3 *Calcium Signaling*

Among the ubiquitous signaling mechanisms, Ca^{2+} plays critical roles in contractile and neuronal processes including exocytosis, for example [52]. Irregularities in signaling associated with Ca^{2+} have been related to multiple diseases such as muscle injury [53], pain [54], and migraines [55]. It is also involved in developmental mechanisms, such as cell differentiation, cell death, and cell proliferation [56]. A classic second messenger role for Ca^{2+} involves GPCRs. When phosphatidylinositol 4,5 bisphosphate (PIP_2) is cleaved by the G-protein-activated PLC-β into 1,4,5-inositol trisphosphate (IP_3) and DAG, the IP_3 binds to a receptor at the endoplasmic reticulum that opens a calcium channel, and the concentration of Ca^{2+} in the cytosol increases. Calcium can then bind to PKC and initiate movement toward the membrane once PKC has been activated by DAG. Additionally, through the higher concentration of calcium ions, calcium-sensitive DAG kinases can convert DAG into phosphatidic acid, or the DAG lipase yields arachidonic acid [56].

2.3.1.5 Amplification

One other implication of the two-state switches and signal transduction cascades described above is the profound amplification that can occur in signaling. There are many examples of this phenomenon in signal transduction, and enzymatic generation of second messengers plays a key role in this phenomenon. In the vertebrate visual system, the rhodopsin molecule in rod outer segments is activated by the light-induced isomerization of cis-retinal within the membrane-spanning region of rhodopsin. The physiological response is exquisitely sensitive in that a single photon of light can cause a detectable electrophysiological event in the rod cells that underlie night vision. The light-initiated active rhodopsin conformation (R*) leads to activation of the heterotrimeric G-protein transducin. As shown in Figure 2.4a, the activated transducin then binds to an "effector" protein, PDE, to stimulate its activity. PDE is an enzyme that degrades cGMP. This turns off a Ca^{2+} current in the rod outer segment membrane initiating neural signaling. This signaling cascade provides tremendous signal sensitivity by amplification through the sequential signaling steps ([57] and Table 2.1). Each amplification step results in a gain of 10–100-fold or more, so sequential steps in signal transduction provide exquisite sensitivity.

Table 2.1 Signal amplification in the visual system.

Step	Mammal	Frog
Transducins/R*/s	400	150
Transducins/R*	16	60
cGMP degraded /active GC/s	630	630
cGMP degraded/R*	2000	75000
Reduced Ca^{++} influx (Ca^{2+} ions)/R*	6×10^8	2×10^{11}

Rhodopsin can activate 100–500 transducins per second, but the lifetime of a single activated rhodopsin molecule (R*) is short, 40 ms in mammals and 400 ms in frogs. Even so, each R* can activate 16–60 transducin molecules. They, in turn activate the highly efficient guanylyl cyclase (GC) enzyme, which rapidly degrades cGMP. Binding of 2–3 cGMP to the cyclic nucleotide gated ion channel (CNGA1) causes the influx of over 3 million Ca^{2+} ions per second. Thus, a single photon of light can reduce the number of Ca^{2+} ions traversing into the rod outer segment by a billion or more.

This large amplification is essential since signaling receptors are generally present in vanishingly small numbers (femtomolar concentrations), yet they remain capable of influencing biochemical processes that often involve micromolar to millimolar concentrations of substrates. Another familiar example of this is the adrenaline-mediated release of glucose in the liver in which amplification happens first via a receptor-mediated G-protein (G_s) that impacts the effector enzyme (AC), leading to kinase cascades (PKA and phosphorylase kinase) that ultimately drive phosphorylase enzyme to degrade glycogen, resulting in sufficient glucose release to provide energy for the body in times of stress.

2.3.1.6 Turn-Off Mechanisms

It has become clear that mechanisms that turn off signal transduction are very important and are often tightly regulated. These systems include the phosphatases that remove phosphates from proteins that have been phosphorylated by kinases [58] and enzymes that break down second messengers such as the cAMP and cGMP PDEs [59]. For activated G-proteins (either small or large), the GTPase-activating proteins (GAPs) are critical in the timely deactivation of the active G-proteins. These include the Ras GAPs [36] and, for the heterotrimeric G-proteins, the RGS proteins [37, 38]. Many of these turn-off mechanisms have been considered viable drug targets [36, 38, 58, 59].

2.3.1.7 Localization

The early concepts of second messengers and signaling considered the cell to be a homogenous and well-mixed environment separated largely into cytoplasm, nucleus, and organelles. We now know that signaling is much more complex and that there is a tremendous amount of localized control. This not only includes scaffold proteins holding receptor and effector complexes together [60, 61] but also allows for localized production or degradation of signaling molecules that generates signaling compartments, even within the cytosol. There have been many advances in this area with proteins such as A kinase anchoring proteins (AKAPs) serving to control cAMP signaling in remarkable ways [41]. A variety of other molecular mechanisms also lead to localized signaling [62–65].

2.3.1.8 Biased Signaling/Functional Selectivity

As seen above, GPCRs are involved in visual and olfactory recognition, which are only a fraction of their versatile functions. Taste also relies on GPCRs, and myriads of cellular signaling cascades are initiated at these receptors. Since these receptors are involved in nearly every physiological function, they are a major target for pharmaceutical compounds [66]. Generally, the proteins are imbedded in the plasma membrane due to seven hydrophobic α-helices and can bind agonists, antagonists, and allosteric modulators at the extracellular face. In the case of a binding event, the receptor becomes activated in the form of a conformational change that enables it to interact with a molecule of the heterotrimeric G-proteins, which consists of three different subunits α, β, and γ, as the name suggests, and has a molecule of GDP bound to the α unit. When the activated GPCR binds to the G-protein, GDP dissociates and is replaced by GTP. This exchange of GDP with GTP is the reason that GPCRs are also ascribed to have a GEF activity [67] and causes the G-protein to dissociate from the receptor either in one piece with weakened association of the α- and the βγ-subunit or as two molecules. The subunits then move along the plasma membrane and can trigger release of secondary messengers such as cAMP at adenylyl cyclase or DAG and inositol trisphosphate (IP_3) at PLC.

GPCRs in an active state, when a ligand is bound, can also be phosphorylated by G-protein receptor kinases (GRKs) and then interact with β-arrestins. This pathway was long thought to desensitize the receptor and cease signaling, since the receptor-β-arrestin complex is internalized via endocytosis in clathrin-coated pits and then either degraded or recycled. However, recent research suggests that β-arrestins initiate G-protein-independent signaling cascades, thereby broadening the importance and involvement of GPCRs in cellular activity, such as the activation of ERKs or all three MAPK cascades [66, 68].

Binding of ligands to GPCR does not create the same active GPCR state for all ligands, such that depending on the ligand or even a complex of ligands, the active conformation of the receptor varies and so does the recruitment of different G-proteins and downstream signaling. The functional selectivity called biased agonism is a rapidly growing field of study, which did not just reveal extracellular biased ligand binding but also that binding of specific complexes in the cytoplasm can modify the receptors affinity for specific binding.

2.4 Receptor Tyrosine Kinases

Another major family of receptors is the RTKs that catalyze the intrinsic transfer of a phosphate from ATP onto a tyrosine residue in targeted proteins. These receptors are structurally similar to GPCR in that they have an extracellular face to bind ligands outside the cell and an intracellular domain. However, these two domains are only connected by one membrane-spanning helix instead of seven (with GPCRs). The intracellular cytoplasmic part of the peptide is where the protein TK core sequence is found as well as other sequences where the receptors activity can be regulated through phosphorylation via other protein kinases or autophosphorylation [69].

RTKs often reside in the cellular membrane as monomeric receptors that dimerize upon ligand binding. The monomers then phosphorylate their own or their dimer partner's cytoplasmic domain. This can either activate the RTK's kinase activity or can serve to recruit intracellular signaling molecules (Figure 2.7). The RTKs can also phosphorylate downstream effector proteins (see Chapters 3 and 4). The patterns of dimerization and phosphorylation can

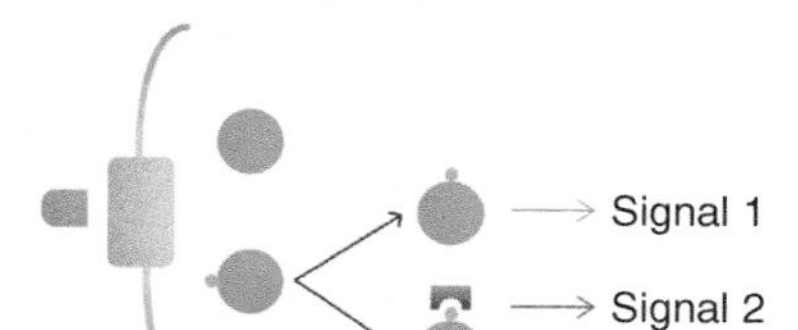

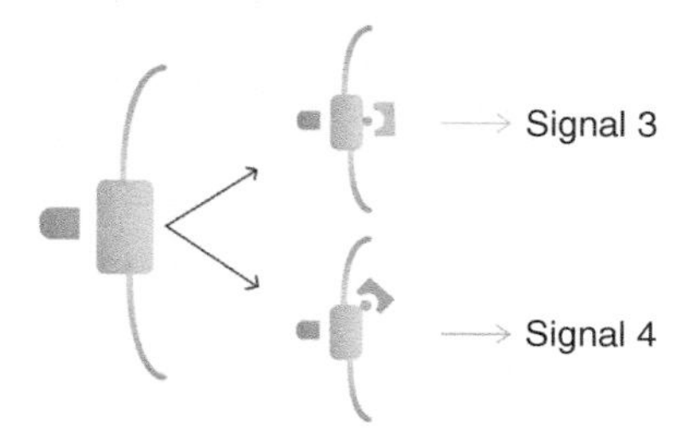

Figure 2.7 Role of posttranslational modifications in signaling. Many signaling processes are mediated by posttranslational modifications of protein, DNA, or lipids. There are two general mechanisms by which this can occur. The modification can change the activity of the protein being modified (effector activation). A good example of this would be the increase in activity of many kinase enzymes upon phosphorylation in their "activation loop." Alternatively, receptors that are autophosphorylated upon ligand binding often function by recruiting the next signaling mediator to the receptor and membrane (effector recruitment). Depending on which amino acid in the receptor is phosphorylated, different effectors may be recruited. The PDGF receptor is an excellent example of this phenomenon.

vary with the kind of ligand bound, also introducing biased agonism with respect to the initiation of signaling cascades. Like GPCRs, activated RTKs exchange of GTP for GDP on G-proteins, as in the case of G-protein Ras, the Ras family is made up of monomeric small GTPases that are structurally similar to the α-subunit of G-proteins associated with the signaling of GPCRs (known as large GTPases) and are often used to initiate signaling cascades involved in cellular processes to control growth, differentiation, and cell death through apoptosis [70]. For example, once Ras is active in the GTP-bound state, interaction with Raf activates Raf, and in turn the activated Raf stimulates MAPK pathways. These not only affect cytoplasmic and membrane-bound substrates, but also translocate into the nucleus where transcription factors can be activated to increase gene transcription. RTKs are also associated with the activation of phosphoinositol as second messengers, as well as DAG and calcium through phospholipid kinases and PDEs [69].

2.5 Steroid Receptors

Steroid hormones take an alternative form of signaling, as the classical steroid receptors are located in the cytosol and closely associated with the nuclear membrane and, upon hormone binding, these receptors undergo a conformational change that allows interaction with DNA to influence transcription. A central cysteine-rich domain in the receptor protein is assumed to be responsible for the interplay of DNA and activated receptor. Second messenger production has been linked to steroid receptors as well, connecting steroid hormones to other non-genomic cellular signaling cascades [71, 72].

2.6 Reactive Oxygen Species (ROS)

Reactive oxygen species (ROS) have long been thought to merely be a by-product of the electron transport chain, generated in the endoplasmic reticulum by nicotinamide adenine dinucleotide phosphate (NADPH) cytochrome p450 reductase [73, 74] or the result of UV irradiation. Now one of these, H_2O_2, fits the criteria of effector specificity in time and space to qualify as a second messenger [75]: it can be generated, among other ways, enzymatically through the action of superoxide dismutase and can be easily degraded. In contrast to hydroxyl radicals, which react indiscriminately, hydrogen peroxide can oxidize cysteine residues to sulfenic acid [73], and its actions have the potential to act as a redox switch [74, 75]. The role of ROS in signaling and pathophysiology will be explored in more detail in Chapters 3 and 4.

2.7 Summary

Tremendous advances have been made to enhance our understanding of receptors and signal transduction over the last 50 years. We have moved from receptors being simply a philosophical construct to having real substance and detailed structures available. These advances have driven the field to the exciting and complex description we have today. The remaining chapters will expand on these general concepts and show how they are applied in complex regulatory and disease mechanisms. Also, some of the experimental approaches used to study these systems will be outlined to give readers a handle on how to dissect these complex mechanisms.

References

1 Duan, G. and Walther, D. (2015). The roles of post-translational modifications in the context of protein interaction networks. *PLoS Comput. Biol.* 11 (2): e1004049.
2 Ardito, F., Giuliani, M., Perrone, D. et al. (2017). The crucial role of protein phosphorylation in cell signaling and its use as a targeted therapy (review). *Int. J. Mol. Med.* 40 (2): 271–280.
3 Cross, T.G., Scheel-Toellner, D., Henriquez, N.V. et al. (2000). Serine/threonine protein kinases and apoptosis. *Exp. Cell Res.* 256 (1): 34–41.
4 Paul, M.K. and Mukhopadhyay, A.K. (2004). Tyrosine kinase – role and significance in cancer. *Int. J. Med. Sci.* 1 (2): 101–115.
5 Barford, D. (1996). Molecular mechanisms of the protein serine/threonine phosphatases. *Trends Biochem. Sci.* 21 (11): 407–412.
6 Sasco, F., Perfetto, L., Castagnoli, L., and Cesareni, G. (2012). The human phosphatase interactome: an intricate family portrait. *FEBS Lett.* 586 (17): 2732–2739.
7 Lin, H., Su, X., and He, B. (2012). Protein lysine acylation and cysteine succination by intermediates of energy metabolism. *ACS Chem. Biol.* 7 (6): 947–960.
8 Lee, S. (2013). Post-translational modification of proteins in toxicological research: focus on lysine acylation. *Toxicol. Res.* 29 (2): 81–86.
9 Wellen, K.E., Hatzivassiliou, G., Sachdeva, U.M. et al. (2009). ATP-citrate lyase links cellular metabolism to histone acetylation. *Science* 324 (5930): 1076–1080.
10 Kim, S.C., Sprung, R., Chen, Y. et al. (2006). Substrate and functional diversity of lysine acetylation revealed by a proteomics survey. *Mol. Cell.* 23 (4): 607–618.
11 Bannister, A.J. and Kouzarides, T. (2011). Regulation of chromatin by histone modifications. *Cell Res.* 21 (3): 381–395.
12 Eberharter, A. and Becker, P.B. (2002). Histone acetylation: a switch between repressive and permissive chromatin: second in review series on chromatin dynamics. *EMBO Rep.* 3 (3): 224–229.

13 Müller, M.M. (2018). Post-translational modifications of protein backbones: unique functions, mechanisms, and challenges. *Biochemistry* 57 (2): 177–185.
14 Walsh, C.T., Garneau-Tsodikova, S., and Gatto, G.J. Jr. (2005). Protein posttranslational modifications: the chemistry of proteome diversifications. *Angew. Chem. Int. Ed.* 44 (45): 7342–7372.
15 Chiu, V.K., Silletti, J., Dinsell, V. et al. (2004). Carboxyl methylation of Ras regulates membrane targeting and effector engagement. *J. Biol. Chem.* 279 (8): 7346–7352.
16 Cox, A.D., Der, C.J., and Philips, M.R. (2015). Targeting RAS membrane association: Back to the future for anti-RAS drug discovery? *Clin. Canc. Res.* 21 (8): 1819–1827.
17 Theillet, F., Smet-Nocca, C., Liokatis, S. et al. (2012). Cell signaling, post-translational protein modifications and NMR spectroscopy. *J. Biomol. NMR* 54 (3): 217–236.
18 Arey, B.J. (2011). The role of glycosylation in receptor signaling. *Intech Open*. http://cdn.intechopen.com/pdfs-wm/39461.pdf (accessed 20 September 2018).
19 Opdenakker, G., Rudd, P.M., Wormald, M. et al. (1995). Cells regulate the activities of cytokines by glycosylation. *FASEB J.* 9 (5): 453–457.
20 Waetzig, G.H., Chalaris, A., Rosenstiel, P. et al. (2010). N-linked glycosylation is essential for the stability but not the signaling function of interleukin-6 signal transducer glycoprotein 130. *J. Biol. Chem.* 285 (3): 1781–1789.
21 Markiv, A., Rambaruth, N.D., and Dwek, M.V. (2012). Beyond the genome and proteome: targeting protein modifications in cancer. *Curr. Opin. Pharmacol.* 12 (4): 408–413.
22 Resh, M.D. (2013). Covalent lipid modifications of proteins. *Curr. Biol.* 23 (10): R431–R435.
23 Xu, G. and Jaffrey, S.R. (2012). The new landscape of protein ubiquitination. *Nat. Biotechnol.* 29 (12): 1098–1100.
24 Maehle, A.H., Prull, C.R., and Halliwell, R.F. (2002). The emergence of the drug receptor theory. *Nat. Rev. Drug Discov.* 1 (8): 637–641.
25 Neubig, R.R., Spedding, M., Kenakin, T. et al. (2003). International Union of Pharmacology Committee on receptor nomenclature and drug classification. XXXVIII. Update on terms and symbols in quantitative pharmacology. *Pharmacol. Rev.* 55 (4): 597–606.
26 Pasquale, E.B. (2010). Eph receptors and ephrins in cancer: bidirectional signalling and beyond. *Nat. Rev. Cancer* 10 (3): 165–180.
27 Qin, J., Vinogradova, O., and Plow, E.F. (2004). Integrin bidirectional signaling: a molecular view. *PLoS Biol.* 2 (6): e169.
28 Coughlin, S.R. (1994). Expanding horizons for receptors coupled to G proteins: diversity and disease. *Curr. Opin. Cell Biol.* 6 (2): 191–197.
29 Gilman, A.G. (1987). G proteins: transducers of receptor-generated signals. *Annu. Rev. Biochem.* 56: 615–649.

30 Rosenbaum, D.M., Rasmussen, S.G., and Kobilka, B.K. (2009). The structure and function of G-protein-coupled receptors. *Nature* 459 (7245): 356–363.

31 Wettschureck, N. and Offermanns, S. (2005). Mammalian G proteins and their cell type specific functions. *Physiol. Rev.* 85 (4): 1159–1204.

32 Gschwind, A., Fischer, O.M., and Ullrich, A. (2004). The discovery of receptor tyrosine kinases: targets for cancer therapy. *Nat. Rev. Cancer* 4 (5): 361–370.

33 Yarden, Y. (2001). Biology of HER2 and its importance in breast cancer. *Oncology* 61 (Suppl 2): 1–13.

34 Pawson, T. (2004). Specificity in signal transduction: from phosphotyrosine-SH2 domain interactions to complex cellular systems. *Cell* 116 (2): 191–203.

35 Pawson, T. and Gish, G.D. (1992). SH2 and SH3 domains: from structure to function. *Cell* 71 (3): 359–362.

36 Zhou, B., Der, C.J., and Cox, A.D. (2016). The role of wild type RAS isoforms in cancer. *Semin. Cell Dev. Biol.* 58: 60–69.

37 Hollinger, S. and Hepler, J.R. (2002). Cellular regulation of RGS proteins: modulators and integrators of G protein signaling. *Pharmacol. Rev.* 54 (3): 527–559.

38 Neubig, R.R. and Siderovski, D.P. (2002). Regulators of G-protein signalling as new central nervous system drug targets. *Nat. Rev. Drug Discov.* 1 (3): 187–197.

39 Rodbell, M. (1995). Signal transduction: evolution of an idea. *Environ. Health Perspect.* 103 (4): 338–345.

40 Tasken, K. and Aandahl, E.M. (2004). Localized effects of cAMP mediated by distinct routes of protein kinase a. *Physiol. Rev.* 84 (1): 137–167.

41 Wong, W. and Scott, J.D. (2004). AKAP signalling complexes: focal points in space and time. *Nat. Rev. Mol. Cell. Biol.* 5 (12): 959–970.

42 Wright, P.T., Schobesberger, S., and Gorelik, J. (2015). Studying GPCR/cAMP pharmacology from the perspective of cellular structure. *Front. Pharmacol.* 6: 148.

43 Gloerich, M. and Bos, J.L. (2010). Epac: defining a new mechanism for cAMP action. *Annu. Rev. Pharmacol. Toxicol.* 50: 355–375.

44 Corrigan, R.M. and Grundling, A. (2013). Cyclic di-AMP: another second messenger enters the fray. *Nat. Rev. Microbiol.* 11 (8): 513–524.

45 Gries, C.M., Bruger, E.L., Moormeier, D.E. et al. (2016). Cyclic di-AMP released from *Staphylococcus aureus* biofilm induces a macrophage type I interferon response. *Infect. Immun.* 84 (12): 3564–3574.

46 Waters, C.M., Lu, W., Rabinowitz, J.D., and Bassler, B.L. (2008). Quorum sensing controls biofilm formation in Vibrio cholerae through modulation of cyclic di-GMP levels and repression of vpsT. *J. Bacteriol.* 190 (7): 2527–2536.

47 Barker, J.R., Koestler, B.J., Carpenter, V.K. et al. (2013). STING-dependent recognition of cyclic di-AMP mediates type I interferon responses during *Chlamydia trachomatis* infection. *MBio* 4 (3): e00018–e00013.

48 Liscovitch, M. and Cantley, L.C. (1994). Lipid second messengers. *Cell* 77 (3): 329–334.

49 Di Paolo, G. and De Camilli, P. (2006). Phosphoinositides in cell regulation and membrane dynamics. *Nature* 443 (7112): 651–657.

50 Gallegos, L.L., Kunkel, M.T., and Newton, A.C. (2006). Targeting protein kinase C activity reporter to discrete intracellular regions reveals spatiotemporal differences in agonist-dependent signaling. *J. Biol. Chem.* 281 (41): 30947–30956.

51 Galadari, S., Rahman, A., Pallichankandy, S., and Thayyullathil, F. (2015). Tumor suppressive functions of ceramide: evidence and mechanisms. *Apoptosis* 20 (5): 689–711.

52 Berridge, M.J. (2004). Calcium signal transduction and cellular control mechanisms. *Biochim. Biophys. Acta.* 1742 (1–3): 3–7.

53 Cheung, J.Y., Bonventre, J.V., Malis, C.D., and Leaf, A. (1986). Calcium and ischemic injury. *N. Engl. J. Med.* 314 (26): 1670–1676.

54 Snutch, T.P. and Zamponi, G.W. (2018). Recent advances in the development of T-type calcium channel blockers for pain intervention. *Br. J. Pharmacol.* 175 (12): 2375–2383.

55 Pietrobon, D. (2013). Calcium channels and migraine. *Biochim. Biophys. Acta.* 1828 (7): 1655–1665.

56 Clapham, D.E. (2007). Calcium signaling. *Cell* 131 (6): 1047–1058.

57 Arshavsky, V.Y. and Burns, M.E. Current understanding of signal amplification in phototransduction. *Cell. Logist.* 4: e29390.

58 Frankson, R., Yu, Z.H., Bai, Y. et al. (2017). Therapeutic targeting of oncogenic tyrosine phosphatases. *Cancer Res.* 77 (21): 5701–5705.

59 Bender, A.T. and Beavo, J.A. (2006). Cyclic nucleotide phosphodiesterases: molecular regulation to clinical use. *Pharmacol. Rev.* 58 (3): 488–520.

60 Good, M.C., Zalatan, J.G., and Lim, W.A. (2011). Scaffold proteins: hubs for controlling the flow of cellular information. *Science* 332 (6030): 680–686.

61 Magalhaes, A.C., Dunn, H., and Ferguson, S.S. (2012). Regulation of GPCR activity, trafficking and localization by GPCR-interacting proteins. *Br. J. Pharmacol.* 165 (6): 1717–1736.

62 Neubig, R.R. (1994). Membrane organization in G-protein mechanisms. *FASEB J.* 8 (12): 939–946.

63 Dunn, H.A. and Ferguson, S.S. (2015). PDZ protein regulation of G protein-coupled receptor trafficking and signaling pathways. *Mol. Pharmacol.* 88 (4): 624–639.

64 Fritschy, J.M., Panzanelli, P., and Tyagarajan, S.K. (2012). Molecular and functional heterogeneity of GABAergic synapses. *Cell. Mol. Life Sci.* 69 (15): 2485–2499.

65 Sarrouilhe, D., di Tommaso, A., Metaye, T., and Ladeveze, V. (2006). Spinophilin: from partners to functions. *Biochimie.* 88 (9): 1099–1113.

66 Heitzler, D., Durand, G., Gallay, N. et al. (2012). Competing G protein-coupled receptor kinases balance G protein and beta-arrestin signaling. *Mol. Syst. Biol.* 8: 590.

67 Vischer, H.F., Castro, M., and Pin, J.P. (2015). G protein-coupled receptor Multimers: a question still open despite the use of novel approaches. *Mol. Pharmacol.* 88 (3): 561–571.
68 DeFea, K.A., Vaughn, Z.D., O'Bryan, E.M. et al. (2000). The proliferative and antiapoptotic effects of substance P are facilitated by formation of a beta-arrestin-dependent scaffolding complex. *Proc. Natl. Acad. Sci. U.S.A.* 97 (20): 11086–11091.
69 Schlessinger, J. (2000). Cell signaling by receptor tyrosine kinases. *Cell* 103 (2): 211–225.
70 Rocks, O., Peyker, A., and Bastiaens, P.I. (2006). Spatio-temporal segregation of Ras signals: one ship, three anchors, many harbors. *Curr. Opin. Cell. Biol.* 18 (4): 351–357.
71 Wehling, M. (1994). Nongenomic actions of steroid hormones. *Trends Endocrinol. Metab.* 5 (8): 347–353.
72 Barton, M., Filardo, E.J., Lolait, S.J. et al. (2018). Twenty years of the G protein-coupled estrogen receptor GPER: historical and personal perspectives. *J. Steroid Biochem. Mol. Biol.* 176: 4–15.
73 Reth, M. (2002). Hydrogen peroxide as second messenger in lymphocyte activation. *Nat. Immunol.* 3 (12): 1129–1134.
74 Sauer, H., Wartenberg, M., and Hescheler, J. (2001). Reactive oxygen species as intracellular messengers during cell growth and differentiation. *Cell Physiol. Biochem.* 11 (4): 173–186.
75 Forman, H.J., Maiorino, M., and Ursini, F. (2010). Signaling functions of reactive oxygen species. *Biochemistry* 49 (5): 835–842.

3

From Cellular Mechanisms to Physiological Responses

Functional Signal Integration Across Multiple Biological Levels

Robert H. Newman

Department of Biology, North Carolina A&T State University, Greensboro, NC, USA

3.1 Introduction

The environmental and nutritional landscapes encountered by organisms are in constant flux. To survive in such an environment, organisms must continuously sense changes in their surroundings and mount a coordinated functional response that spans several levels of biological organization, including the cellular level, the tissue level, and the organismal level. Failure to properly integrate functional responses at any of these levels – due to mutation and/or disruption of endogenous signaling processes by pharmacological/toxicological agents – underlies pathological disorders such as diabetes, cancer, and cardiovascular disease. In the previous chapter, we examined the molecular properties of key signaling molecules, including receptors, signaling enzymes, and small molecule second messengers. We then discussed some of the ways in which select signaling molecules can be organized into discrete signaling modules inside the cell. In this chapter, we will build on this discussion by exploring specific mechanisms by which cellular signaling modules – and, ultimately, the larger signaling networks of which they are a part – convert environmental cues into integrated functional responses. While our discussion will focus primarily on signal integration at the cellular level, we will also touch on some of the ways in which signaling processes can be further coordinated at both the tissue and the organismal levels. Our discussion will be centered on three fundamental questions: (i) how do individual cells sense and respond to changes in their environment (i.e. cellular information processing); (ii) how are individual signaling modules and, by extension, cellular signaling networks regulated in cellular time and space; and (iii) why do different cell types respond differently to the same signal? To help illustrate these concepts, we

Cellular Signal Transduction in Toxicology and Pharmacology: Data Collection, Analysis, and Interpretation, First Edition. Edited by Jonathan W. Boyd and Richard R. Neubig.

will examine the mechanisms by which various types of cells respond to elevations in blood glucose and how these responses are integrated across multiple tissues and organs to elicit a coordinated, systemic response. In this context, we will also discuss how dysregulation of these mechanisms contributes to the etiology of type 2 diabetes mellitus (T2D) and its complications.

3.2 Cellular Information Flow: Mechanisms of Cellular Signal Integration and Regulation

As discussed in Chapter 2, cells contain a diverse set of signaling molecules that collectively regulate nearly all aspects of cellular physiology. Therefore, to better understand how cells sense and respond to different environmental cues, we must first understand how the activities of these signaling molecules are modulated under various cellular conditions. Moreover, we must understand how changes in these activities lead to a specific functional response. Traditionally, researchers have taken a reductionist approach to the study of cellular signal transduction. These studies, which often focus on the behavior of a purified signaling enzyme (or a set of signaling enzymes and/or their cofactors) *in vitro*, have offered important insights into the molecular mechanisms underlying the regulation of key players within many cellular signaling pathways. For example, focused biochemical and structural analyses of the cyclic adenosine monophosphate (AMP)-dependent protein kinase, PKA, and other kinase family members have uncovered exquisite details about the molecular architecture of the kinase domain and its dynamic reorganization following activation [1]. Similarly, careful kinetic analyses have helped characterize each step in the activation cycle of small G-proteins and how these steps are regulated by their respective guanine nucleotide exchange factors (GEFs) and GTPase-accelerating proteins (GAPs) [2, 3]. However, despite the great strides that have been made in understanding the biochemical regulation of cellular signaling enzymes, it is clear that biochemical regulation alone cannot explain the diverse set of signaling activities and the exquisite specificity characteristic of cellular signal transduction. This is because, in addition to biochemical regulation, there exists another level of "contextual" control that plays an important role in cellular information flow. This level of control is dependent on many factors, including the relative levels of various signaling enzymes, their subcellular distribution, and their activity profiles which, in turn, are a function of the activity of their endogenous regulatory factors [4]. Therefore, to understand how information is processed within the cell, it is important not only to consider how individual signaling enzymes function in isolation but also to understand how they are regulated in the context of their endogenous signaling pathways. Indeed, in many cases, it is the complex, systems-level behaviors of many signaling molecules acting in concert that lead to a given functional response. To illustrate this notion, we will begin with a relatively simple model describing the linear, unidirectional flow of information from the extracellular environment to the nucleus via a canonical signaling pathway. Specifically, we will examine how changes in extracellular insulin levels are sensed by the insulin receptor (InsR) in hepatocytes and how this information is transmitted through atypical PKC (aPKC) family members and other intracellular signaling molecules to drive changes in nuclear factor kappa B (NF-κB)-dependent gene expression. To better understand how signaling processes are coordinated inside the cell, we will then layer in additional levels of regulation that more accurately represent the cellular context in which these signaling processes actually occur. However, even as the number of interactions grows and the complexity of the system increases, it is important to remember that each additional node still involves a series of simple steps (e.g. protein–protein interactions, posttranslational modification, translocation, transcriptional activation, etc.) that together lead to the emergence of the observed systems-level properties.

3.2.1 The InsR-aPKC-NF-κB Signaling Axis

In the fed state, increases in blood glucose levels stimulate the release of the hormone insulin from pancreatic beta cells into the bloodstream. Circulating insulin is distributed throughout the body to various tissues, including hepatocytes, where it binds dimeric InsR molecules located on the surface of target cells (Figure 3.1). Association of insulin with the extracellular ectodomain of the InsR induces conformational changes in the receptor that lead to activation of its intracellular tyrosine kinase (TK) domain (Figure 3.2, step 1) [5, 6]. Receptor activation promotes autophosphorylation of Tyr residues within the TK domain, which in turn facilitate the binding of adaptor proteins, such as the insulin receptor substrate (IRS) isoforms, IRS-1 and IRS-2, via their phosphotyrosine binding (PTB) domains (step 2). In hepatocytes, both IRS isoforms are expressed at appreciable levels, but, as we will see, each isoform is involved in regulating distinct aspects of the insulin response [7]. Here, we will first focus on IRS-2 since it is the predominant isoform involved in the regulation of signaling by aPKC family members (which include PKCι, PKCλ, and PKCζ in hepatocytes [7, 8].

Once bound to the phosphorylated InsR, IRS-2 is then phosphorylated on conserved tyrosine residues by the receptor's kinase domain, creating binding sites for Src homology domain 2 (SH2)-containing effector proteins, such as phosphoinositide-3-kinase (PI3K) (step 3). PI3K catalyzes the conversion of the membrane phospholipid, phosphoinositide (4,5)-bisphosphate (PIP_2), to phosphoinositide (3,4,5)-trisphosphate (PIP_3), leading to the recruitment of pleckstrin homology (PH) domain-containing proteins, such as phosphoinositide-dependent protein kinase 1 (PDK1) and Akt, to the membrane (step 4).

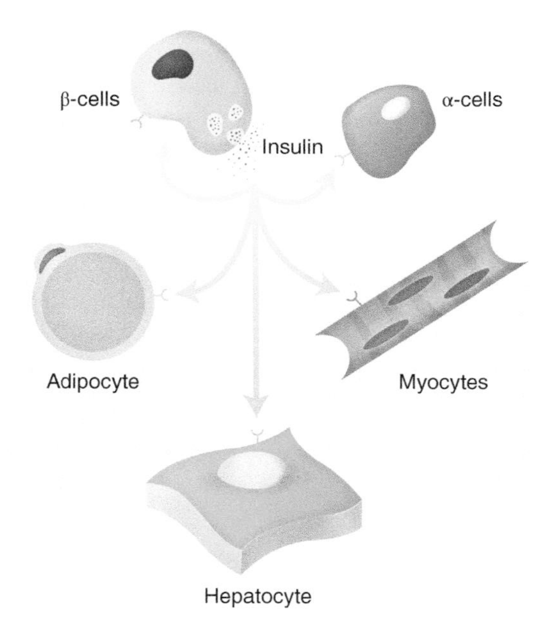

Figure 3.1 **Insulin-responsive cells.** In response to elevated blood glucose levels, insulin is secreted from pancreatic β-cells and distributed to various insulin-sensitive cells, including pancreatic α-cells, skeletal muscle cells (myocytes), liver (hepatocytes), and adipose tissue (adipocytes). Insulin-responsive cells express the insulin receptor on their cell surface.

PDK1, which phosphorylates several members of the PKA-PKG-PKC (AGC) family of serine/threonine kinases on a conserved threonine residue on the activation loop located within the kinase active site, functions as a central regulator of AGC kinase activity in the cell. As we will see later, PDK1 also phosphorylates several other important AGC family members involved in insulin-dependent signaling (including Akt). For simplicity, we will initially focus only on the impact of PDK1-mediated phosphorylation of aPKC as it relates to NF-κB-dependent transcriptional regulation (step 5).

Once activated by PDK1, aPKC catalyzes the phosphorylation of the beta subunit of the inhibitor of NF-κB kinase (IKKβ) on Ser177 and Ser181 (step 6) [9]. IKKβ, which is part of a heterotrimeric complex composed of IKKα, IKKβ, and a regulatory subunit known as NF-κB essential modulator (NEMO), then phosphorylates the alpha isoform of the inhibitor of NF-κB (IκBα) on Ser32 and Ser36 (step 7) [10]. These residues are part of a phosphorylation-dependent destruction motif (i.e. a phosphodegron) that is recognized by a member of the Skip1-Cullin-F-box (SCF) family of ubiquitin E3 ligase complexes. In the case of phosphorylated IκBα, the SCF complex is recruited to the site of ubiquitination via interactions between pSer32 and pSer36 on IκBα and the

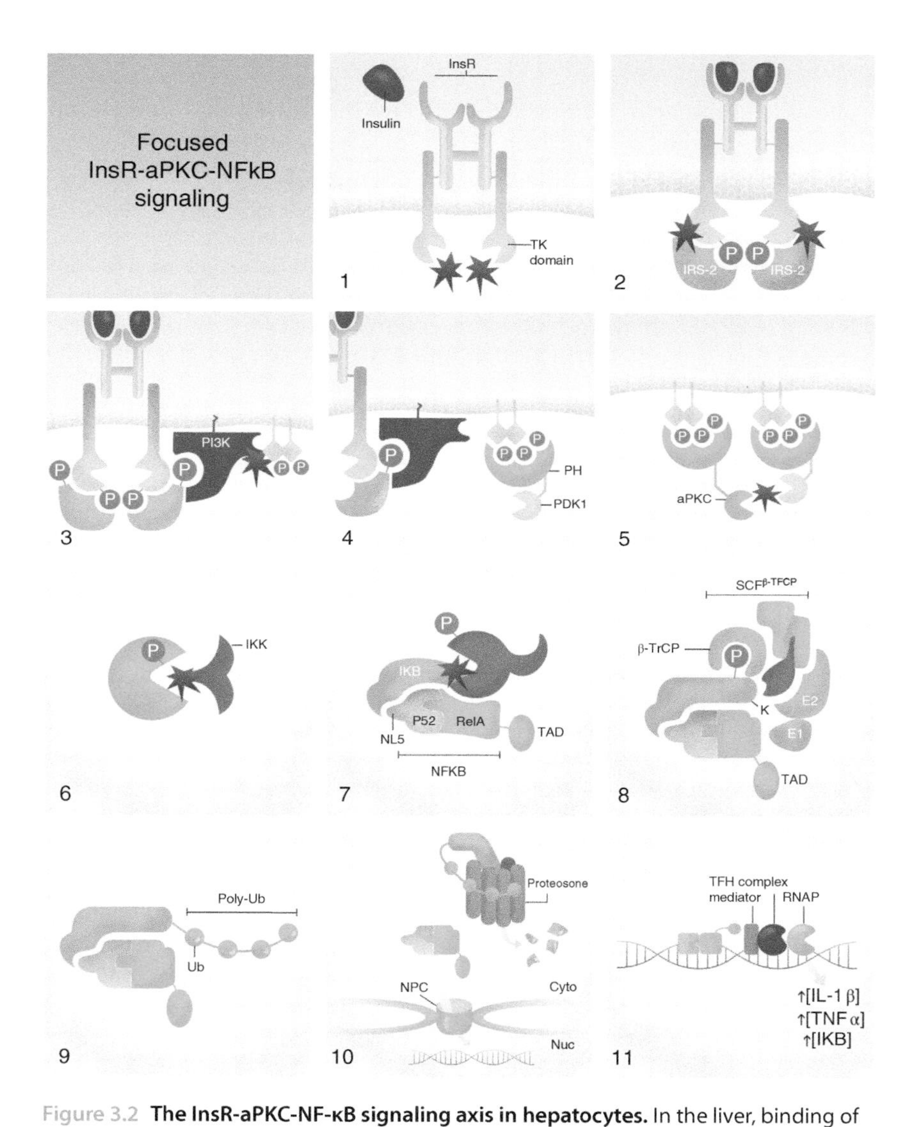

Figure 3.2 **The InsR-aPKC-NF-κB signaling axis in hepatocytes.** In the liver, binding of insulin to dimeric insulin receptors (InsR) on the cell surface initiates several intracellular signaling pathways, including insulin receptor substrate 2 (IRS-2)/atypical protein kinase C (aPKC)-mediated activation of the transcription factor, nuclear factor kappa b (NF-κB), and subsequent expression of NF-κB-dependent genes, such as inhibitor of kappa B (IκB), and the pro-inflammatory cytokines, interleukin 1 beta (IL-1β) and tumor necrosis factor alpha (TNF-α). See text for details. *trans*-phosphorylation (P) reactions are depicted as starbursts. Abbreviations: TK, tyrosine kinase; PI3K, phosphoinositide 3-kinase; PDK1, phosphoinositide-dependent kinase 1; PH, pleckstrin homology domain; IκB, inhibitor of NF-κB; IKK, IκB kinase; NLS, nuclear localization signal; p52, NF-κB p52 subunit; RelA, v-rel avian reticuloendotheliosis viral oncogene homolog A; TAD, transactivation domain; Ub, ubiquitin; $SCF^{\beta\text{-TrCP}}$, Skip-Cullin-F-box Ub E3 ligase with the β-TrCP substrate binding subunit; E1, Ub ligase enzyme 1; E2, Ub ligase enzyme 2; NPC, nuclear pore complex; TFII, general transcription factor II complex; RNAP, RNA polymerase.

WW phosphoamino acid binding domain (PAABD) of the SCF complex's β-TrCP substrate binding subunit (step 8) [11, 12]. The formation of K48-linked polyubiquitin chains by $SCF^{\beta\text{-TrCP}}$ marks IκBα for degradation by the proteasome (steps 9–10). This is important because, prior to its degradation, interactions between IκBα and NF-κB mask NF-κB's nuclear localization sequence (NLS), thereby sequestering NF-κB in the cytoplasm away from its target sites in the nucleus. Once released from IκB, NF-κB translocates into the nucleus where it binds NF-κB response elements and drives expression of target genes, such as the cytokines interleukin 1B (IL-1B) and tumor necrosis factor alpha (TNF-α) (step 11) [13].

3.2.2 Modes of Regulation in InsR-PKC-NF-κB Signaling Axis

The signaling pathway described above illustrates some of the ways in which cellular signaling molecules must work together to convert information about changes in the extracellular environment (i.e. an increase in circulating insulin) into a functional intracellular response (i.e. changes in gene expression). However, such linear models of cellular information flow do not adequately capture many important aspects of cellular signal transduction, including negative or positive feedback regulation. For instance, the model described above does not account for potential differences in the expression levels of different NF-κB-regulated genes. Likewise, it does not describe how attenuation of the signal occurs when insulin levels decrease, let alone how graded or oscillatory responses can be achieved. Therefore, to better understand how these events occur and how they modulate information flow inside the cell, we must examine how various steps in the pathway are regulated. For illustrative purposes, we will begin at the transcriptional level and work our way "back up" the signaling pathway to InsR regulation. However, it is important to remember that, at any given time, many (if not all) of the regulatory steps that we discuss could be occurring to varying extents within the cellular environment to produce a dynamic, integrated response.

3.2.3 Transcriptional Regulation

When considering the extent to which the expression of a particular target gene is regulated by its cognate transcription factor (TF), several parameters must be taken into account. For example, we must consider (i) the rate of induction, (ii) the duration of functional promoter-bound complexes, and (iii) the relative stability of the mRNA products [13, 14]. The first of these, the rate of induction, is a function of several parameters intrinsic to chromosomal structure, including (i) promoter architecture (e.g. the number of response elements and their distribution relative to each other and to the transcriptional start site), (ii) the accessibility of the promoter (e.g. due to the chromatin state), and (iii) the presence or absence of co-regulatory TFs and/or DNA-binding proteins at nearby sites. For example, over 400 NF-κB target genes have been identified in humans [15]. These can be divided into three groups based on their expression profiles, with "early" genes showing maximal expression approximately one hour after NF-κB activation followed by "middle" genes (approximately three hours) and "late" genes (approximately six hours) [14]. Bioinformatics analysis of the promoter regions demonstrated that genes in the early group contained a substantially higher density of NF-κB binding sites than those in the middle and late groups. Both the number of sites and their proximity to one another are greater in the early group (Figure 3.3a) [16]. This suggests that clustering TF binding sites near the transcriptional start site of genes in the early group enhances the recruitment of activated NF-κB to their promoters compared with genes in the late group. These contain fewer NF-κB binding sites arranged in a more dispersed manner [16]. Indeed, clustering of TF binding sites appears to be a general mechanism of regulation employed by eukaryotic TFs, which typically exhibit weaker affinity for their sites than prokaryotic TFs [17].

In addition to promoter architecture, posttranslational modifications, either to the TF itself or to auxiliary proteins involved in transcriptional activation at a particular locus, can also dramatically influence the rate of induction. For example, a subset of NF-κB-regulated genes in the middle and late groups requires that the chromatin structure near their start sites is actively remodeled for transcription to proceed efficiently. To accomplish this, NF-κB recruits the CREB binding protein (CBP)/p300 transcriptional coactivator to these loci. CBP/p300, which exhibits histone acetyltransferase (HAT) activity, then acetylates lysine residues on nearby histone tails, promoting their transition from a transcriptionally silent heterochromatic state to a transcriptionally competent euchromatic state. Interactions between NF-κB and CBP/p300 are dependent on phosphorylation of NF-κB on Ser276 of the p65 (RelA) subunit [18, 19]. Interestingly, phosphorylation at this site is believed to be mediated by IκB-NF-κB-associated PKA catalytic subunits (PKAc) that are activated in a cyclic AMP (cAMP)-independent manner [19]. Initially, PKAc bound to the IκB-NF-κB complex in the cytoplasm is held in an inactive state by the complex. Upon IκBα degradation, PKAc is released from the complex, relieving the inhibition and allowing it to phosphorylate the RelA subunit of the NF-κB heterodimer [19]. In this way, the signal triggering the movement of NF-κB into the nucleus (i.e. IκB degradation) is linked to the signal to recruit the CBP/p300 coactivator important for its function (i.e. PKAc-mediated phosphorylation of RelA[S276]).

Posttranslational modification of RelA can also differentially affect the duration of NF-κB-mediated transcription once it has begun. For instance, in addition to the IL-1β and TNF-α cytokines mentioned above, NF-κB also strongly upregulates the expression of its negative regulator, IκBα. Once synthesized,

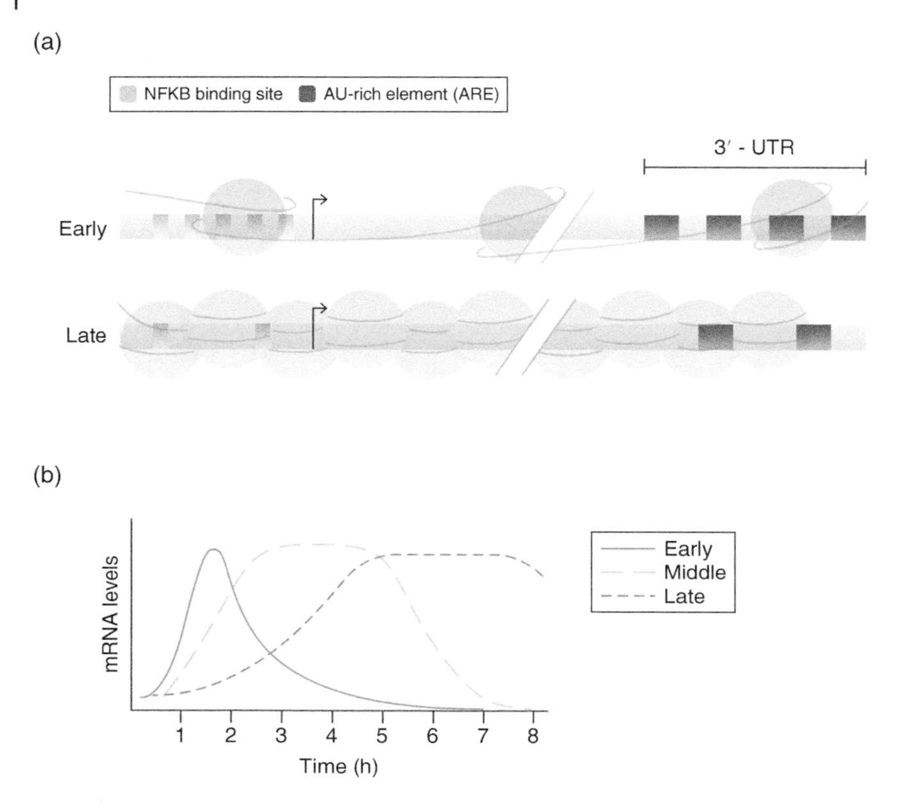

Figure 3.3 **Genetic elements affecting gene expression profiles.** (a) Genes that are expressed in the early group (i.e. maximal expression approximately one hour after stimulation) tend to be located in euchromatic regions and contain a relatively high density of (1) NF-κB binding sites in their promoters and (2) AU-rich elements (AREs) in their 3′-untranslated regions (3′-UTR). In contrast, genes in the late group (i.e. maximal expression approximately six hours after stimulation) tend to be located in heterochromatic regions and contain a relatively low density of NF-κB binding sites and AREs. Finally, genes in the middle group combine some elements characteristic of the late group (e.g. similar promoter architecture) with elements of early genes (e.g. a large number of AREs in the 3′-UTR), leading to an intermediate half-life ($t_{1/2}$). (b) Expression profiles of genes in the early (solid line), middle (long dashes), and late (short dashes) groups.

IκBα proteins translocate into the nucleus, where they associate with DNA-bound NF-κB molecules [13]. The newly formed NF-κB–IκBα complexes are then exported back to the cytoplasm via IκBα's nuclear export signal (NES), where they are sequestered until the next round of induction. In this way, IκBα and NF-κB form an integrated circuit that first enables activation and then a coordinated attenuation of NF-κB-regulated gene expression. However, at select NF-κB target gene loci, this circuit can be disrupted via posttranslational modification of NF-κB. For example, acetylation of RelA by CBP/p300 present at certain genomic loci decreases the affinity of IκBα for NF-κB [20]. As a consequence, NF-κB remains bound to these promoters for a longer period of time, allowing transcription to persist longer.

The IκB–NF-κB circuit can also be modulated in other interesting ways. For example, as we saw previously, NF-κB activation increases the expression of TNF-α. In hepatocytes, secreted TNF-α protein can feed back on the NF-κB pathway through autocrine signaling mechanisms involving the TNF receptor 1 (TNFR1) (Figure 3.4). Accordingly, secreted TNF-α binds a preassembled TNFR1 homo-trimer on the cell surface. Reorganization of the TNFR1 intracellular domain promotes the association of the adapter protein, TNF receptor-associated protein with a death domain (TRADD), and the Ser/Thr kinase, receptor-interacting protein 1 (RIP1) [21]. TRADD then recruits the adapter protein, TNFR-associated factor 2 (TRAF2), to the complex which, in turn, recruits the cellular inhibitor of apoptosis proteins, cIAP-1 and cIAP-2. cIAP-1 and cIAP-2, which function as E3 ubiquitin ligases, then catalyze the formation of K63-linked polyubiquitin chains on RIP1, TRAF2, and themselves [21].

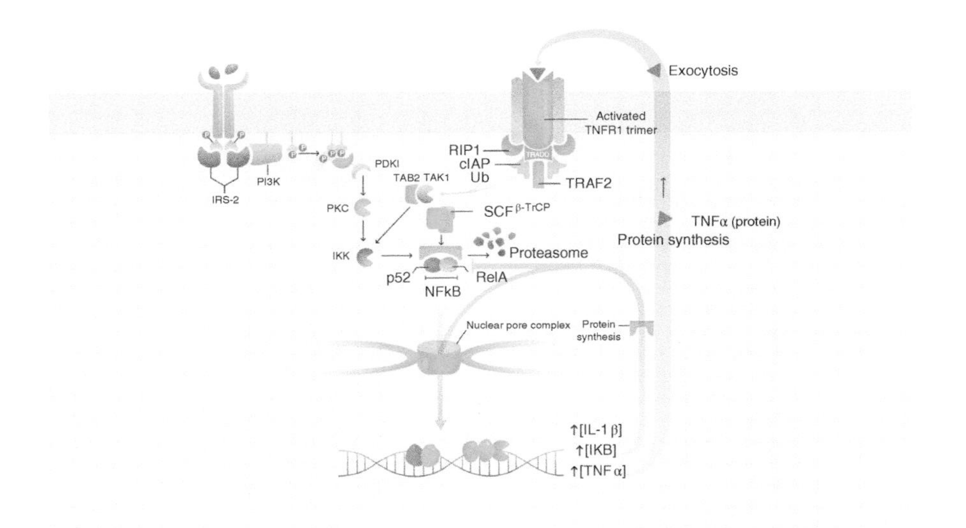

Figure 3.4 **Regulation of the IRS-2/PI3K/aPKC signaling axis in hepatocytes by positive and negative feedback.** The IRS-2/PI3K/aPKC signaling axis is regulated by both positive and negative feedback loops involving TNF-α (pink triangles) and IκB (green cradle), respectively. See text for details. Abbreviations are the same as those in Figure 3.2 and, additionally, TNFR1, TNF-α-receptor 1; TRAF2, TNFR-associated factor 2; cIAP, cellular inhibitor of apoptosis proteins 1 and 2; RIP1, receptor-interacting protein 1; TAK1, TNF-associated kinase; TAB2, TAK1-binding protein 2; TRADD, TNF receptor-associated protein with a death domain.

Unlike the K48-linked polyubiquitin chains involved in proteasomal degradation, K63-linked chains do not lead to degradation. Rather, they serve as binding sites for regulatory proteins, such as the linear ubiquitin chain assembly complex (LUBAC) E3 ligase complex. Once bound, the LUBAC complex promotes the K63-linked polyubiquitinylation of the NEMO subunit of the IKK complex, bringing it into close proximity to activated TNF-associated kinase 1 (TAK1) and its binding partner, TAB2. TAK1 phosphorylates IKKβ, leading to its activation. Finally, as we discussed above, activated IKKβ phosphorylates IκBα within its phosphodegron, marking it for $SCF^{\beta\text{-TrCP}}$-mediated ubiquitinylation and subsequent degradation [11, 12]. Thus, NF-κB regulates genes involved in both its inhibition (IκBα) and its activation (TNF-α). Together, these components form an oscillatory circuit whose amplitude and frequency are modulated by the relative levels of IκBα and TNF-α (Figure 3.5).

Finally, the expression patterns of TF-regulated genes can be further refined based on the decay rates of their mRNA products. For instance, aside from containing a large number of NF-κB binding sites in close proximity to the transcriptional start site, many of the NF-κB-regulated genes in the early group have an abundance of AU-rich elements (AREs) within their 3′-untranslated regions (3′-UTR) (Figure 3.3a) [16, 22]. AREs within the 3′-UTR are recognized by a family of proteins known as ARE-specific binding proteins (AUBPs) that alter the stability and localization of their mRNA targets [23]. As a

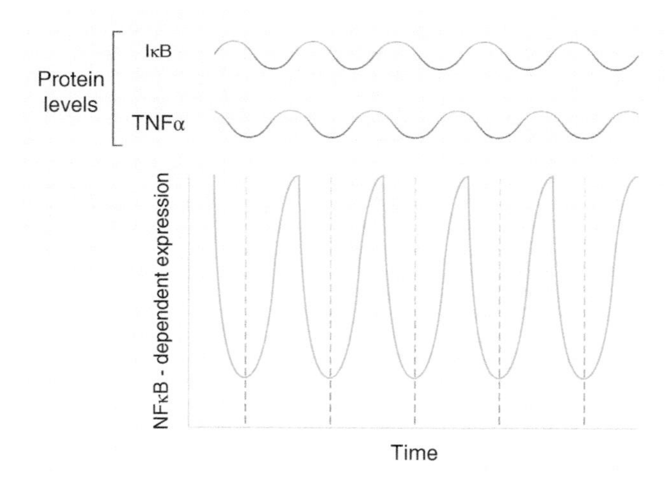

Figure 3.5 **Opposing actions of TNF-α and IκB can lead to oscillations in NF-κB-dependent gene expression.** Though the expression of both TNF-α and IκB is regulated by NF-κB, subsequent NF-κB-dependent gene expression is enhanced by TNF-α and inhibited by IκB. As a consequence, these factors can form an integrated circuit that drives oscillatory patterns of NF-κB gene expression.

consequence, early genes rich in AREs tend to decay more rapidly than those in the late group, which contain markedly fewer AREs in their 3′-UTRs [16, 22]. Interestingly, recent evidence suggests that posttranslational modification of AUBPs (e.g. via phosphorylation, acetylation, methylation, ubiquitination, and proline isomerization) plays an important role in modulating their interactions with ARE-containing elements within their target transcripts [23].

Thus, by considering both the rate of induction and the rate of decay of a given transcript, it becomes clear how differential expression patterns can be achieved between NF-κB-dependent genes in the early, middle, and late groups (Figure 3.3b). For instance, though genes in the early group are expected to undergo rapid induction due to a high density of NF-κB binding sites within their promoter regions, they are also expected to decay quickly due to the large number of AREs in their 3′-UTRs. Meanwhile, genes in the late group may be activated more slowly due to a relatively sparse distribution of NF-κB binding sites and/or the presence of heterochromatin at their promoters but their gene products are likely to persist longer due to the lower abundance of AREs within their 3′-UTRs. Finally, genes in the middle group would combine some elements characteristic of late genes (e.g. similar promoter architecture) with elements of early genes (e.g. a large number of AREs in the 3′-UTR), leading to an intermediate half-life ($t_{1/2}$). Interestingly, genes within each group are often related to one another with respect to their cellular function [13]. For example, cytokines tend to be expressed in the early group, while cell surface adhesion proteins and adaptor molecules are generally expressed in the late group [16].

3.2.4 Regulating the Regulators: Phosphatase-Mediated Regulation of Signaling Molecules

In addition to NF-κB, aPKC family members also regulate the activity of other TFs important for the cellular insulin response. In some cases, the products of the genes regulated by these TFs can feed back on the InsR-aPKC signaling pathway to attenuate the signal. For instance, aPKC-mediated phosphorylation of the TF, Sp1, leads to an increase in the expression of the *PTPN1* gene [24]. This gene codes for protein tyrosine phosphatase, non-receptor type 1 (PTPN1), which is the founding member of the protein tyrosine phosphatase family of enzymes. Once expressed, PTPN1 (also known as PTP1B) dephosphorylates pTyr residues on the InsR that are required for IRS-2 docking interactions (Figure 3.6). As a consequence, PI3K fails to associate with IRS-2 at the membrane, thereby preventing further production of PIP_3. Meanwhile, existing PIP_3 molecules are actively converted back to PIP_2 via dephosphorylation by the lipid phosphatases, SH2 domain-containing inositol-5-phosphatase 1 (SHIP1) and phosphatase and tensin homolog (PTEN). Therefore, the equilibrium between PIP_3 and PIP_2 is shifted toward the latter, preventing activation of additional PDK1 and, by extension, phosphorylation and activation of

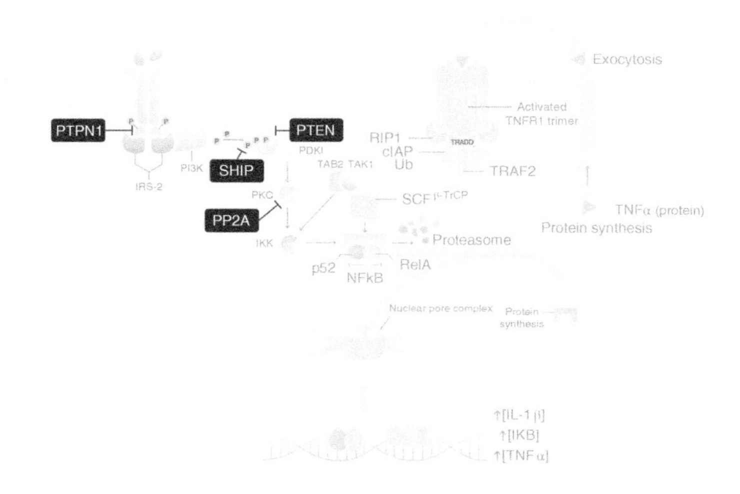

Figure 3.6 **Phosphatase-mediated regulation of the IRS-2/PI3K/aPKC signaling axis.** Insulin-dependent signaling through the IRS-2/PI3K/aPKC signaling axis is negatively regulated by both protein (i.e. PTPN1 and PP2B) and lipid (i.e. SHIP1 and PTEN) phosphatases that catalyze the dephosphorylation of key signaling molecules, such as IRS-2 (via PTPN1), aPKC (via PP2B), and PIP_3 (via SHIP1 and PTEN). Abbreviations are the same as in Figure 3.4.

additional aPKC molecules. This is important because activated aPKCs are believed to be dephosphorylated by the constitutively active Ser/Thr protein phosphatase, protein phosphatase 2A (PP2A) (Figure 3.6). Thus, attenuation of PDK1 activity shifts aPKC family members back toward the inactive state.

3.3 Crosstalk and Functional Signal Integration in Response to Insulin in Hepatocytes

As the primary site of gluconeogenesis (i.e. synthesis of new glucose molecules from noncarbohydrate precursors), glycogen synthesis (i.e. synthesis of branched polysaccharides used for long-term glucose storage), and lipogenesis (i.e. synthesis of fatty acids and lipids from glucose precursors), the liver plays a central role in glucose homeostasis in vertebrates. To accomplish this, the liver must respond to changes in blood glucose by orchestrating an appropriate functional response that simultaneously activates some metabolic pathways and inactivates others. For instance, in the fed state, elevations in blood glucose lead to increased expression/activity of metabolic enzymes involved in glycogen synthesis and lipogenesis while reducing the expression of those enzymes involved in gluconeogenesis. Amazingly, each of these pathways is directly regulated by InsR signaling. For example, as we saw earlier, insulin triggers autophosphorylation of the InsR, which ultimately leads to aPKC activation via an IRS-2/PI3K-dependent pathway (Figure 3.2, steps 1–5). In parallel, hepatic Akt2 is activated via a similar mechanism involving IRS-1 (and to a lesser extent, IRS-2) and PI3K (Figure 3.7). Like aPKCs, association of Akt2 with PIP_3 at the membrane facilitates PDK1-mediated phosphorylation on Thr308 in its activation loop. This is followed by phosphorylation of Ser473 by the mammalian target of rapamycin complex 2 (mTORC2), leading to the fully activated Akt2 enzyme. Once activated, Akt2 phosphorylates several downstream effectors, many of which are involved in the metabolic pathways alluded to above. For instance, Akt2-mediated phosphorylation inhibits glycogen synthase kinase 3-beta (GSK3-β) activity by promoting the formation of an intramolecular pseudosubstrate that blocks the GSK3-β active site, thereby preventing substrate binding. As a consequence, GSK3-β is unable to catalyze the phosphorylation of its downstream target, glycogen synthase. Since GSK3-β-mediated phosphorylation normally inhibits glycogen synthase activity, Akt2-induced inhibition of GSK3-β activity has the overall effect of activating glycogen synthase, leading to increased glycogen synthesis.

At the same time, insulin-dependent activation of Akt2 also promotes downregulation of genes involved in gluconeogenesis, such as *Pck1* (encoding phosphoenolpyruvate carboxykinase [PEPCK], which increases hepatic glucose production) and *G6pc* (encoding glucose-6-phosphatase [G6Pase], which facilitates the transport of newly synthesized glucose out of the liver). Akt2 accomplishes this via phosphorylation of the TF, forkhead box O1 (Foxo1). In the fasted state, Foxo1 typically shuttles between the cytoplasm and the nucleus. Inside the nucleus, it binds regions of DNA containing the insulin response element (IRE: 5′-CAAAACAA-3′), including the *Pck1* and *G6pc* promoters, driving transcription at these sites. In contrast, during the fed state, InsR-dependent activation of Akt2 promotes phosphorylation of Foxo1 on three residues (T24, S256, and S319). The most important of these sites with respect to the regulation of Foxo1 function is pS256. Indeed, Akt2-mediated phosphorylation of S256 promotes interactions between Foxo1 and SKP2, a subunit of the SCF ubiquitin E3 ligase complex. Similar to IκB in the NF-κB pathway discussed earlier, the SCF complex catalyzes the conjugation of K48-linked ubiquitin chains to Foxo1, leading to its sequestration in the cytoplasm and eventual degradation by the 26S proteasome. This, in turn, reduces expression of PEPCK and G6Pase enzymes and slows the rate of gluconeogenesis.

Interestingly, Akt2–Foxo1 interactions appear to be modulated by crosstalk between the IRS-1/PI3K/Akt2 and IRS-2/PI3K/aPKC pathways. For instance, Akt2 and Foxo1 are brought into close proximity with one another in the cytoplasm via interactions with the scaffold protein, ProF [25]. ProF, which contains seven WD40 repeats and a Fab1/YOTB/Vac1/EEA1 (FYVE) zinc finger domain, also

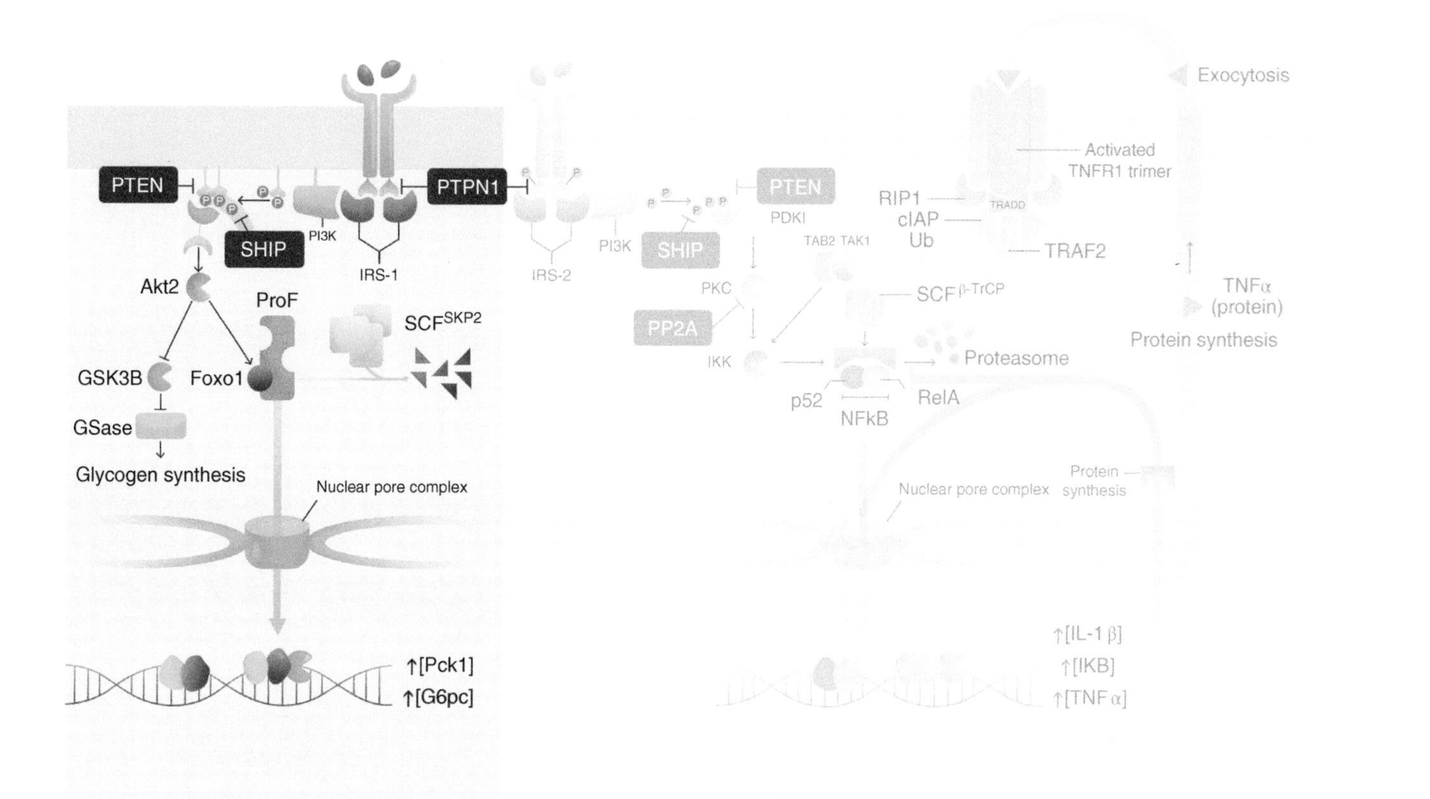

Figure 3.7 **Regulation of gluconeogenesis and glycogen synthesis via the IRS-1/PI3K/Akt2 signaling axis in hepatocytes.** In hepatocytes, in addition to the IRS-2/PI3K/aPKC signaling axis (background), insulin also stimulates the IRS-1/PI3K/Akt2 pathway, which regulates (1) gluconeogenesis via Akt2-mediated phosphorylation/inhibition of Foxo1 and (2) glycogen synthesis by Akt2-mediated phosphorylation/inhibition of glycogen synthase kinase 3-beta (GSK3-β). Abbreviations are the same as those in Figure 3.6 and, additionally, GSase, glycogen synthase; ProF, ProF/WW scaffold protein; SCF^{SKP2}, Skip-Cullin-F-box ubiquitin E3 ligase with the SKP2 substrate binding subunit; Foxo1, forkhead box O1; Pck1, phosphoenolpyruvate

interacts with aPKC. As we will discuss in more detail below, during periods of hyperinsulinemia, activated aPKC disrupts interactions between Akt2 and ProF [26, 27]. As a result, Akt2 fails to phosphorylate Foxo1, leading to continued expression of gluconeogenic enzymes and subsequent glucose synthesis despite the presence of activated Akt2. Though it is not currently known whether aPKC displaces Akt2 from ProF by phosphorylation of the ProF scaffold or by direct phosphorylation of Akt2, itself, it is known that pharmacological inhibition of aPKC restores Akt2/ProF interactions and promotes phosphorylation of Foxo1 and attenuation of gluconeogenesis [27].

Likewise, crosstalk between the IRS-1/PI3K/Akt2 and IRS-2/PI3K/aPKC signaling pathways also appears to be important for the regulation of genes involved in lipogenesis. For instance, sterol response element binding protein 1c (SREBP-1c) is a basic helix–loop–helix (bHLH) TF that controls the expression of key lipogenic enzymes, including fatty acid synthase (FAS) and acetyl-CoA carboxylase (ACC) [28]. In the fasted state, the SREBP-1c precursor protein is anchored in the ER membrane via two transmembrane domains. In the ER, the SREBP-1c precursor associates with the SREBP cleavage activating protein (SCAP) via its C-terminal regulatory domain (Figure 3.8). The SREBP-1c/SCAP complex, which must be shuttled to the Golgi apparatus for proteolytic processing of SREBP-1c prior to its activation, is retained in the ER via interactions with insulin-induced gene 2 (INSIG2). However, in the presence of insulin, Akt2 activation disrupts interactions between INSIG2 and the SREBP-1c/SCAP complex, presumably through an mTORC1- and p70S6K-dependent mechanism [29–32]. Likewise, Akt2-mediated phosphorylation of SREBP-1c promotes association of the SREBP-1c/SCAP complex with COPII-coated vesicles involved in anterograde transport to the Golgi [33]. At the Golgi, the luminal region of the SREBP-1c precursor protein is cleaved in a site-specific manner by the Ser protease, site-1 protease (S1P). Cleavage by S1P uncouples the N-terminal region of SREBP-1c (which contains the bHLH domain) from the C-terminal region (which remains associated with SCAP). This is followed by a second cleavage event mediated by the matrix metalloprotease, site-2 protease (S2P). S2P cleavage releases the mature N-terminal SREBP-1c TF from the membrane, allowing it to translocate into the nucleus and associate with DNA regions containing the sterol response element (SRE; 5′-TCACNCCAC-3′). Interestingly, GSK3-mediated phosphorylation of SREBP-1c has been shown to negatively regulate its activity. Therefore, like glycogen synthase discussed above, Akt-dependent inactivation of GSK3 serves to increase SREBP-1c transcriptional activity [34].

However, despite the presence of active SREBP-1c, many SRE-regulated genes, including those encoding FAS and ACC, are not expressed efficiently without a second signal. This is because these genes are located in transcriptionally silent heterochromatic regions of the genome. To facilitate expression of these genes, the chromatin in these regions must be actively remodeled

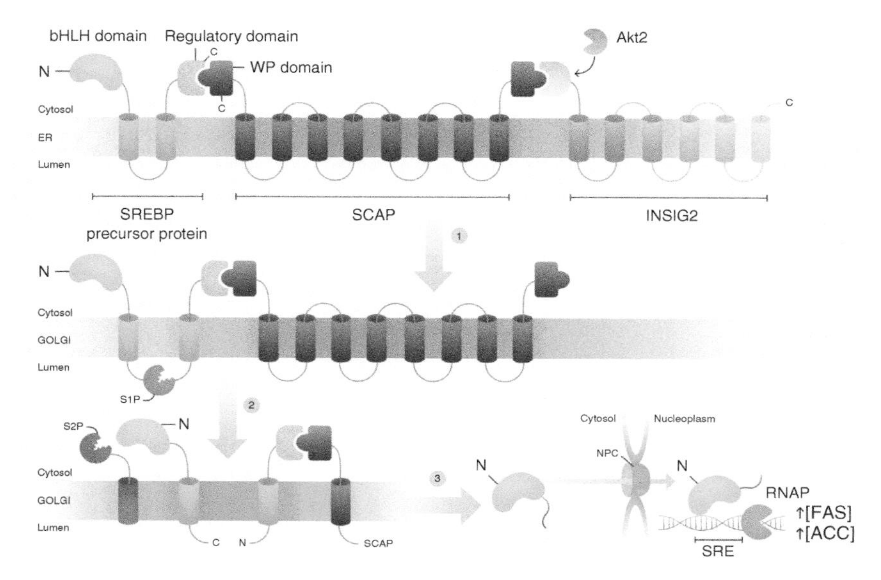

Figure 3.8 **Akt2-mediated activation of SREBP1 and lipogenesis.** The sterol response element binding protein (SREBP) precursor protein, containing an N-terminal basic helix–loop–helix (bHLH) DNA-binding domain and a C-terminal regulatory domain flanking two transmembrane regions, is initially retained in the endoplasmic reticulum through interactions with SREBP cleavage-associated protein (SCAP) and insulin-induced gene 2 (INSIG2). Insulin-dependent activation of the IRS-1/PI3K/Akt2 signaling axis. Akt2-mediated phosphorylation of INSIG2 disrupts SCAP–INSIG2 interactions, facilitating anterograde transport of the SCAP/SREBP precursor protein ternary complex to the Golgi apparatus via coat protein II (COPII)-coated vesicles (step 1). At the Golgi, the site-1 protease (S1P) cleaves the linker region in the lumen, releasing the N-terminal region of the SREBP precursor protein from SCAP (step 2). This is followed by a second cleavage event mediated by the matrix metalloprotease, site-2 protease (S2P), releasing the mature SREBP into the cytosol (step 3). The mature SREBP protein then translocates into the nucleus through the nuclear pore complex (NPC), where it drives expression of lipogenic genes containing sterol response elements (SRE) in their promoter regions, such as fatty acid synthase (FAS) and acetyl-CoA carboxylase (ACC). See text for further details.

through the action of the mammalian SWItch/Sucrose Non-Fermentable (SWI/SNF) complex. Mammalian SWI/SNF complexes are composed of several Brg1/Brm-associated factor (BAF) family members, including the core ATPase, BAF190, and BAF60, which is believed to recruit the SWI/SNF complex to its site of action though interactions with TFs. During insulin signaling, cytoplasmic BAF60 is specifically phosphorylated by aPKC on S247, leading to its redistribution from the cytoplasm to the nucleus. Once inside the nucleus, BAF60 associates with upstream stimulatory factor 1 (USF1) which, interestingly, also plays a role in recruiting SREBP-1c to SREs in a

DNA-dependent protein kinase (DNA-PK)-dependent manner [33]. Subsequent interactions between BAF60 and the SWI/SNF complex then promote remodeling of the chromatin in these regions, converting them to transcriptionally active euchromatic regions.

3.4 Systemic Signal Integration

To this point, we have focused primarily on signaling pathways involved in the hepatic response to insulin. However, it is important to note that insulin also simultaneously stimulates other cells and tissues that are critical for an appropriate physiological response to glucose during feeding (i.e. the efficient uptake and utilization of glucose and its derivatives). Here, we will briefly discuss how different tissues respond to elevations in glucose and how the specific responses from each of these tissues are coordinated throughout the system. During this discussion, we will highlight three cell types whose activities are tightly linked to those that we examined previously in the liver. Specifically, we will focus on the mechanisms by which insulin- and InsR-dependent signaling processes regulate glucose metabolism in skeletal muscle cells, adipocytes, and pancreatic β-cells.

3.4.1 Pancreatic β-Cells

The systemic insulin response begins in pancreatic β-cells, where the insulin precursor protein, preproinsulin, is initially synthesized from the *INS* gene. Preproinsulin is then processed to the mature insulin protein by a series of proteolytic processing steps, first in the rough ER and then in the Golgi apparatus, before being packaged into vesicles for eventual secretion [35, 36]. Insulin secretion, which is primarily regulated by elevations in blood glucose but can also be triggered by other stimuli, including various carbohydrates, amino acids, and vagal nerve stimulation, is initiated by depolarization of electrically sensitive β-cells [37, 38]. Accordingly, in the fed state, blood glucose is transported across the β-cell's plasma membrane via the low-affinity glucose transporter, GLUT2 [39]. Intracellular glucose is then shuttled into the glycolytic pathway, leading to the production of ATP. As ATP levels increase, so too does the intracellular ATP/ADP ratio, causing closure of ATP-sensitive sulfonylurea receptor 1 (SUR1)/K_{ir}6.2 inward rectifying potassium channels [40]. This, in turn, leads to membrane depolarization, which activates voltage-gated Ca^{2+} channels, resulting in an influx of extracellular Ca^{2+} and an increase in the intracellular Ca^{2+} concentration, both from the aforementioned influx of extracellular Ca^{2+} and the subsequent release of internal Ca^{2+} stores from the ER via stimulation of Ca^{2+}-sensitive ryanodine receptors, trigger vesicle fusion with the membrane – potentially via a signaling mechanism involving the classical and/or novel PKC family members, PKCα and PKCε, respectively – and subsequent secretion of insulin into the hepatic portal vein. Interestingly, β-cells express the InsR at appreciable levels. Moreover, exogenous addition of insulin leads to a decrease in *INS* gene expression [41]. As a consequence, insulin secretion has long been thought to feed back on β-cells to regulate its own expression through an autocrine signaling pathway involving the IRS-2/Akt/Foxo1 signaling axis [42]. However, the relevance of this autocrine signaling pathway under normal physiological conditions has recently been called into question [43].

3.4.2 Skeletal Muscles

After insulin is secreted from β-cells, it first encounters neighboring α- and β-cells in the local islet milieu (modulating the synthesis/secretion of glucagon and somatostatin, respectively, in an InsR-dependent manner) before feeding into the hepatic portal vein. In the liver, insulin-mediated activation of InsR regulates a series of metabolic pathways, including gluconeogenesis, glycogen synthesis, and lipogenesis, as discussed in detail in Section 3.2. Interestingly, due to the high concentration of InsR on the surface of hepatic cells, roughly half of the insulin secreted from β-cells is removed from circulation via receptor-mediated endocytosis [39]. The remaining insulin is retained in the systemic circulatory system, where it is distributed to various tissues throughout the body. One of its primary targets is skeletal muscles. In myocytes, InsR activation stimulates glucose uptake via the high affinity glucose transporter, GLUT4, and subsequent glycogen synthesis. Interestingly, these processes are regulated by both aPKC and Akt in an exclusively IRS-1/PI3K-dependent manner. Indeed, unlike the liver – where the IRS-1/PI3K and IRS-2/PI3K signaling axes are predominately associated with Akt- and aPKC-mediated signaling, respectively – in myocytes, the IRS-1/PI3K pathway directly activates both Akt and aPKC (although IRS-2 is expressed in myocytes, the function of the IRS-2/PI3K pathway in muscle is currently unknown) [8]. For instance, insulin-dependent activation of aPKC promotes the phosphorylation of vesicle-associated membrane protein 2 (VAMP2) on the surface of GLUT4 vesicles. Phosphorylation of VAMP2, which is a key component of the soluble *N*-ethylmaleimide-sensitive attachment receptor (SNARE) complex, promotes interactions with syntaxin 4 on the inner surface of the plasma membrane, facilitating vesicle fusion to the membrane. Moreover, PKC-mediated phosphorylation of the syntaxin 4-associated protein, Munc18c, promotes its dissociation from syntaxin 4 [44]. This is important because Munc18c normally inhibits interactions between VAMP2 and proteins required for vesicle docking and subsequent fusion of GLUT4 to the membrane. Integration of GLUT4 into the membrane results in highly efficient transport of glucose into the cell via facilitated diffusion. In fact, due to the high affinity of GLUT4 for glucose and the relatively large biomass of skeletal muscle, myocytes in skeletal muscle account for the clearance of nearly 70% of postprandial glucose [39, 45]. Once inside the cell, glucose monomers are integrated into glycogen for storage. Much like we saw in hepatocytes, the rate of glycogen synthesis is regulated in an insulin-dependent manner through Akt-mediated inhibition of GSK3, leading to an increase in glycogen synthase activity (e.g. see Figure 3.7).

3.4.3 Adipose Tissue

In addition to the liver and muscle, adipose is another major insulin-responsive tissue. Adipocytes, which play an essential role in metabolic regulation through fat storage and the secretion of adipokines such as leptin, adiponectin, and resistin, express the InsR at fairly high levels. Like skeletal muscles, InsR activation promotes the integration of high-affinity GLUT4 glucose transporters into the plasma membrane via SNARE-mediated vesicle fusion. This process, which is also governed by an IRS-1/PI3K/aPKC-dependent signaling mechanism [13, 46], leads to the efficient transport of glucose into the cytoplasm of adipocytes. As a consequence, adipocytes account for between 5 and 15% of total glucose clearance after a meal [39, 47]. Once inside the cell, glucose is rapidly converted to acetyl-CoA, which is then incorporated into fatty acids through the action of FAS. In adipocytes, these fatty acids are ultimately conjugated to glycerol to form triglycerides for long-term energy storage. Similar to the liver, expression of FAS and other enzymes involved in lipogenesis is regulated in an insulin-dependent manner through Akt-mediated activation of SREBP-1c (Figure 3.8). In parallel, InsR activation represses lipolysis in adipocytes. This process is also believed to be mediated by Akt, which phosphorylates and activates the phosphodiesterase, PDE3B. PDE3B degrades the second messenger, cAMP, inhibiting PKA activity and, by extension, PKA-mediated activation of hormone-sensitive lipase (HSL). Therefore, through the action of the IRS-1/PI3K/Akt signaling axis, insulin simultaneously stimulates the metabolic pathways involved in fatty acid synthesis (i.e. lipogenesis) and inhibits those involved in breaking down fatty acids (i.e. lipolysis).

3.5 Dysregulation of Insulin Signaling in the Etiology of Type 2 Diabetes

Given the central importance of insulin-dependent signaling to glucose clearance and metabolic regulation, perhaps it is not surprising that dysregulation of this system underlies several of the disorders associated with the metabolic syndrome, including obesity, cardiovascular disease, and T2D. The latter, which is characterized by elevated blood glucose levels stemming from decreased insulin sensitivity, has reached pandemic levels in many "Westernized" societies. For instance, nearly 30 million people in the United States were affected by T2D in 2017 [48]. Moreover, recent estimates suggest that the prevalence of T2D will increase by more than 50% (to almost 55 million Americans) by 2030, making T2D a major health burden in the United States [49].

Though the etiology of T2D is complex, one of the primary causative factors appears to be prolonged periods of caloric excess [8]. Under these conditions, hyper-activation of InsR signaling can lead to insulin insensitivity and diminished glucose clearance and utilization. For instance, as we saw earlier, hyperinsulinemia leads to disruption of the Akt2–ProF complex in hepatocytes via an IRS-2/PI3K/aPKC-dependent signaling mechanism. As a result, Akt2 is unable to phosphorylate/inhibit Foxo1, leading to sustained gluconeogenesis in spite of elevated systemic blood glucose levels. The *de novo* synthesis of glucose leads to further increases in blood glucose, stimulating further insulin secretion from pancreatic β-cells and continued hyperinsulinemia. Under normal conditions, a significant proportion of the excess glucose produced during gluconeogenesis would be cleared by skeletal muscles and, to a lesser extent, adipose tissue. However, during hyperinsulinemia, several factors related to over-activation of hepatic aPKC and Akt lead to diminished systemic glucose clearance. For instance, concomitant increases in hepatic lipogenesis (mediated by Akt2/SREBP1) produce fatty acids and other lipids that reduce InsR signaling mediated through the IRS-1/PI3K pathway in myocytes. Since both aPKC and Akt are activated by this pathway in myocytes, aPKC-mediated redistribution of GLUT4 to the membrane is severely diminished, resulting in impaired glucose transport. Moreover, recall that in hepatocytes, IRS-2/PI3K/aPKC also leads to increased expression of pro-inflammatory cytokines, such as IL-1B and TNF-α, in an NF-κB-dependent manner (Figure 3.2). Not only does TNF-α feed back on hepatocytes to activate aPKC via TNFR1 and TAK (Figure 3.4), but, together with IL-1B, it also induces system-wide inflammation that further diminishes InsR signaling in myocytes and adipocytes [50]. To make matters worse, the inflammatory response is reinforced and intensified by fat expansion in adipocytes driven by hyperlipidemia. Specifically, increases in circulating lipids caused by hepatic lipogenesis lead to adipocytic secretion of the pro-inflammatory cytokines TNF-α, IL-6, and leptin while simultaneously reducing secretion of the anti-inflammatory hormone, adiponectin. Together, the resulting inflammatory response leads to systemic insulin resistance that results in elevated blood glucose levels characteristic of T2D. Interestingly, recent evidence suggests that ubiquitous environmental contaminants, such as the plasticizer, bisphenol A (BPA), may further contribute to dysregulation of the systemic insulin response. Indeed, environmentally relevant doses of BPA, which is an endocrine-disrupting chemical (EDC) present in reusable plastic bottles and the internal coatings of many food and beverage cans, have been shown to inhibit adiponectin secretion and to promote the release of TNF-α and IL-6 from adipocytes, even in the absence of overnutrition [51, 52]. Likewise, similar concentrations of BPA (approximately 1 nM) were shown to dramatically decrease the activity of SUR1/K_{ir}6.2 potassium channels in pancreatic β-cells, thereby potentiating glucose-induced insulin secretion and hyperinsulinemia [53]. Similar effects have also been reported for other EDCs, including dioxins and diethylstilbestrol (DES) [54]. Thus, dysregulation of insulin-dependent signaling mechanisms, either by caloric excess or environmental agents (or a combination thereof), can reduce insulin sensitivity and increase blood glucose levels, leading to the development of T2D and its complications.

References

1 Taylor, S.S. and Kornev, A.P. (2011). Protein kinases: evolution of dynamic regulatory proteins. *Trends Biochem. Sci.* 36 (2): 65–77.

2 Iversen, L., Tu, H.L., Lin, W.C. et al. (2014). Molecular kinetics. Ras activation by SOS: allosteric regulation by altered fluctuation dynamics. *Science* 345 (6192): 50–54.

3 Zeeh, J.C., Antonny, B., Cherfils, J., and Zeghouf, M. (2008). In vitro assays to characterize inhibitors of the activation of small G proteins by their guanine nucleotide exchange factors. *Methods Enzymol.* 438: 41–56.

4 Newman, R.H., Fosbrink, M.D., and Zhang, J. (2011). Genetically encodable fluorescent biosensors for tracking signaling dynamics in living cells. *Chem. Rev.* 111 (5): 3614–3666.

5 Hubbard, S.R. (2013). The insulin receptor: both a prototypical and atypical receptor tyrosine kinase. *Cold Spring Harb. Perspect. Biol.* 5 (3): a008946.

6 Tatulian, S.A. (2015). Structural dynamics of insulin receptor and transmembrane signaling. *Biochemistry* 54 (36): 5523–5532.

7 Farese, R.V. and Sajan, M.P. (2010). Metabolic functions of atypical protein kinase C: "good" and "bad" as defined by nutritional status. *Am. J. Physiol. Endocrinol. Metab.* 298 (3): E385–E394.

8 Farese, R.V., Lee, M.C., and Sajan, M.P. (2014). Hepatic atypical protein kinase C: an inherited survival-longevity gene that now fuels insulin-resistant syndromes of obesity, the metabolic syndrome and type 2 diabetes mellitus. *J. Clin. Med.* 3 (3): 724–740.

9 Lallena, M.J., Diaz-Meco, M.T., Bren, G. et al. (1999). Activation of IkappaB kinase beta by protein kinase C isoforms. *Mol. Cell Biol.* 19 (3): 2180–2188.

10 Clifton, D.R., Rydkina, E., Freeman, R.S., and Sahni, S.K. (2005). NF-kappaB activation during rickettsia rickettsii infection of endothelial cells involves the activation of catalytic IkappaB kinases IKKalpha and IKKbeta and phosphorylation-proteolysis of the inhibitor protein IkappaBalpha. *Infect. Immun.* 73 (1): 155–165.

11 Kanarek, N. and Ben-Neriah, Y. (2012). Regulation of NF-kappaB by ubiquitination and degradation of the IkappaBs. *Immunol. Rev.* 246 (1): 77–94.
12 Suzuki, H., Chiba, T., Kobayashi, M. et al. (1999). IkappaBalpha ubiquitination is catalyzed by an SCF-like complex containing Skp1, cullin-1, and two F-box/WD40-repeat proteins, betaTrCP1 and betaTrCP2. *Biochem. Biophys. Res. Commun.* 256 (1): 127–132.
13 Sen, R. and Smale, S.T. (2010). Selectivity of the NF-{kappa}B response. *Cold Spring Harb. Perspect. Biol.* 2 (4): a000257.
14 Paszek, P., Lipniacki, T., Brasier, A.R. et al. (2005). Stochastic effects of multiple regulators on expression profiles in eukaryotes. *J. Theor. Biol.* 233 (3): 423–433.
15 Boston University (2011). NF-kB Target Genes. http://www.bu.edu/nf-kb/gene-resources/target-genes (accessed 1 February 2018).
16 Iwanaszko, M., Brasier, A.R., and Kimmel, M. (2012). The dependence of expression of NF-kappaB-dependent genes: statistics and evolutionary conservation of control sequences in the promoter and in the 3′ UTR. *BMC Genomics* 13: 182.
17 Wunderlich, Z. and Mirny, L.A. Different gene regulation strategies revealed by analysis of binding motifs. *Trends Genet.* 25 (10): 434–440.
18 Zhong, H., May, M.J., Jimi, E., and Ghosh, S. (2002). The phosphorylation status of nuclear NF-kappa B determines its association with CBP/p300 or HDAC-1. *Mol. Cell* 9 (3): 625–636.
19 Zhong, H., Voll, R.E., and Ghosh, S. (1998). Phosphorylation of NF-kappa B p65 by PKA stimulates transcriptional activity by promoting a novel bivalent interaction with the coactivator CBP/p300. *Mol. Cell* 1 (5): 661–671.
20 Chen, L.F., Mu, Y., and Greene, W.C. (2002). Acetylation of RelA at discrete sites regulates distinct nuclear functions of NF-kappaB. *EMBO J.* 21 (23): 6539–6548.
21 Wajant, H. and Scheurich, P. (2011). TNFR1-induced activation of the classical NF-kappaB pathway. *FEBS J.* 278 (6): 862–876.
22 Hao, S. and Baltimore, D. (2009). The stability of mRNA influences the temporal order of the induction of genes encoding inflammatory molecules. *Nat. Immunol.* 10 (3): 281–288.
23 Shen, Z.J. and Malter, J.S. Regulation of AU-rich element RNA binding proteins by phosphorylation and the Prolyl Isomerase Pin1. *Biomolecules* 5 (2): 412–434.
24 Inada, S., Ikeda, Y., Suehiro, T. et al. (2007). Glucose enhances protein tyrosine phosphatase 1B gene transcription in hepatocytes. *Mol. Cell. Endocrinol.* 271 (1–2): 64–70.
25 Fritzius, T. and Moelling, K. (2008). Akt- and Foxo1-interacting WD-repeat-FYVE protein promotes adipogenesis. *EMBO J.* 27 (9): 1399–1410.
26 Sajan, M.P., Ivey, R.A. 3rd, and Farese, R.V. (2015). BMI-related progression of atypical PKC-dependent aberrations in insulin signaling through IRS-1, Akt,

FoxO1 and PGC-1alpha in livers of obese and type 2 diabetic humans. *Metabolism* 64 (11): 1454–1465.

27 Sajan, M.P., Ivey, R.A., Lee, M.C., and Farese, R.V. (2015). Hepatic insulin resistance in Ob/Ob mice involves increases in ceramide, aPKC activity, and selective impairment of Akt-dependent FoxO1 phosphorylation. *J. Lipid Res.* 56 (1): 70–80.

28 Jeon, T.I. and Osborne, T.F. (2012). SREBPs: metabolic integrators in physiology and metabolism. *Trends Endocrinol. Metab.* 23 (2): 65–72.

29 Bakan, I. and Laplante, M. (2012). Connecting mTORC1 signaling to SREBP-1 activation. *Curr. Opin. Lipidol.* 23 (3): 226–234.

30 Guo, S. (2014). Insulin signaling, resistance, and the metabolic syndrome: insights from mouse models into disease mechanisms. *J. Endocrinol.* 220 (2): T1–T23.

31 Li, S., Brown, M.S., and Goldstein, J.L. (2010). Bifurcation of insulin signaling pathway in rat liver: mTORC1 required for stimulation of lipogenesis, but not inhibition of gluconeogenesis. *Proc. Natl. Acad. Sci. U.S.A.* 107 (8): 3441–3446.

32 Owen, J.L., Zhang, Y., Bae, S.H. et al. (2012). Insulin stimulation of SREBP-1c processing in transgenic rat hepatocytes requires p70 S6-kinase. *Proc. Natl. Acad. Sci. U.S.A.* 109 (40): 16184–16189.

33 Wang, Y., Viscarra, J., Kim, S.J., and Sul, H.S. (2015). Transcriptional regulation of hepatic lipogenesis. *Nat. Rev. Mol. Cell. Biol.* 16 (11): 678–689.

34 Kim, K.H., Song, M.J., Yoo, E.J. et al. (2004). Regulatory role of glycogen synthase kinase 3 for transcriptional activity of ADD1/SREBP1c. *J. Biol. Chem.* 279 (50): 51999–52006.

35 Du, X., Kristiana, I., Wong, J., and Brown, A.J. (2006). Involvement of Akt in ER-to-Golgi transport of SCAP/SREBP: a link between a key cell proliferative pathway and membrane synthesis. *Mol. Biol. Cell* 17 (6): 2735–2745.

36 Yellaturu, C.R., Deng, X., Cagen, L.M. et al. (2009). Insulin enhances post-translational processing of nascent SREBP-1c by promoting its phosphorylation and association with COPII vesicles. *J. Biol. Chem.* 284 (12): 7518–7532.

37 Chandra, R. and Liddle, R.A. (2014). Recent advances in the regulation of pancreatic secretion. *Curr. Opin. Gastroenterol.* 30 (5): 490–494.

38 Molina, J., Rodriguez-Diaz, R., Fachado, A. et al. (2014). Control of insulin secretion by cholinergic signaling in the human pancreatic islet. *Diabetes* 63 (8): 2714–2726.

39 Bedinger, D.H. and Adams, S.H. (2015). Metabolic, anabolic, and mitogenic insulin responses: a tissue-specific perspective for insulin receptor activators. *Mol. Cell. Endocrinol.* 415: 143–156.

40 Koo, B.K., Cho, Y.M., Park, B.L. et al. (2007). Polymorphisms of KCNJ11 (Kir6.2 gene) are associated with type 2 diabetes and hypertension in the Korean population. *Diabet. Med.* 24 (2): 178–186.

41 Leibiger, B., Leibiger, I.B., Moede, T. et al. (2001). Selective insulin signaling through A and B insulin receptors regulates transcription of insulin and glucokinase genes in pancreatic beta cells. *Mol. Cell.* 7 (3): 559–570.

42 White, M.F. (2006). Regulating insulin signaling and beta-cell function through IRS proteins. *Can. J. Physiol. Pharmacol.* 84 (7): 725–737.

43 Rhodes, C.J., White, M.F., Leahy, J.L., and Kahn, S.E. (2013). Direct autocrine action of insulin on beta-cells: does it make physiological sense? *Diabetes* 62 (7): 2157–2163.

44 Smithers, N.P., Hodgkinson, C.P., Cuttle, M., and Sale, G.J. (2008). 80K-H acts as a signaling bridge in intact living cells between PKCzeta and the GLUT4 translocation regulator Munc18c. *J. Recept. Signal Transduct. Res.* 28 (6): 581–589.

45 DeFronzo, R.A. and Tripathy, D. (2009). Skeletal muscle insulin resistance is the primary defect in type 2 diabetes. *Diabetes Care* 32 (Suppl 2): S157–S163.

46 Sajan, M.P., Rivas, J., Li, P. et al. (2006). Repletion of atypical protein kinase C following RNA interference-mediated depletion restores insulin-stimulated glucose transport. *J. Biol. Chem.* 281 (25): 17466–17473.

47 Biddinger, S.B. and Kahn, C.R. (2006). From mice to men: insights into the insulin resistance syndromes. *Annu. Rev. Physiol.* 68: 123–158.

48 Center for Disease Control (2017). National Diabetes Statistics Report, Atlanta.

49 Rowley, W.R., Bezold, C., Arikan, Y. et al. (2017). Diabetes 2030: insights from yesterday, today, and future trends. *Popul. Health Manag.* 20 (1): 6–12.

50 Cai, D., Yuan, M., Frantz, D.F. et al. (2005). Local and systemic insulin resistance resulting from hepatic activation of IKK-beta and NF-kappaB. *Nat. Med.* 11 (2): 183–190.

51 Ben-Jonathan, N., Hugo, E.R., and Brandebourg, T.D. (2009). Effects of bisphenol a on adipokine release from human adipose tissue: implications for the metabolic syndrome. *Mol. Cell. Endocrinol.* 304 (1–2): 49–54.

52 Hugo, E.R., Brandebourg, T.D., Woo, J.G. et al. (2008). Bisphenol a at environmentally relevant doses inhibits adiponectin release from human adipose tissue explants and adipocytes. *Environ. Health Perspect.* 116 (12): 1642–1647.

53 Soriano, S., Alonso-Magdalena, P., Garcia-Arevalo, M. et al. (2012). Rapid insulinotropic action of low doses of bisphenol-a on mouse and human islets of Langerhans: role of estrogen receptor beta. *PLoS One* 7 (2): e31109.

54 Alonso-Magdalena, P., Quesada, I., and Nadal, A. (2011). Endocrine disruptors in the etiology of type 2 diabetes mellitus. *Nat. Rev. Endocrinol.* 7 (6): 346–353.

4

Signal Transduction in Disease

Relating Cell Signaling to Morbidity and Mortality

Patricia E. Ganey[1] *and Sean A. Misek*[2]

[1] *Department of Pharmacology and Toxicology, Michigan State University, East Lansing, MI, USA*
[2] *Department of Physiology, Michigan State University, East Lansing, MI, USA*

4.1 Introduction

Disease pathogenesis can often arise from disruptions in normal signaling pathways, which are often complex and integrated. For example, perturbations in intracellular calcium homeostasis have been implicated in the development of neurodegenerative diseases such as Parkinson's, Huntington's, and Alzheimer's disease [1–3]. Additionally, simple dysregulation of signal transduction cascades in response to alterations in the local cellular environment (e.g. oxidative stress, endoplasmic reticulum stress, mitogen-activated protein kinase (MAPK) activity, and alterations in calcineurin, to name a few [1, 4–7]) has also been shown to contribute to disease pathogenesis. Further, perturbations in signal transduction networks can lead to altered physiological states, such as asthma and diabetes. Asthma arises from disruption of inter- and intracellular signaling in inflammatory cells, as well as alterations in the molecular machinery for synthesis of mucus; these changes involve lipid mediators, cytokines, T cells, and innate lymphoid cells [8–11]. Disruption of pathways signaling through Notch, MAPKs, phosphoinositide-3-kinase (PI3K), and protein tyrosine kinases (PTKs) has been shown to contribute to the pathogenesis of asthma [12–14]. Diabetes, which was discussed in detail in Chapter 3, is associated with alterations in Wnt and bone morphogenetic protein 4 (BMP4) signaling [15], as well as disturbance of the function of glucose transporter 4 [16, 17] and peroxisome proliferator-activated receptors (PPARs) [18].

One of the most well-researched examples of altered signal transduction in disease pathogenesis is the study of cancer. Large efforts in the field have been dedicated to understanding the imbalances that lead to the development of this

Cellular Signal Transduction in Toxicology and Pharmacology: Data Collection, Analysis, and Interpretation, First Edition. Edited by Jonathan W. Boyd and Richard R. Neubig.

disease. From a holistic point of view, inhibition of apoptotic pathways (that are initiated via signal transduction), which would normally eliminate mutated or otherwise damaged cells, plays a critical permissive role in carcinogenesis [19, 20]. Further, imbalances in cellular homeostasis can result in adaptations at the cellular level that facilitate disease development (e.g. loss of pathways that degrade hypoxia-inducible factor-1 alpha [HIF-1α] leads to its stabilization, affording a survival advantage to oxygen-deprived tumor cells [21]). Additionally, changes in DNA methylation, microRNAs, and disrupted Notch signaling are among the other pathways for which perturbation can contribute to the development of cancer [22–25].

From a genetic perspective, one of the most common oncogenes – or mutated genes that drives aberrant cancer growth and survival – is the small GTPase protein Ras. Mutations that prevent GTP hydrolysis and "turn-off" of the protein are found commonly in the KRAS gene in pancreatic, colorectal, and lung cancers, as well as the NRAS gene in melanoma, both of which participate in Ras signaling. Protein kinases are another well-studied species in the field of signal transduction and cancer, as they are often mutated or overexpressed. Kinases are affected by the activity of oncogenes as well as tumor suppressor genes. The first definitively identified oncogene was src from the Rous sarcoma virus. Both this viral protein and its mammalian cellular homolog, Src, are PTKs encoded by the src gene; other known examples of oncogenes associated with tyrosine kinases include the epidermal growth factor receptor (EGFR), the EGFR family protein HER2 in breast cancer, platelet-derived growth factor (PDGF), vascular endothelial growth factor (VEGF), and anaplastic lymphoma kinase (ALK), which is important in lung cancer as well as lymphoma. A number of serine/threonine kinases are also mutated in melanoma, lung, and other cancers. These include the v-Raf murine sarcoma viral oncogene homolog B (BRAF) and the kinases that activate mitogen-activated protein kinase (MAPK) and extracellular-regulated kinase (ERK) kinase (MEK1/2). The activity of tumor suppressor genes that code for important signaling proteins can be inactivated or lost in cancer, also resulting in dysregulation of kinase activity. For example, one of the most common tumor suppressor genes is phosphatase and tensin homolog (PTEN), which encodes phosphatidylinositol-3,4,5-trisphosphate 3-phosphatase. This lipid phosphatase breaks down phosphatidylinositol (3,4,5)-trisphosphate (PIP_3) to bisphosphate product phosphatidylinositol (4,5)-bisphosphate (PIP_2). Since PIP_3 is the primary activator of the pro-survival Akt kinases, loss of a key degradation pathway for PIP_3 provides an important stimulus for cancer development and progression.

The above examples merely scratch the surface of intracellular signaling pathways that contribute to disease. In this chapter, we will focus on changes that go beyond single cells to understand the complex interactions between cell types that often contribute to disease. Rather than attempt complete coverage, this chapter will focus on two examples: (i) First, diseases in which tissue fibrosis develops will illustrate the complexity and extensive interaction of cells and signaling events that participate in the development of fibrotic diseases. (ii) Second, the development of cancer resistance to the new "targeted" drugs that modulate signal transduction can involve cell–cell interactions, and it is important to note that both heterogeneity and intercellular interactions among the large population of cancer cells will contribute to the complexity of response. With these two topics as the focus of the chapter, we aim to illustrate general principles that are broadly applicable to signal transduction in disease.

4.2 Fibrosis as an Example of Complex Signaling

Fibrosis is a chronic condition characterized by excess deposition of extracellular matrix (ECM) proteins and thickening and stiffening of tissue, resulting in impairment of function and alterations in tissue architecture. It is considered a wound healing process of last resort, one that is initiated when repair processes fail. Fibrosis occurs in a variety of tissues including liver, lung, and kidney, and it is a result of persistent damage and inflammation. Fibrosis can range from a benign condition to one that is quite serious, and it can even be a precursor to neoplastic disease [26]. However, recent efforts have brought to light that even advanced changes are reversible, and resolution can be achieved [27]. This section will focus on intracellular signaling events and cellular interactions that lead to fibrosis, using hepatic fibrosis as an example. Hepatic fibrosis is the example of choice, as many of the signaling pathways involved in fibrosis of the liver are relevant to other tissues.

4.2.1 Development of Liver Fibrosis

Liver fibrosis can occur as a result of chronic injury due to viral infection, chemical insult (including chronic alcohol consumption), nonalcoholic steatohepatitis (fatty liver), autoimmune reactions, or cholestatic liver disease, among others [28]. Necrosis, accumulation of inflammatory cells, increased numbers of myofibroblasts, and excessive deposition of ECM are all evident in liver fibrosis [29]. Development of fibrosis occurs due to interactions among multiple resident liver cell types, inflammatory cells, and mediators, as well as ECM components.

Hepatocellular damage is the initiator of fibrosis. In response to cell damage, reactive oxygen species (ROS) and other signal transduction mediators are released from dying hepatocytes and Kupffer cells (KCs), the resident macrophages of the liver (Figure 4.1). These mediators attract other macrophages and inflammatory cells from the blood, leading to increased production and

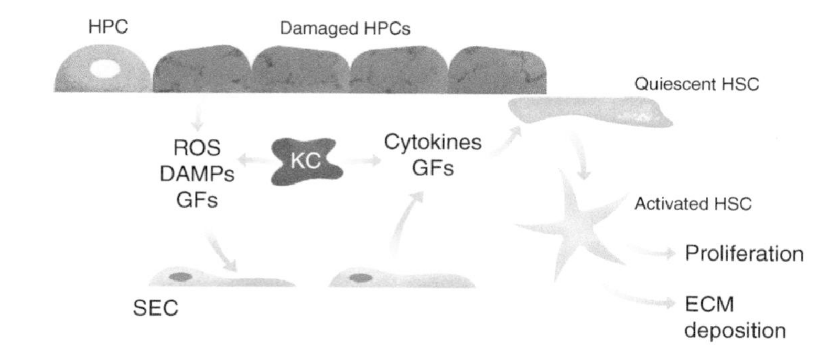

Figure 4.1 Initiation of fibrosis. Damaged hepatocytes (HPCs) release mediators that stimulate sinusoidal endothelial cells (SECs) and Kupffer cells (KCs). KCs phagocytose dead and dying cells and release mediators that attract more macrophages. Some mediators (cytokines, growth factors) released by KCs and SECs activate quiescent hepatic stellate cells (HSCs) to change phenotype to a myofibroblast-like cell that proliferates and secretes extracellular matrix (ECM) components. ROS, reactive oxygen species; DAMPs, damage-associated molecular pattern molecules; GFs, growth factors.

release of pro-inflammatory mediators [27]. Cytokines, such as tumor necrosis factor-α (TNF-α), interleukin (IL)-6, IL-1β, and transforming growth factor-β1 (TGF-β1), are released by macrophages to promote inflammation and recruit other immune cells, including T lymphocytes (T cells). Growth factors, including PDGF and epidermal growth factor (EGF), are also generated during this response. TGF-β1 released by hepatocytes is a major activator of hepatic stellate cells (HSCs), which then release more TGF-β1 to act in an autocrine manner for further activation [30]. Consequences of HSC activation are discussed in the following sections.

4.2.2 Animal Models of Hepatic Fibrosis

Animal models have been instrumental in our understanding of the pathogenesis of liver fibrosis. In general, the histologic changes that occur in animal models are similar to those seen in humans and have shed light on potential molecular mechanisms. The most common models involve ligation of the bile duct or repeated treatment of mice or rats with carbon tetrachloride (CCl_4) over a period of weeks [31]. Other less common but useful models include treatment with thioacetamide or chronic exposure to alcohol [32]. Dietary modifications, such as methionine/choline deficiency or high cholesterol diet, have also been used. In addition, a variety of genetic models, alone or in combination with nutritional changes, have been employed (reviewed in [33]). Through the use of animal models, much has been learned about the mechanism of fibrosis, which will be discussed throughout the remainder of the chapter.

4.2.3 Activation of Hepatic Stellate Cells

Myofibroblasts are critical to the development of liver fibrosis. They can arise from resident fibroblasts, such as portal fibroblasts, which are few in number in the liver, or from HSCs. Lineage tracing studies have revealed that activated HSCs represent the majority of myofibroblasts in livers of mice that had undergone bile duct ligation (BDL) or treatment with CCl_4 [34]. HSCs reside in the space of Disse, between the endothelium and hepatic parenchymal cells. Quiescent HSCs store lipids, especially vitamin A, which explains why they were formerly called "fat-storing" cells or Ito cells. These cells take on a myofibroblast phenotype upon activation.

Myofibroblasts differ from quiescent HSCs in that they contain less lipid, are contractile, and proliferate at a greater rate. The myofibroblast phenotype of activated HSCs is also characterized by expression of α-smooth muscle actin (αSMA), and this has been used *in vitro* and *in vivo* as a marker of HSC activation. The percentage of HSCs that express αSMA increases in patients with liver fibrosis, and hepatic expression of αSMA is elevated in livers of animals that have undergone BDL or treatment with CCl_4 [35].

Activated HSCs also express the gene encoding collagen type I, alpha 1 (COL1a1), and they subsequently release large amounts of collagen, elastin, and other ECM proteins [36–39]. The resulting change in composition and amount of ECM results in loss of fenestrae in sinusoidal endothelial cells (SECs) and changes in hepatic parenchymal cells, ultimately leading to disruption of liver architecture. Under normal circumstances, the amount of ECM present is maintained by a balance between its synthesis and its degradation by matrix metalloproteinases (MMPs). Activated HSCs upregulate and release tissue inhibitor of metalloproteinase 1 (Timp1), which inhibits MMP-mediated degradation of ECM [40, 41]. In hepatic tissue taken from patients with chronic fibrotic liver disease, the level of expression of Timp1 correlated positively with the severity of disease [42, 43].

Another consequence of activation of HSCs is the release of cytokines and growth factors (e.g. TGF-β1, PDGF, hepatocyte growth factor) and the upregulation of growth factor receptors [44–49]. The upregulation and release of TGF-β1 contributes to the development of fibrosis through further activation of HSCs and stimulation of their production of collagen, so it is not surprising that the level of TGF-β1 was positively related to the severity of fibrosis in human tissue [42, 43]. Intracellular signaling initiated by TGF-β1 is critical to development of fibrosis.

Activation of HSCs is thought to involve both an initiation phase and a perpetuation phase [50]. The initiation phase is mediated by products released from neighboring cells, including cells that are damaged. Mediators include cytokines such as TGF-β1, TNF-α, and IL-6, and growth factors such as PDGF and EGF, as well as lipid peroxides, and endothelin [51]. The perpetuation phase involves proliferation of HSCs and replacement of normal ECM with collagen type 1. Although a variety of mediators participate in the perpetuation phase, a primary stimulus for collagen production is TGF-β1 [52]. The role of matrix stiffness in the HSC phenotype has subsequently been investigated. *In vitro* experiments using hydrogels, for which stiffness was controlled, have demonstrated that HSCs remained quiescent in a soft matrix, whereas in a stiff matrix, their lipid droplet content was reduced, and they expressed collagen type 1 and αSMA [53, 54]. Furthermore, softening the matrix led to return to a quiescent phenotype. Thus, the stiffening of tissue that accompanies fibrosis likely contributes to HSC activation. Interestingly, mice treated with CCl_4 exhibited increased liver stiffness prior to evidence of fibrosis [55]. Clearly, the complex cell–cell and cell–environment interactions can be investigated to help elucidate the progression of disease states, like fibrosis.

4.2.4 Epithelial-to-Mesenchymal Transition (EMT)

Other cell types may also give rise to myofibroblasts. One potential mechanism to facilitate this is epithelial-to-mesenchymal transition (EMT). Fibrocytes are inactive mesenchymal circulating cells derived from bone marrow, and they can be found in liver after BDL injury or exposure to CCl_4. Studies have suggested that secretions from fibrocytes may have the potential to induce EMT. While they have been shown to express collagen and αSMA, their contribution to the pathology has been debated [56–60].

Myofibroblast-like cells can also arise from hepatic parenchymal cells (hepatocytes) and cholangiocytes though the EMT process [61–64]. EMT is effected through several transcription factors, including Snail1 and Snail2. These transcription factors repress transcription of genes associated with cell adhesion, such as cadherins. As a result, cells dissociate from the basement membrane, lose intercellular adhesion complexes, and take on mesenchymal properties, including motility. Exposure of hepatocytes to TGF-β1 was accompanied by a loss of E-cadherin and an increase in αSMA and vimentin [65, 66]. Hepatocytes isolated from CCl_4-treated animals displayed a myofibroblast phenotype [64]. Furthermore, hepatocytes that had undergone EMT contributed to liver fibrosis in CCl_4-treated mice [67]. On the other hand, the hypothesis that EMT contributes to the population of myofibroblasts in fibrotic liver has been challenged by studies that demonstrated a lack of conversion of hepatocytes or cholangiocytes to collagen-expressing cells in livers of CCl_4-treated mice [68, 69]. Additional research on this topic is needed to clarify this disparity.

4.2.5 Other Cellular Interactions in Fibrosis

Although HSCs play a critical function in the development of liver fibrosis, other cell types, including KCs, SECs, natural killer (NK) cells, and hepatocytes, participate. There is evidence of extensive crosstalk among cell types [70]. KCs, or resident tissue macrophages, actually play a dual role in liver fibrosis. Early in the development of fibrosis, they release pro-inflammatory and pro-fibrotic mediators, such as ROS, ILs, and TGF-β1, contributing to the pathogenesis. Later, macrophages produce mediators, such as MMPs, that promote resolution of fibrosis. A population of macrophages that contributes to resolution of fibrosis was observed in livers of CCl_4-treated mice, and deficiency in this population inhibited tissue remodeling [71]. Clearance of necrotic hepatocytes is critical to restoration of normal tissue after fibrosis develops. One hypothesis is that macrophages involved in the restoration of normal tissue are those that have phagocytosed dying cells. Macrophages exposed *in vitro* to debris from dead hepatocytes expressed a phenotype similar to that of macrophages that mediate resolution of liver fibrosis. In addition, the ability of macrophages to clear dead cells after exposure to CCl_4 was impaired in mice with reduced levels of HIF-1α in HSCs [72]. This defect leads to hepatic deposition of collagen. These results suggest that HIF-1α signaling in HSCs is important in determining the phenotype of macrophages during fibrosis.

Hepatic SECs are characterized by many fenestrae and a discontinuous basement membrane. This allows blood-borne substances access to the hepatocytes. Like other endothelial cells, SECs express nitric oxide (NO) synthase; SEC-derived NO contributes to changes in blood flow and pressure. Fibrosis causes loss of fenestrae and thickening of the basement membrane in a process termed "capillarization," which is accompanied by a reduction in NO release from SECs. In the absence of liver disease, SECs are thought to help maintain HSCs in a quiescent state, which is supported by the observation that HSCs are resistant to activation when cocultured with SECs. The loss of SEC-derived NO might contribute to HSC activation during fibrosis, although there is no direct evidence for that relationship [73, 74]. NO signals through soluble guanylate cyclase and protein kinase G. Activation of soluble guanylate cyclase slows progression of fibrosis in thioacetamide-treated rats. Activation of soluble guanylate cyclase inhibited activation of HSCs *in vitro*, and this effect was greater in the presence of SECs. However, the progression of fibrosis through this signaling pathway was not completely dependent on NO [75, 76].

As mentioned above, fibrosis is associated with tissue hypoxia and activation of HIF-1α. VEGF is a HIF-1α-regulated gene that is expressed by hypoxic hepatocytes and HSCs. VEGF from hepatocytes and HSCs contributes to the maintenance of SEC fenestrae through NO-dependent and NO-independent mechanisms [75]. NO-dependent VEGF signaling might underlie some of the actions mediated by guanylate cyclase, which is not surprising considering its role in angiogenic responses during fibrosis.

NK cells are innate lymphoid cells that reside in several tissues, including liver. As their name implies, NK cells kill tumor and other cells, in addition to other functions. In effecting these functions, they produce various cytokines, including interferon-γ (IFN-γ), and they interact with other cells within the liver. In general, NK cells are thought to play an anti-fibrotic role, primarily by killing activated HSCs through release of IFN-γ and signal transduction involving TNF-related apoptosis-inducing ligand (TRAIL). Cytotoxic NK cells accumulate in livers of patients with fibrosis [77]. When HSCs and NK cells isolated from patients with liver fibrosis were cocultured, NK cells became activated and induced apoptosis in the HSCs [78, 79]. The cytotoxic potential of NK cells toward HSCs decreased with increasing disease severity [78], suggesting that the more effective these cells are at eliminating HSCs, the more fibrosis is limited. Mice lacking NK cells developed more hepatic fibrosis in response to CCl_4 than control mice [80]. Furthermore, activation of NK cells in a model of hepatic fibrosis was protective. When mice were fed a diet containing 3,5-diethoxycarbonyl-1,4-dihydrocollidine to induce cholestatic liver fibrosis, treatment with polyinosinic–polycytidylic acid (poly I:C) to activate NK cells resulted in death of HSCs and a reduction of liver fibrosis. This result depended on NK cells and IFN-γ [81]. Treatment with poly I:C also reduced CCl_4-induced liver fibrosis. It should be noted, however, that poly I:C also affects function of dendritic cells and B cells [82, 83], so the effect may be through more cell types than just NK cells.

4.2.6 Intracellular Signaling Pathways Critical to Liver Fibrosis

4.2.6.1 TGF-β1

As previously mentioned, the activation of HSCs is affected primarily by TGF-β1. TGF-β1 is a member of the TGF-β superfamily of proteins that includes activins, inhibins, BMPs, and other proteins. TGF-β1 is synthesized as a propeptide. The latent form is bound to latency-associated peptide (LAP) and latent TGF-β binding proteins (LTBPs), forming a large complex. TGF-β1 can be activated by integrins, MMPs, plasmin, ROS, and other substances, as well as by myofibroblast contraction [27, 84–87].

Activated TGF-β1 binds a heterotetrameric complex containing two types of cell surface serine/threonine-specific protein kinase receptors, the TGF-β receptors type I and type II (TBRI and TBRII). Binding of TGF-β1 induces phosphorylation of TBRI by TBRII, which initiates intracellular signaling through phosphorylation of intracellular regulators (Figure 4.2).

The TGF-β1 signaling response involves Smad proteins, which are the main signal transducers for the TGF-β family. Smad proteins are either receptor-regulated (R-Smads; Smad1, Smad2, Smad3, Smad5, and Smad8), inhibitory (Smad6 and Smad7), or common (Smad4). They contain two Mad Homology (MH) domains separated by a linker region. MH1 binds DNA, whereas MH2 binds to TGF receptors or other MH2 domains in coactivators and corepressors. When TGF-β1 binds to its receptor, phosphorylation of TBRI reveals its Smad binding site, allowing association of R-Smads and their phosphorylation

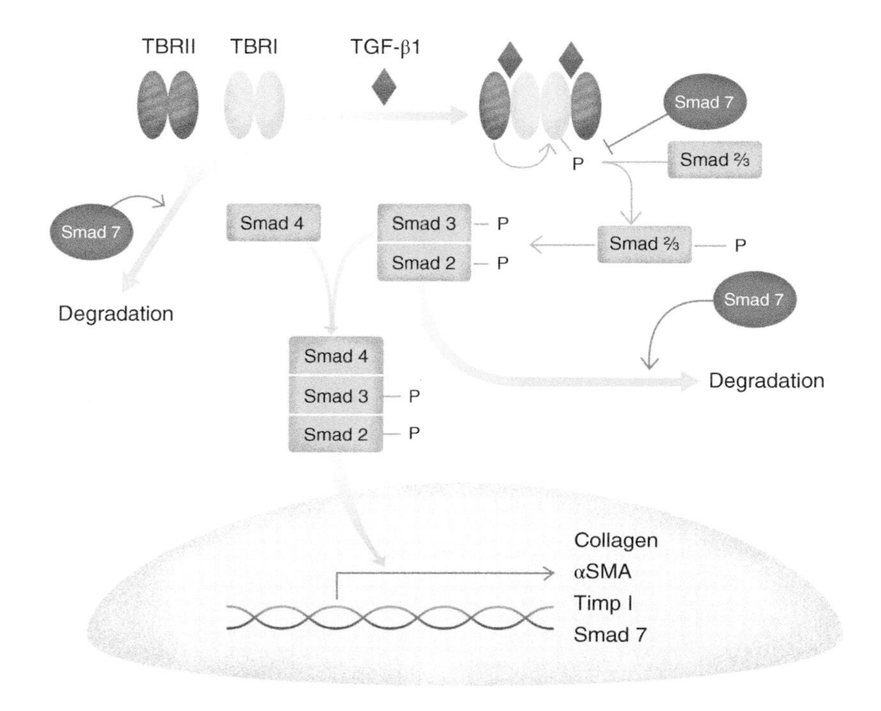

Figure 4.2 Smad-dependent TGF-β1 signaling. TGF-β1 binds to two receptors, TBRI and TBRII. These two receptors exist as dimers, and, when ligated by TGF-β1, they form a complex capable of phosphorylating Smad proteins. Phosphorylated Smad2 and Smad3 dimerize and associate with Smad4 to form a complex that translocates to the nucleus and acts as a transcription factor to regulate transcription of several genes involved in fibrosis. Smad7 negatively regulates the signaling process by inhibiting binding of Smad proteins to TBRI and by enhancing degradation of TBRI and Smad2 proteins. αSMA, α-smooth muscle actin; Timp1, tissue inhibitor of metalloproteinase 1.

at C-terminal serine domains. Phosphorylation creates a motif capable of binding MH2 domains on R-Smads or on Smad4 [88, 89]. Once phosphorylated, Smad2 and Smad3 associate with Smad4 and translocate to the nucleus where the complex functions as a transcription factor [88–91]. Although free Smads move easily between the cytoplasm and the nucleus using nuclear porins, phosphorylated Smads use a variety of nuclear transport factors [92, 93].

A conserved domain known as MH1 facilitates sequence-specific Smad binding to DNA. Phosphorylation of the MH1 domain of Smad2 or Smad3 inhibits DNA binding [94], illustrating one method by which signal transduction can regulate the activity of Smads. Smad-regulated genes include those encoding collagen, αSMA, E-cadherin, and Timp1 [40, 41]. The gene for inhibitory Smad7 is also regulated through this pathway; its transcription and translation lead to downregulation of Smad signaling.

Interaction among Smad proteins determines the outcome of signaling. For example, Smad2 and Smad3 might not contribute equally to fibrosis. In human fibrotic tissue and cells, both Smad2 and Smad3 are phosphorylated [29]. In Smad3 knockout mice, treatment with CCl_4 elicited a smaller increase in expression of collagen mRNA, although αSMA expression was increased relative to wild-type (WT) mice [95]. Results were similar in HSCs isolated from Smad3-deficient mice. Levels of collagen and Timp1 were decreased in TGF-β1-treated human HSCs deficient in Smad3. Conversely, overexpression of Smad3 increased the expression of collagen and Timp1 and decreased expression of MMP-2 *in vitro* [29]. Manipulation of Smad2 *in vitro* had the opposite effect of that described for Smad3 [29, 95]. In extrahepatic tissues, deficiency of Smad3 blocked EMT and fibrosis, while Smad2 deletion increased fibrosis [96, 97]. Both functioning Smad2 and Smad3 are required for transcriptional activity, as evidenced by a lack of nuclear translocation in the absence of either [29, 95]. These results suggest a degree of complexity in Smad signaling at the level of interaction of Smads. The essential role of Smad4 in fibrosis is illustrated by the observation that deletion of Smad4 inhibits TGF-β1-induced ECM deposition by mesangial cells [98].

Smad7 downregulates Smad signaling by regulating binding of Smads to TBRI, thereby preventing their phosphorylation. In addition, Smad7 facilitates the degradation of Smad2 and TBRI by recruiting ubiquitin-specific processing protease 15 and Smad ubiquitination regulatory factors (SMURFs) [99, 100], which target the proteins for degradation. In CCl_4-induced liver fibrosis, loss of Smad7 enhanced HSC activation and development of fibrosis, but overexpression of Smad7 diminished fibrosis [101, 102].

4.2.6.2 Kinase Pathways Involved in Fibrotic Responses

Activation of protein kinases also plays a role in development of liver fibrosis. TGF-β1 increases phosphorylation of ERK, c-Jun N-terminal kinase (JNK), and p38 MAPKs [103–105]. Thioacetamide- or CCl_4-induced fibrosis in mice was associated with activation of ERK [106], and interference with ERK signaling reduced expression of collagen [107]. Furthermore, transcriptomic analysis of human keratinocytes revealed a role for ERK-regulated pathways in cell adhesion and motility; inhibition of MAPK signaling inhibited EMT [103, 108]. Conversely, activation of ERK can lead to phosphorylation of Smad proteins in the linker region that causes their retention in the cytoplasm [109], thereby inhibiting their transcriptional activity.

PI3 kinase/Akt signaling is also involved in fibrosis, especially through effects on EMT. Loss of Akt is associated with loss of EMT phenotype and increased expression of E-cadherin [110]. Snail1 and Snail2 are transcription factors that are upregulated through TGF-β1 signaling (Figure 4.3). Their overexpression

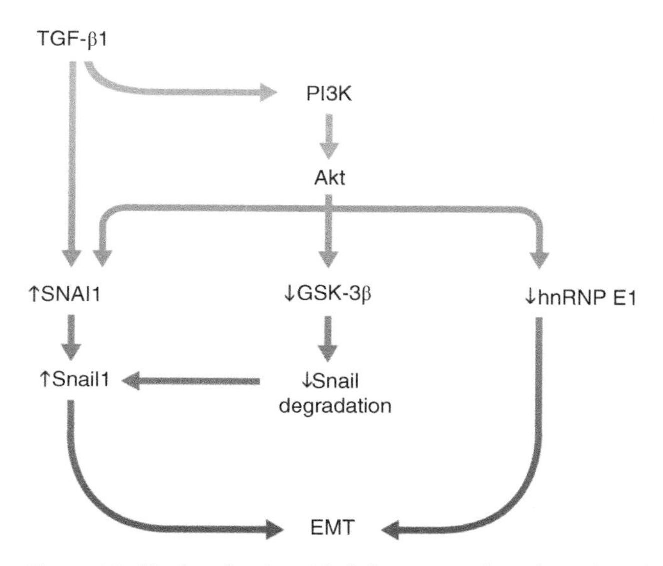

Figure 4.3 Akt signaling in epithelial-to-mesenchymal transition (EMT). TGF-β1 increases expression of SNAI1, the gene encoding Snail1, and also activates PI3 kinase (PI3K). Signaling through PI3K activates Akt, which also upregulates SNAI1, leading to increased Snail1 that promotes EMT. Akt also inhibits glycogen synthase kinase-3β (GSK-3β), which normally targets Snail for degradation. In addition, Akt inhibits a number of proteins such as heterogeneous nuclear ribonucleoprotein E1 (hnRNP E1) that repress EMT.

induces EMT [111, 112]. They complex with corepressors and histone deacetylase to reduce expression of E-cadherin. Akt signals through nuclear factor κB (NF-κB) to induce transcription of SNAI1, the gene for Snail1 [113]. Akt also inhibits glycogen synthase kinase-3β (GSK-3β), which targets Snail1 protein for degradation [114]. Accordingly, Akt both increases Snail1 protein synthesis and inhibits its degradation, permitting Snail-mediated EMT. In addition, Akt inactivates proteins (e.g. heterogeneous nuclear ribonucleoprotein E1 [hnRNP E1]) that suppress EMT [115]. Furthermore, in tumor cells, Akt activates mammalian target of rapamycin (mTOR) complexes 1 and 2, both of which contribute to EMT [116, 117].

4.2.6.3 HIF-1α

As mentioned previously, tissue hypoxia occurs during fibrosis, likely from diminished hepatic blood flow as well as deposition of fibrin in sinusoids [118, 119]. Hypoxia leads to activation of HIF-1α. HIF-1α was increased in livers of mice subjected to BDL [120]. This was accompanied by an increase in mRNA for collagen, αSMA, and PDGF. These responses were reduced in livers of mice deficient in HIF. In patients with fibrotic liver disease, HIF-1α protein was detected in macrophages, hepatocytes, and fibroblasts [121], and experimental evidence suggests that activation of HIF-1α in KCs contributes to hepatic fibrosis. In mice with myeloid cell-specific deficiency in HIF-1α or HIF-1β, BDL led to smaller increases in expression of collagen and αSMA [122]. Furthermore, exposure of isolated KCs to hypoxia led to increased mRNA for PDGF and VEGF; this response was absent in KCs isolated from HIF-1β-deficient mice [122].

4.2.6.4 miRNA

MicroRNAs (miRNAs) are small noncoding RNAs that regulate the translation and transcription of genes. The role of miRNAs in disease has been a relatively recent subject of research exploration. Several miRNAs have been suggested to play different roles in the pathogenesis of hepatic fibrosis, although no causality has been demonstrated directly. In human tissue, expression of miR-199a, miR-199a*, miR-200a, and miR-200b tracked positively with severity of fibrosis [123]. Similarly, miR-21 has a positive effect on activation of HSCs as well as EMT [124, 125].

On the other hand, some miRNAs correlate negatively with fibrosis. Microarray analysis of livers from CCl_4-treated mice detected 31 differentially expressed miRNAs, of which the majority were decreased in expression compared to control mice [126]. Among these, decreased expression of miR-29b correlated with decreased liver fibrosis. Expression of miR-29 family members was also decreased in livers and plasma from patients with fibrosis/cirrhosis [126]. miR-29 family members were highly expressed in HSCs isolated from livers of control mice, and expression was decreased by exposure to TGF-β1 *in vitro*. Overexpression of miR29b in HSC led to decreased expression of Col1a1 but did not affect Timp1 or αSMA expression. Similar results were observed with miR-122 [127].

4.2.6.5 Toll-Like Receptors (TLRs)

Toll-like receptors (TLRs) are receptors for pathogen-associated and damage-associated molecular pattern molecules. They mediate responses to inflammagens, such as bacterial products (e.g. lipopolysaccharide) and molecules released by dying cells (e.g. high-mobility group box 1 protein). Both TLR2 and TLR4 have been implicated in fibrogenesis. Activation of TLR4 signaling pathways leads to decreased expression of the BMP and activin membrane-bound inhibitor homolog (Bambi), which is a pseudoreceptor for TGF-β1 that lacks intracellular kinase activity. Reductions in Bambi play a permissive role for profibrogenic actions of TGF-β1 on HSCs [128]. In TLR2-deficient mice treated with CCl_4, hepatic fibrosis, HSC activation, and collagen deposition were reduced, as were inflammation and necrosis [129]. Similar results were observed in TLR5-deficient mice [130]. These latter two studies demonstrate the role of inflammation in the development of fibrosis. Taken together, these results suggest that inflammatory signaling through TLRs can contribute to the pathogenesis of liver fibrosis.

4.3 Cancer Drug Resistance: Complex Cellular and Population Changes

Modern cancer research is based on a deep understanding of signal transduction mechanisms. Many of the genetic mutations, amplifications, or chromosomal translocations that cause cancer (i.e. oncogenes or tumor suppressors) are in signal transduction proteins as described in Chapter 2. Oncogenic mutations occur in a broad array of receptors (e.g. receptor tyrosine kinases [RTKs]), many kinase pathways such as the mitogen-activated kinase (MAPK) and cell cycle-dependent kinase (CDK) mechanisms, small GTPases (e.g. KRAS), and transcription factors (e.g. MYC). Tumor suppressors include phosphatases (e.g. PTEN) and GTPase deactivators (e.g. neurofibromatosis 1 [NF1]). As might be expected, the mutated oncoproteins are excellent drug targets when inhibitors can be found. EGFR is commonly mutated in non-small cell lung cancer (NSCLC), and EGFR tyrosine kinase inhibitors (TKIs) have become a valuable treatment strategy for these tumors [131–133]. In melanoma, the MAPK pathway is frequently mutated with nearly 60% of melanomas having mutations in the serine/threonine kinase BRAF. Inhibitors of BRAF (BRAFi) produce excellent clinical responses in BRAF-mutant melanomas [134].

Drug resistance, however, remains a major hurdle in achieving durable therapeutic responses for EGFR and BRAFi and for virtually every "targeted" cancer therapy that is based on signal transduction mechanisms. Drug resistance can be divided into two broad classes: (i) intrinsic resistance and (ii) acquired resistance. The exact definition of intrinsic versus acquired resistance is still debated and depends on whether the cancer is being discussed on a tumor level or on a cellular level. For example, a point mutation found in 0.01% of tumor cells may promote intrinsic resistance to a drug on the cellular level for those rare cells with a resistance-causing mutation. For the tumor, resistance to the drug would be acquired as the mutant cell clonally expands. Due to this confusion, another way to stratify resistance mechanisms is to consider genomic resistance and non-genomic resistance mechanisms.

4.3.1 Genomic Resistance Mechanisms

Genomic resistance mechanisms are typically caused by mutations, copy number alterations (CNAs), or chromosomal translocations. Cells that harbor genomic resistance mechanisms may only comprise a small subpopulation of cells in drug-naïve tumors (Figure 4.4). For lung cancers treated with EGFR

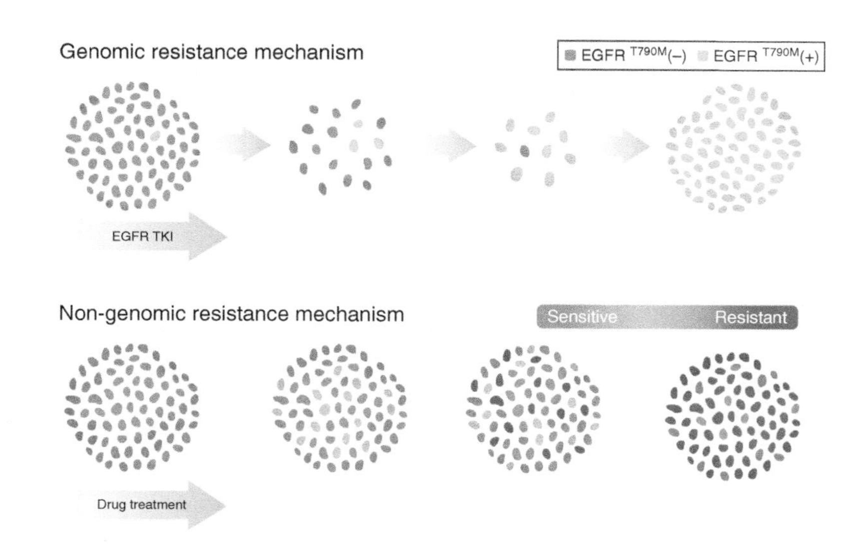

Figure 4.4 Comparison of genomic and non-genomic resistance mechanisms. *Genomic* – rare cells within a non-small cell lung cancer (NSCLC) tumor contain $EGFR^{T790M}$ mutations. Upon treatment with an EGFR TKI, such as erlotinib, cells with $EGFR^{T790M}$ mutations can continue proliferating, leading to the selection for these cells within the bulk tumor population. Ultimately this results in tumor relapse. *Non-genomic* – upon drug treatment, the signaling circuitry within cancer cells is altered. These alterations allow for cancer cells to retain their ability to proliferate even in the presence of drug treatment. This potentially can affect all cells in the tumor at once.

TKIs, one frequently observed resistance mechanism is the so-called "gatekeeper" mutation in EGFR, in which a single amino acid is altered from threonine to methionine (i.e. $EGFR^{T790M}$) [135–137]. This prevents drug binding and causes loss of effectiveness to first-generation EGFR inhibitors. Finally, activation of parallel signaling pathways can promote drug resistance, since activation of the parallel pathway may phenocopy the role of the drug target pathway. An example of this is deletion or mutation of PTEN, which promotes resistance to a wide range of targeted therapies via activation of the PI3K/Akt pathway [138–140].

While drug-naïve tumors may have only a few cells with this mutation, drug treatment kills the cells with WT EGFR, providing selective pressure that gives a growth advantage to the $EGFR^{T790M}$ cells, allowing for their clonal expansion. Similarly, various estrogen receptor (ER) mutations in breast cancer can promote resistance to ER antagonists [141]. In both cases, drug resistance results from direct genomic modification of the drug target to result in signaling pathway reactivation.

Pathway reactivation can also result from mutation or CNAs in genes which encode proteins upstream or downstream of the drug target (Figure 4.5).

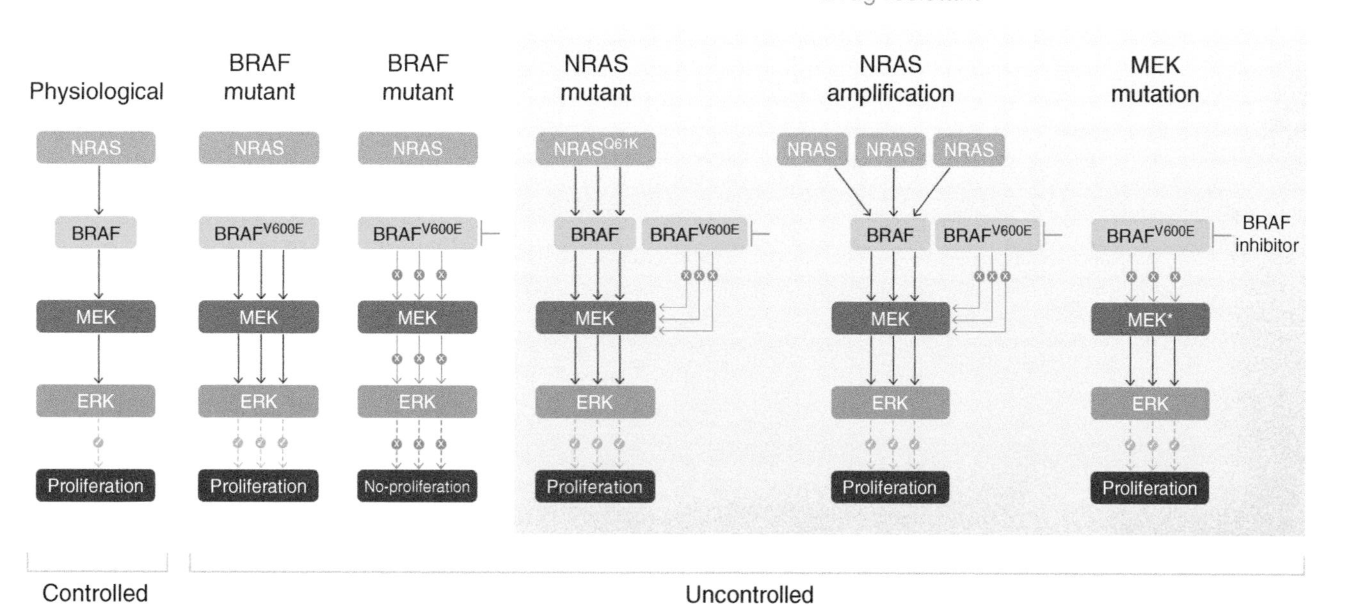

Figure 4.5 MAPK pathway reactivation results in BRAF inhibitor resistance in melanoma tumors. *Physiological* – signaling diagram detailing the MAPK pathway under physiological conditions. *BRAF-mutant* – BRAF^{V600E} mutations result in constitutive MAPK activation regardless of upstream stimulus. Inhibition of BRAF^{V600E} results in MAPK pathway inactivation and suppression of proliferation. *MAPK-reactivating mutations* – selection for cells with NRAS^{Q61K} mutations, NRAS amplification, and MEK mutations results in tumor recurrence since these genomic alterations result in BRAF^{V600E}-independent MAPK pathway reactivation. MAPK-reactivating mutations are located within the shaded region of the diagram.

BRAFi initially produce excellent responses in BRAF-mutant melanomas [134], but resistance develops rapidly. For example, one common mechanism of resistance to BRAFi is mutations or CNAs in NRAS or MEK, which flank BRAF in the pathway [142, 143]. Because of this, MEK inhibitors (MEKi) have also been used (see below) with significant benefit.

4.3.2 Non-genomic Mechanisms

Non-genomic mechanisms, such as epigenetic remodeling or rewiring of signaling circuitry, can also result in drug resistance. Like genomic resistance mechanisms, non-genomic resistance mechanisms can also result in pathway reactivation or activation of parallel signaling mechanisms. Inhibition of one RTK may lead to compensatory upregulation of other RTKs, resulting in reactivation of downstream signaling pathways. In other cases, specific epigenetic states poise cancer cells to upregulate a resistance gene program in response to drug treatment [144]. A common thread among many non-genomic resistance mechanisms is that drug treatment induces specific alterations in the cell's signaling circuitry. These mechanisms most commonly fall in the category of acquired resistance mechanisms.

4.3.3 Non-cancer Drug Resistance Paradigms

Drug resistance is not a problem that is unique to cancer. For example, bacterial drug resistance poses a major problem in both livestock management [145] and human health [146]. These non-cancer systems can provide some guidance to future targeted cancer therapies. One success story in combatting drug resistance is illustrated by the development of modern HIV therapeutic strategies. In this case, the use of drug combinations provides dramatically prolonged disease suppression, sometimes for the remainder of the patient's lifetime [147, 148]. While such an approach works well for bacterial or viral drug targets, success using these types of approaches has remained elusive in cancer. This is because cancer cells exploit the same signaling pathways that nonmalignant cells depend on; many of these drug combinations can have significant toxicity. This issue can be partially blunted by developing mutation-specific therapeutic strategies. For example, the BRAFi vemurafenib has approximately 10–100-fold selectivity for the cancer-specific mutant $BRAF^{V600E/K}$ over the normal protein $BRAF^{WT}$ [149], boosting the therapeutic index.

One mechanism of bacterial antibiotic resistance is expression of multidrug efflux pumps. Similar proteins, such as the multidrug resistance protein 1 (MDR1) (gene name ABCB1), are expressed on the surface of cancer cells and promote resistance to a wide range of therapeutics by preventing the accumulation of drug within the cell [150]. Unfortunately, in humans MDR1 inhibitors display significant toxicities at therapeutically relevant doses, which highlights yet another challenge in targeting cancer drug resistance mechanisms [151].

4.3.4 Tumor Heterogeneity as a Driver of Drug Resistance

As was first proposed by Nowell in 1976, tumors become substantially more heterogeneous during disease progression as different cellular clones acquire new genomic and non-genomic alterations [152]. These tumor cell clones are spatially heterogeneous, meaning that clonally similar tumor cells are located in close proximity to each other within a tumor [153]. This observation is not surprising, given what we now know about how sequential acquisition of new alterations in cancer cells contributes to oncogenesis and tumor progression. In most cases, the targeted therapies used in the clinic were designed to target the "driver oncogene." These approaches fail to account for the considerable intratumor heterogeneity, ultimately leading to selection for clones with drug-resistant characteristics.

Advances in sequencing technology, especially the ability to determine gene expression in an individual cell (i.e. single-cell transcriptomic profiling), have led to a revolution in our understanding of tumor heterogeneity and the clonal evolution of cancer. Most major tumor types have been transcriptomically characterized on a single-cell level [154–158]. In all cases, tumor cells can be stratified into clusters based on their transcriptional signatures. In melanoma, two cell states were identified, which show different gene expression programs. One has high expression of the protein kinase AXL (AXL^{High}), and the other has high expression of MITF, a transcription factor that promotes melanocyte differentiation ($MITF^{High}$) [156]. Cells in the AXL^{High} cell state show resistance to MAPK inhibitor therapy (i.e. MAPKi, which consists of BRAF and/or MEKi alone or in combination) [159]. While each tumor could be classified as predominantly AXL^{High} or $MITF^{High}$ through bulk RNA-Seq analysis, drug-resistant AXL^{High} cells were present in all tumors to some degree. This was found even in drug-naïve tumors that had a $MITF^{High}$ signature on a bulk transcriptomic level. This point is important, since bulk RNA-Seq analysis would suggest that a $MITF^{High}$ tumor would likely be sensitive to MAPKi therapy; however, owing to the presence of a subpopulation of resistant cells, tumor relapse would occur eventually.

Single-cell analysis of the tumor microenvironment is also important for studying drug resistance. It has become clear that cancer progression, metastasis, and treatment response depend on other cells in the vicinity of the cancer as well as the cancer cells themselves [160–163]. Cells in the tumor microenvironment, or stroma, that may affect these processes can include fibroblasts, macrophages, T cells, etc. [164–168]. In many contexts, including response to BRAFi, tumor-killing T lymphocytes ($CD8^+$ T cells) are pivotal for maintaining the drug response. Depletion or exhaustion of CD8+ T cells can be recognized by reduced cell numbers or an altered transcriptional program including changes in expression of T-bet, eomesodermin (EOMES), and B lymphocyte-induced maturation protein 1 (BLIMP1). This results in drug resistance [169, 170]. As with cancer cells, there is a high degree of heterogeneity in $CD8^+$ T cells, especially within the T-cell receptor (TCR), which determines the selectivity of immune recognition. Characterizing the heterogeneity and clonal expansion of these cells can provide important information on mechanism and prediction of prolonged drug responses. Using single-cell sequencing to identify tumors with an abundance of exhausted T cells, or simply a dearth of anti-tumor T cells, will help identify tumors that may be intrinsically resistant to select targeted therapies.

4.3.5 Mutational Drivers of Drug Resistance

As discussed above, drug resistance of NSCLC to EGFR TKIs is common due to the $EGFR^{T790M}$ mutation [135–137]. This mutation is located within the catalytic cleft of the tyrosine kinase domain of EGFR. Co-crystallization studies have demonstrated that the T790 residue is critical for inhibitor binding. Increased steric hindrance, due to methionine's bulkier side chain, prevents drug binding and also increases the affinity of EGFR for its substrate ATP [171]. What is interesting is that this mutation is often present in only a small subpopulation of cells. In one of the original studies of the $EGFR^{T790M}$ mutation, it was only detected in ~1% of cells [137]. Other studies have found this mutation to be even rarer. Despite this, 50–60% of patients treated with EGFR TKIs will relapse with $EGFR^{T790M}$ mutations [172]. What makes the identification of $EGFR^{T790M}$ such a success story is the development of second-line therapeutics strategies to combat this drug resistance. Tumors that developed resistance to first-line EGFR inhibitors such as erlotinib or gefitinib could be treated with second-generation $EGFR^{T790M}$ inhibitors such as rociletinib and osimertinib. Most $EGFR^{T790M}$-positive tumors respond strongly to one of these second-generation inhibitors [173, 174]. These findings are an excellent example of how identification of drug resistance mechanisms can lead to the development of new therapeutic options that substantially improve patient outcome.

Another success story in combatting drug resistance clinically is the development of BRAF + MEKi dual therapy. The most common mechanism of resistance to BRAFi is the emergence of secondary mutations that reactivate the MAPK pathway ([175] and see Figure 4.5). However, unlike the $EGFR^{T790M}$ mutations in lung cancer, there is a wide range of MAPK-reactivating mutations that emerge in melanoma. The most common alterations include NRAS or BRAF amplification, $NRAS^{Q61K}$ mutations, NF1 mutation/deletion, or MEK mutations. In all cases, these genomic alterations result in MAPK pathway activation. Due to this, MEKi were developed and used in combination with

BRAFi [176, 177]. It may initially appear counterintuitive to concurrently use two different inhibitors that target the same signaling pathway. However, the MEKi delays or prevents MAPK-reactivating resistance mechanisms from arising, since all these mutations ultimately signal downstream to activate MEK/ERK. Unfortunately, tumors still become resistant to BRAF + MEKi. One possibility is that MEKi are typically used at low concentrations due to on-target toxicity issues, resulting in incomplete MAPK pathway inactivation *in vivo*. Alternatively, it could be due to the emergence of MAPK-independent resistance mechanisms.

The mutations that drive EGFR TKI resistance or BRAFi resistance are just two examples of mutationally driven resistance mechanisms. However, there are mutationally driven resistance mechanisms for virtually every targeted cancer therapy that blocks a signal transduction mechanism. For example, mutations in the ER promote resistance to ER antagonists [141], and mutations in or deletions of the retinoblastoma tumor suppressor render cells nonresponsive to cell cycle checkpoint inhibitors such as blockers of CDK4/6 [178]. Thus far, only the most prominent mutations in well-studied genes have been characterized for their role in resistance to targeted therapies. However, tumors can have hundreds or thousands of mutations, and few are understood in terms of functional consequences. To make matters worse, these mutations may only be present in a fraction of tumor cells and may consequently be undetectable by bulk sequencing. Even if a mutation is identified and a causal link is established between that mutation and drug resistance, it is unlikely that there is an FDA-approved therapeutic avenue to target that resistance mechanism. From these examples, it is evident that while we have made enormous progress in understanding how genomic alterations contribute to drug resistance, there is still a long way to go before we can transform such knowledge into clinical benefit for patients.

4.3.6 Drug-Induced Rewiring of Signaling Networks as a Mechanism of Drug Resistance

Until now, we have discussed mutationally driven resistance mechanisms. On a cellular level, these mechanisms are mostly static, in the sense that drug treatment does not specifically induce the mutations; rather the drug simply selects for resistant cells. In many cases, drug treatment can specifically and dynamically induce changes in signaling pathway activation and can rewire the signaling networks within a cell without inducing genomic alterations. The time scales of these alterations can vary greatly from seconds to days or weeks. This may involve epigenetic mechanisms or stable alterations in gene transcription programs. Since this broad class of drug resistance mechanisms is induced by drug treatment, these mechanisms fall under the umbrella of acquired resistance mechanisms.

Most BRAF-mutant melanoma tumors respond to the BRAFi vemurafenib, but this response is not observed in BRAF-mutant colorectal cancer (CRC) patients even though both tumor types harbor the same mutation. One reason for this lack of response is because CRC tumors adaptively activate EGFR in response to BRAF inhibition [179]. This compensatory EGFR activation results in activation of the PI3K/Akt pathway and reactivation of the MAPK pathway, which subsequently promotes vemurafenib resistance (Figure 4.6). One might then assume that simultaneous inhibition of BRAF and EGFR may be enough to prevent drug resistance. This strategy works initially, but, ultimately, patients still relapse on these therapies. In some cases, relapse on anti-EGFR + BRAF therapy results from compensatory activation of other RTKs such as c-Met.

Resistance to targeted therapies through activation of RTKs appears to be a recurrent theme in cancer biology. In addition to CRC, melanoma tumors can also develop resistance to BRAFi through activation of RTKs such as EGFR, PDGFR, IGF-1R, and AXL. Ultimately, this results in compensatory reactivation of the MAPK pathway through a BRAF-independent mechanism or via activation of parallel pathways such as the PI3K/Akt pathway [180]. Due to the degree of similarity in RTK effector pathways, compensatory activation of

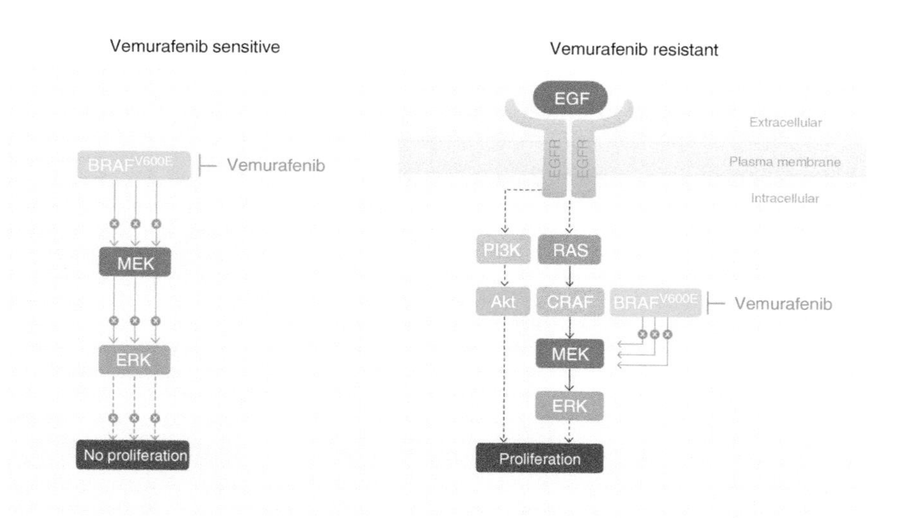

Figure 4.6 Compensatory EGFR activation results in BRAF inhibitor resistance. As in melanoma tumors, BRAF^{V600E} mutations in colorectal cancer result in constitutive MAPK pathway activation and constitutive pro-proliferative signals. Inhibition of mutant BRAF with vemurafenib suppresses proliferation. However, colorectal cancer cells develop resistance to these therapies via feedback activation of EGFR. Ultimately EGFR signals by activating the BRAF-related kinase CRAF to reactivate the MAPK pathway. EGFR also activates the PI3K/Akt pathway. Due to these actions, the cancer cells can proliferate even in the presence of vemurafenib. This ultimately results in drug resistance and tumor relapse.

other RTKs is also a common bypass mechanism of resistance to anti-RTK targeted therapies [181]. Earlier we discussed how selection for the rare NSCLC cell that harbors an $EGFR^{T790M}$ mutation promotes resistance to EGFR inhibitors. However, this resistance mechanism is not present in all tumors, and some cancers develop resistance through compensatory activation of other signaling pathways. For example, in lung cancer, activation of the RTK AXL results in resistance to the EGFR inhibitor erlotinib [182]. Also, signaling by fibroblast growth factor 2 (FGF2) acting through the fibroblast growth factor receptor 1 (FGFR1) promotes resistance to the EGFR inhibitor gefitinib through an autocrine/paracrine loop [183]. There are other mechanisms of compensatory RTK activation that are not necessarily drug induced, such as selection for NSCLC cells that harbor MET amplification [184]. Ultimately, activation of any of these RTKs results in activation of signaling pathways involved in cell proliferation such as the MAPK pathway or the PI3K pathway.

4.3.7 Parallel Pathways and Combination Treatments

As one looks at the problem of cancer drug resistance, patterns begin to emerge. Non-genomic drug resistance does not always occur simply by stochastic activation of other signaling pathways. Rather, drug treatment can induce specific alterations in the cellular signaling circuitry. Some of these are found repeatedly in different types of tumors. Cancer is a complex disease, and, to simplify it, we often think of each signaling pathway in isolation; the reality is that signaling pathways within a cell are generally linked to each other in complex networks. By learning how these signaling pathways interact, we can begin to understand how cancer cells adaptively respond to drug treatments. This information can be exploited to develop new therapeutic approaches to prevent or reverse drug resistance. For example, inhibition of one signaling pathway may render cells exquisitely sensitive to inhibition of another, a phenomenon known as "synthetic lethality." This can be due to effects of two different drugs on the same pathway or different pathways. The latter is frequently used in antiviral or antibacterial therapies where drug combinations target multiple pathways with mutually exclusive resistance mechanisms.

Alternatively, consider a situation in which two distinct signaling pathways can promote proliferation of a cancer cell (Figure 4.7). These pathways function as a logical OR gate where either one being "on" results in cancer. Specifically, we could have a situation in which Pathway 1 includes proteins "A," "B," "C," and "D" and Pathway 2 includes proteins "E," "F," "G," and "H." Pathway 1 could be the oncogenic driver in the cancer cell through an activating mutation in any of the four proteins, A–D. Activation of this pathway results in uncontrolled proliferation and tumor growth. Inhibiting signal transduction by this pathway with a small molecule inhibitor prevents the cell from proliferating. Activation of a second "parallel" signaling pathway may also be able to

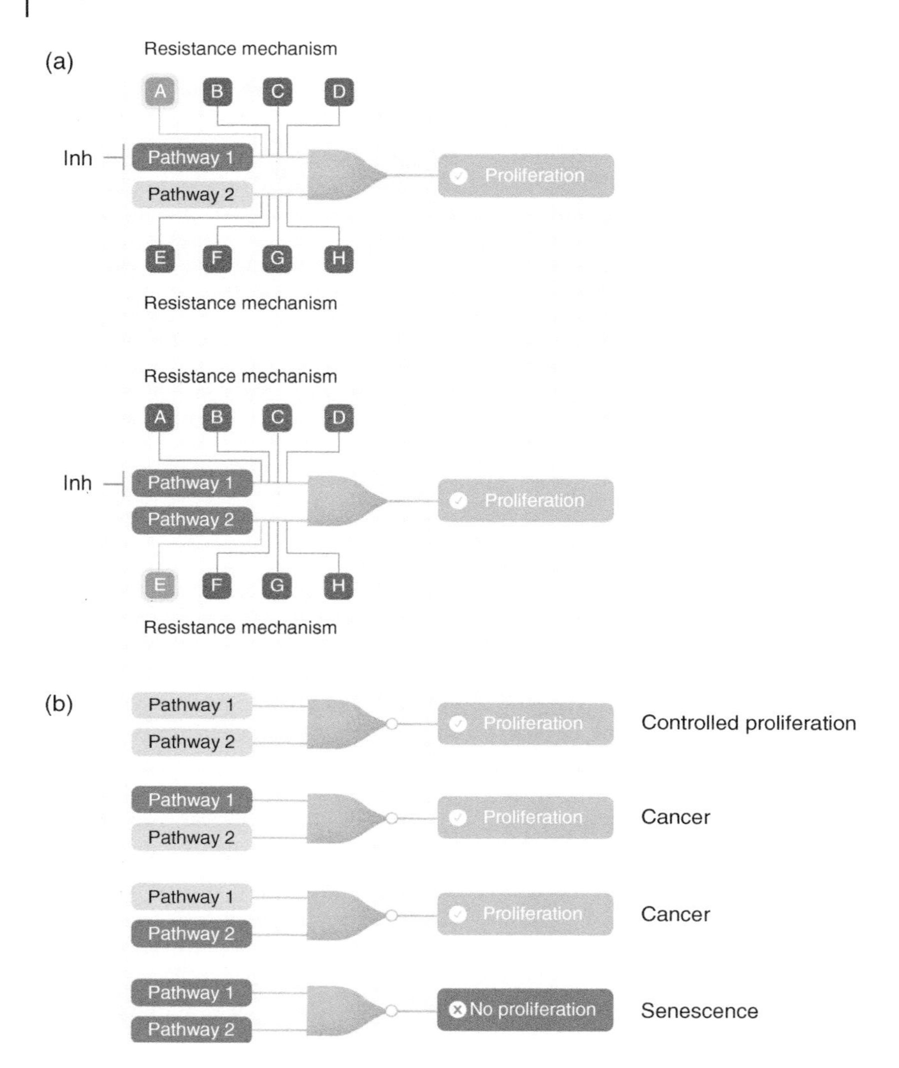

Figure 4.7 Parallel pathways can lead to resistance. (a) In a cancer cell driven by a mutant oncogene in Pathway 1, there are two primary mechanisms of genomic resistance. A new mutation can arise in a protein (e.g. A) downstream of the oncogene (and its inhibitor) to reactivate the original oncogenic pathway. Alternatively, a parallel Pathway 2 could substitute for Pathway 1 in maintaining the oncogenic proliferation. A driving mutation in Pathway 2 could also result in drug resistance. Approaches to therapy would differ based on which of these two resistance mechanisms occurred. (b) In some cases, two distinct pathways that each can drive proliferation and oncogenesis on their own, will interact to cause senescence when activated together.

promote proliferation without the need for activation of Pathway 1. If this occurs when a cancer is treated with an inhibitor of Pathway 1, this would represent drug resistance. Such resistance mechanisms could occur through an adaptive response, wherein drug treatment actively induces the compensatory activation of the other signaling pathway. Alternatively, cells could have activation of Pathway 2 through a mutational mechanism. In this case, rare cells with a mutation in protein "E" in Pathway 2 would also be resistant to Pathway 1 inhibitors.

If there are multiple signaling pathways that control cellular proliferation, then why do cancer cells not simply acquire activation of *all* these pathways during the clonal evolution of a tumor? Certainly, one would assume that maximizing proliferative signaling is something cancer cells would want. Counterintuitively, excessive activation of pro-proliferative signaling pathways results in oncogene-induced senescence [185, 186]. In this case, an overload of pro-proliferative signaling causes the cell to exit from the cell cycle and prevents proliferation. In some cases, concurrent activation of two pro-proliferative signaling pathways results in suppressed proliferation, but when one of the pathways is inhibited (for example, by drug treatment), it would allow for that cell to proliferate. In this case, this signaling circuitry functions as an XOR gate rather than an OR gate (Figure 4.7B). These signaling mechanisms may explain why many tumors relapse during drug treatment with no apparent genomic resistance mechanisms. One example of this phenomenon is in melanoma where EGFR expression or TGF-β treatment suppresses proliferation of $BRAF^{V600E}$-mutant melanoma cells. However, in the presence of a BRAFi, TGF-β-treated cells or cells with EGFR expression are conferred a growth advantage.

4.3.8 Epigenetic Mechanisms of Drug Resistance

It is becoming better appreciated that stable changes in the function of a cell do not require alterations in DNA sequence. Epigenetics refers to the creation of stable functional alterations by long-lasting, but potentially reversible, modification of DNA, RNA, or proteins. This often involves methylation of DNA or methylation, acetylation, or phosphorylation of histones or other proteins [187, 188]. Epigenetic changes in cancer cells clearly contribute to drug resistance. We discussed above how EGFR activation in melanoma cells promotes resistance to BRAFi; however, the mechanism was not described. As melanoma cells develop resistance to BRAFi, the landscape of DNA methylation marks in the genome, which largely serve to suppress gene expression, is dynamically altered [189]. Methylation of adjacent cytosine and guanine bases (i.e. CpG regions) can reduce gene expression. Methylation of CpG in the upstream and downstream enhancer regions of the EGFR gene is inversely correlated with EGFR expression. Methylation at these sites is lost in drug-resistant cells, which should enhance EGFR expression. This prediction is validated by the observation that treatment with the demethylating agent 5-azacytidine results in EGFR upregulation [190]. Of course, this is only one mechanism by which EGFR signaling can be increased, so there are likely other mechanisms at play. Given the importance of epigenetic changes, both as cancer drivers themselves and as mechanisms of resistance, substantial efforts are underway to target epigenetic mechanisms pharmacologically [191].

It is not necessary for epigenetic reprogramming to occur in every cell within a tumor. Melanoma cells can stochastically switch between a "bulk tumor cell" state and a slow cycling "pre-resistant" state; the latter state is very rare before drug treatment due to overgrowth of the faster proliferating cells [192]. Upon treatment with vemurafenib, which is a BRAFi, cells in the "bulk tumor cell state" are unable to proliferate. However, the cells in the "pre-resistant" state continue to proliferate and also undergo epigenetic reprogramming, which results in "burn-in" of a drug-resistant transcriptional state (Figure 4.8). A factor in this transcriptional new state is Sox10, a transcription factor that controls melanocyte differentiation. High Sox10 activity induces melanocyte-specific genes such as another transcription factor, MITF, and an enzyme, tyrosinase. Tyrosinase synthesizes melanin and makes melanocytes dark. The "burn-in" process begins with loss of accessible Sox10 binding sites through epigenetic changes, ultimately resulting in increased active chromatin and more accessible binding sites for the growth-promoting transcription factors activator protein 1 (AP-1) and transcriptional enhancer factor domain family member 1 (TEAD1).

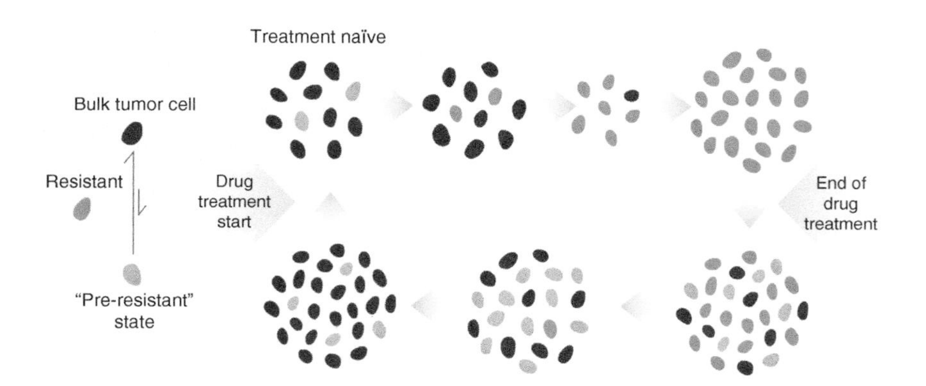

Figure 4.8 Dynamic state changes in non-genomic resistance mechanisms. Melanoma cells can stochastically switch between a bulk tumor cell state (common) and a "pre-resistant" state (rare). Upon drug treatment "pre-resistant" cells are transcriptionally rewired to be resistant to vemurafenib and retain the ability to proliferate in the presence of vemurafenib. This state is reversible and cessation of drug treatment results in the cancer cells returning to a bulk tumor cell state, which is presumably resensitized to the original drug treatment.

Interestingly, Sox10 was previously identified as a mediator of vemurafenib resistance in an shRNA screen of chromatin regulators [193].

Subpopulations of pre-resistant cells are not unique to melanoma. Work from Jeffrey Settelman's group identified a similar subpopulation of resistant cells, or "drug-tolerant persisters (DTPs)," in NSCLC cells treated with EGFR inhibitors [194]. Fitting with the recurring theme of compensatory RTK activation, these DTPs require signaling by the insulin-like growth factor receptor (IGF-1R) for sustained growth and maintenance of an altered chromatin state in the resistant cells. This altered chromatin state is associated with global changes in methylation and acetylation of histones, including changes in di- and tri-methylated histone 3 at lysine 4 ($H3K4me^2$ and $H3K4me^3$) and histone 3 acetylation at lysine 14 ($H3K14^{Ac}$) [194]. Interestingly, this study also demonstrated that cell surface antigens CD133 and CD44 are upregulated in the DTPs. Both CD133 and CD44 are putative cancer stem cell (CSC) markers. CSCs have been implicated in drug resistance in multiple cancer types including pancreatic cancer, glioblastoma, and breast cancer. DTP cells have also been identified in T-cell-derived acute lymphocytic leukemia (T-ALL). These cells are also driven by an epigenetic mechanism [195]. The take-home point from these observations is that the presence of small subpopulations of cells that are primed to become drug resistant is a recurring theme in multiple cancer types. These cells, while genetically similar, differ in their epigenetic and transcriptional landscapes. These differences profoundly alter the signaling circuitry within these cells and represent a major challenge to targeted cancer therapies.

4.3.9 Summary of Cancer Drug Resistance

Throughout this section we have identified several recurring themes in drug resistance. First, compensatory activation of signaling pathways (especially RTKs) through multiple mechanisms can promote drug resistance. This can occur through mutational reactivation of the drug target itself as well as activation of other proteins that share similar downstream effector pathways. Activation of parallel signaling pathways can also result in drug resistance. Signaling pathway plasticity is one of the main hallmarks of cancer and is certainly one of the driving mechanisms behind drug resistance. We also learned how rare subpopulations of cells contribute to drug resistance. These subpopulations can be resistant through genomic mechanisms (e.g. a mutation in a resistance gene) but also through non-genomic or epigenetic mechanisms. As with therapy of HIV/AIDS and tuberculosis, it seems clear that combination therapy will ultimately be needed to keep resistance at bay.

The complexity of drug resistance cannot be overstated; every mechanism that we have discussed here is at play to varying degrees in every single cancer cell. As one might expect, each tumor develops drug resistance through not one but many of these mechanisms. Owing to this complexity, it appears less likely that cancer can be treated as an acute disease, wherein a patient takes an antibiotic-like drug and is completely cured. The recent success of various immunotherapies, though, may challenge this assertion. But even this exciting new immunotherapeutic approach is not without drug resistance issues. Rather, it may be more appropriate to treat cancer as a chronic disease with the expectation that eventually the patient will relapse. Appropriate combinations should delay that outcome, hopefully for the natural life of the patient. Genomic, transcriptomic, and, perhaps, epigenomic profiling of tumors should help define personalized treatment strategies with appropriate combinations that evolve with the patient's disease. Even this vision is still in the future since, in many cases, we lack the arsenal of drugs to combat the various resistance mechanisms. To address this problem, we need to learn more about mechanisms of drug resistance, but we also need to develop new approaches to target those mechanisms.

4.4 Summary

Determining the mechanisms that may lead to morbidity and mortality requires a comprehensive theoretical understanding of cell signaling as well as experimental studies in models that appropriately match the organism (or organ or cell type) of interest. The role of many different interacting cell types and the tremendous heterogeneity even within a single (e.g. cancer) cell type make it challenging to dissect mechanisms of disease. A proper analogy would be trying to decipher the culprit in a murder mystery with no knowledge of who, when, where, how, or why the crime took place. Dissecting both the players and the roles of the multiple cellular constituents that may lead to disease often requires substantial corroborative results that provide independent identification of involvement. To accomplish this, there are many pharmacological, toxicological, and genetic techniques that can provide discriminatory power to aid in building a case (i.e. testing a hypothesis). The remainder of this book will focus on methods and techniques for collecting data, analyzing data, and drawing conclusions, with a final chapter focused on current needs in the field.

References

1 Shah, S.Z., Hussain, T., Zhao, D., and Yang, L. (2017). A central role for calcineurin in protein misfolding neurodegenerative diseases. *Cell Mol. Life Sci.* 74 (6): 1061–1074.

2 Naia, L., Ferreira, I.L., Ferreiro, E., and Rego, A.C. (2017). Mitochondrial Ca^{2+} handling in Huntington's and Alzheimer's diseases – role of ER-mitochondria crosstalk. *Biochem. Biophys. Res. Commun.* 483 (4): 1069–1077.

3 Raymond, L.A. (2017). Striatal synaptic dysfunction and altered calcium regulation in Huntington disease. *Biochem. Biophys. Res. Commun.* 483 (4): 1051–1062.
4 Wang, H., Dharmalingam, P., Vasquez, V. et al. (2017). Chronic oxidative damage together with genome repair deficiency in the neurons is a double whammy for neurodegeneration: is damage response signaling a potential therapeutic target? *Mech. Ageing Dev.* 161 (Pt A): 163–176.
5 Bahar, E., Kim, H., and Yoon, H. (2016). ER stress-mediated signaling: action potential and Ca(2+) as key players. *Int. J. Mol. Sci.* 17 (9): 1558.
6 Segales, J., Perdiguero, E., and Munoz-Canoves, P. (2016). Regulation of muscle stem cell functions: a focus on the p38 MAPK signaling pathway. *Front. Cell Dev. Biol.* 4: 91.
7 Kipanyula, M.J., Kimaro, W.H., and Seke Etet, P.F. (2016). The emerging roles of the calcineurin-nuclear factor of activated T-lymphocytes pathway in nervous system functions and diseases. *J. Aging Res.* 2016 (5081021): 1–20.
8 Sanak, M. (2016). Eicosanoid mediators in the airway inflammation of asthmatic patients: what is new? *Allergy Asthma Immunol. Res.* 8 (6): 481–490.
9 Akdis, M., Aab, A., Altunbulakli, C. et al. (2016). Interleukins (from IL-1 to IL-38), interferons, transforming growth factor B, and TNF-a: receptors, functions, and roles in diseases. *J Allergy Clin. Immunol.* 138 (4): 984–1010.
10 Krishnamurthy, P. and Kaplan, M.H. (2016). STAT6 and PARP family members in the development of T cell-dependent allergic inflammation. *Immune Netw.* 16 (4): 201–210.
11 Kim, H.Y., Umetsu, D.T., and Dekruyff, R.H. (2016). Innate lymphoid cells in asthma: will they take your breath away? *Eur. J. Immunol.* 46 (4): 795–806.
12 Zong, D., Ouyang, R., Li, J. et al. (2016). Notch signaling in lung diseases: focus on Notch1 and Notch3. *Ther. Adv. Respir. Dis.* 10 (5): 468–484.
13 Barnes, P.J. (2016). Kinases as novel therapeutic targets in asthma and chronic obstructive pulmonary disease. *Pharmacol. Rev.* 68 (3): 788–815.
14 Sharma, Y., Bashir, S., Bhardwaj, P. et al. (2016). Protein tyrosine phosphatase SHP-1: resurgence as new drug target for human autoimmune disorders. *Immunol. Res.* 64 (4): 804–819.
15 Smith, U. and Kahn, B.B. (2016). Adipose tissue regulates insulin sensitivity: role of adipogenesis, de novo lipogenesis and novel lipids. *J. Intern. Med.* 280 (5): 465–475.
16 Carnagarin, R., Dharmarajan, A.M., and Dass, C.R. (2015). Molecular aspects of glucose homeostasis in skeletal muscle – a focus on the molecular mechanisms of insulin resistance. *Mol. Cell Endocrincol.* 417: 52–62.
17 Govers, R. (2014). Molecular mechanisms of GLUT4 regulation in adipocytes. *Diabetes Metab.* 40 (6): 400–410.
18 Gross, B., Pawlak, M., Lefebvre, P., and Staels, B. (2017). PPARs in obesity-induced T2DM, dyslipidaemia and NAFLD. *Nature Rev. Endocrinol.* 13 (1): 36–49.

19 Wong, R.S. (2011). Apoptosis in cancer: from pathogenesis to treatment. *J. Exp. Clin. Cancer Res.* 30 (87): 1–14.
20 Croce, C.M. and Reed, J.C. (2016). Finally, an apoptosis-targeting therapeutic for cancer. *Cancer Res.* 76 (20): 5914–5920.
21 Nguyen, T.L. and Duran, R.V. (2016). Prolyl hydroxylase domain enzymes and their role in cell signaling and cancer metabolism. *Int. J. Biochem. Cell Biol.* 80: 71–80.
22 Mohammadi, A., Mansoori, B., and Baradaran, B. (2016). The role of microRNAs in colorectal cancer. *Biomed. Pharmacother.* 84: 705–713.
23 Mahdian-Shakib, A., Dorostkar, R., Tat, M. et al. (2016). Differential role of microRNAs in prognosis, diagnosis, and therapy of ovarian cancer. *Biomed. Pharmacother.* 84: 592–600.
24 Brzozowa-Zasada, M., Piecuch, A., Dittfeld, A. et al. (2016). Notch signalling pathway as an oncogenic factor involved in cancer development. *Contemp. Oncol.* 20 (4): 267–272.
25 Caplakova, V., Babusikova, E., and Blahovcova, E. (2016). DNA methylation machinery in the endometrium and endometrial cancer. *Anticancer Res.* 36 (9): 4407–4420.
26 Dongiovanni, P., Romeo, S., and Valenti, L. (2014). Hepatocellular carcinoma in nonalcoholic fatty liver: role of environmental and genetic factors. *World J. Gastroenterol.* 20 (36): 12945–12955.
27 Pellicoro, A., Ramachandran, P., Iredale, J.P., and Fallowfield, J.A. (2014). Liver fibrosis and repair: immune regulation of wound healing in a solid organ. *Nat. Rev. Immunol.* 14 (3): 181–194.
28 Bataller, R. and Brenner, D.A. (2009). Hepatic fibrosis. In: *The Liver: Biology and Pathobiology*, 5e (ed. I.M. Arias, H.J. Alter, J.L. Boyer, et al.), 433–452. Chichester: Wiley-Blackwell.
29 Zhang, L., Liu, C., Meng, X.M. et al. (2015). Smad2 protects against TGF-B1/Smad3-mediated collagen synthesis in human hepatic stellate cells during hepatic fibrosis. *Mol. Cell Biochem.* 400 (1–2): 17–28.
30 Bissell, D.M., Wang, S.S., Jarnagin, W.R., and Roll, F.J. (1995). Cell-specific expression of transforming growth factor-beta in rat liver. Evidence for autocrine regulation of hepatocyte proliferation. *J Clin. Invest.* 96 (1): 447–455.
31 Wasser, S. and Tan, C.E. (1999). Experimental models of hepatic fibrosis in the rat. *Ann. Acad. Med. Singapore* 28 (1): 109–111.
32 Krishnasamy, Y., Ramshesh, V.K., Gooz, M. et al. (2016). Ethanol and high cholesterol diet causes severe steatohepatitis and early liver fibrosis in mice. *PloS One* 11 (9): 1–20.
33 Sanches, S.C., Ramalho, L.N., Augusto, M.J. et al. (2015). Nonalcoholic steatohepatitis: a search for factual animal models. *BioMed Research International* 2015 (574832): 1–13.
34 Mederacke, I., Hsu, C.C., Troeger, J.S. et al. (2013). Fate tracing reveals hepatic stellate cells as dominant contributors to liver fibrosis independent of its aetiology. *Nat. Commun.* 4 (2823): 1–11.

35 Cassiman, D., Libbrecht, L., Desmet, V. et al. (2002). Hepatic stellate cell/myofibroblast subpopulations in fibrotic human and rat livers. *J. Hepatol.* 36 (2): 200–209.

36 Mannaerts, I., Schroyen, B., Verhulst, S. et al. (2013). Gene expression profiling of early hepatic stellate cell activation reveals a role for Igfbp3 in cell migration. *PloS One* 8 (12): 1–13.

37 Lee, K.S., Buck, M., Houglum, K., and Chojkier, M. (1995). Activation of hepatic stellate cells by TGF alpha and collagen type I is mediated by oxidative stress through c-myb expression. *J Clin. Invest.* 96 (5): 2461–2468.

38 Nieto, N., Friedman, S.L., and Cederbaum, A.I. (2002). Stimulation and proliferation of primary rat hepatic stellate cells by cytochrome P450 2E1-derived reactive oxygen species. *Hepatology* 35 (1): 62–73.

39 Iredale, J.P., Thompson, A., and Henderson, N.C. (2013). Extracellular matrix degradation in liver fibrosis: biochemistry and regulation. *Biochim. Biophys. Acta.* 1832 (7): 876–883.

40 Latella, G., Vetuschi, A., Sferra, R. et al. (2009). Targeted disruption of Smad3 confers resistance to the development of dimethylnitrosamine-induced hepatic fibrosis in mice. *Liver Int.* 29 (7): 997–1009.

41 Masszi, A. and Kapus, A. (2011). Smaddening complexity: the role of Smad3 in epithelial-myofibroblast transition. *Cells Tissues Organs* 193 (1–2): 41–52.

42 Dudas, J., Kovalszky, I., Gallai, M. et al. (2001). Expression of decorin, transforming growth factor-beta 1, tissue inhibitor metalloproteinase 1 and 2, and type IV collagenases in chronic hepatitis. *Am. J. Clin. Pathol.* 115 (5): 725–735.

43 Castilla, A., Prieto, J., and Fausto, N. (1991). Transforming growth factors beta 1 and alpha in chronic liver disease. Effects of interferon alfa therapy. *N. Engl. J. Med.* 324 (14): 933–940.

44 Marra, F., Valente, A.J., Pinzani, M., and Abboud, H.E. (1993). Cultured human liver fat-storing cells produce monocyte chemotactic protein-1. Regulation by proinflammatory cytokines. *J. Clin. Invest.* 92 (4): 1674–1680.

45 Czaja, M.J., Geerts, A., Xu, J. et al. (1994). Monocyte chemoattractant protein 1 (MCP-1) expression occurs in toxic rat liver injury and human liver disease. *J Leukoc. Biol.* 55 (1): 120–126.

46 Pinzani, M., Milani, S., Grappone, C. et al. (1994). Expression of platelet-derived growth factor in a model of acute liver injury. *Hepatology* 19 (3): 701–707.

47 Schirmacher, P., Geerts, A., Pietrangelo, A. et al. (1992). Hepatocyte growth factor/hepatopoietin a is expressed in fat-storing cells from rat liver but not myofibroblast-like cells derived from fat-storing cells. *Hepatology* 15 (1): 5–11.

48 Maher, J.J. (1993). Cell-specific expression of hepatocyte growth factor in liver. Upregulation in sinusoidal endothelial cells after carbon tetrachloride. *J. Clin. Invest.* 91 (5): 2244–2252.

49 Mullhaupt, B., Feren, A., Fodor, E., and Jones, A. (1994). Liver expression of epidermal growth factor RNA. Rapid increases in immediate-early phase of liver regeneration. *J. Biol. Chem.* 269 (31): 19667–19670.

50 Friedman, S.L. (2008). Hepatic stellate cells: protean, multifunctional, and enigmatic cells of the liver. *Physiol. Rev.* 88 (1): 125–172.

51 Reeves, H.L. and Friedman, S.L. (2002). Activation of hepatic stellate cells – a key issue in liver fibrosis. *Front. Biosci.* 7: 808–826.

52 Shah, R., Reyes-Gordillo, K., Arellanes-Robledo, J. et al. (2013). TGF-B1 up-regulates the expression of PDGF-B receptor mRNA and induces a delayed PI3K-, AKT-, and p70(S6K) -dependent proliferative response in activated hepatic stellate cells. *Alcohol. Clin. Exp. Res.* 37 (11): 1838–1848.

53 Caliari, S.R., Perepelyuk, M., Soulas, E.M. et al. (2016). Gradually softening hydrogels for modeling hepatic stellate cell behavior during fibrosis regression. *Integr. Biol. (Camb.)* 8 (6): 720–728.

54 Guvendiren, M., Perepelyuk, M., Wells, R.G., and Burdick, J.A. (2014). Hydrogels with differential and patterned mechanics to study stiffness-mediated myofibroblastic differentiation of hepatic stellate cells. *J. Mech. Behav. Biomed. Mater.* 38: 198–208.

55 Georges, P.C., Hui, J.J., Gombos, Z. et al. (2007). Increased stiffness of the rat liver precedes matrix deposition: implications for fibrosis. *Am. J. Physiol. Gastrointest. Liver Physiol.* 293 (6): 1147–1154.

56 Kisseleva, T., Uchinami, H., Feirt, N. et al. (2006). Bone marrow-derived fibrocytes participate in pathogenesis of liver fibrosis. *J. Hepatol.* 45 (3): 429–438.

57 Scholten, D. and Weiskirchen, R. (2011). Questioning the challenging role of epithelial-to-mesenchymal transition in liver injury. *Hepatology* 53 (3): 1048–1051.

58 Higashiyama, R., Moro, T., Nakao, S. et al. (2009). Negligible contribution of bone marrow-derived cells to collagen production during hepatic fibrogenesis in mice. *Gastroenterology* 137 (4): 1459–1466.

59 Taura, K., Iwaisako, K., Hatano, E., and Uemoto, S. (2016). Controversies over the epithelial-to-mesenchymal transition in liver fibrosis. *J. Clin. Med.* 5 (1): 9.

60 Brenner, D.A., Kisseleva, T., Scholten, D. et al. (2012). Origin of myofibroblasts in liver fibrosis. *Fibrogenesis Tissue Repair* 5 (1): 1–14.

61 Valdes, F., Alvarez, A.M., Locascio, A. et al. (2002). The epithelial mesenchymal transition confers resistance to the apoptotic effects of transforming growth factor Beta in fetal rat hepatocytes. *Mol. Cancer Res.* 1 (1): 68–78.

62 Ikegami, T., Zhang, Y., and Matsuzaki, Y. (2007). Liver fibrosis: possible involvement of EMT. *Cells Tissues Organs* 185 (1–3): 213–221.

63 Omenetti, A., Bass, L.M., Anders, R.A. et al. (2011). Hedgehog activity, epithelial-mesenchymal transitions, and biliary dysmorphogenesis in biliary atresia. *Hepatology* 53 (4): 1246–1258.

64 Nitta, T., Kim, J.S., Mohuczy, D., and Behrns, K.E. (2008). Murine cirrhosis induces hepatocyte epithelial mesenchymal transition and alterations in survival signaling pathways. *Hepatology* 48 (3): 909–919.

65 Kaimori, A., Potter, J., Kaimori, J.Y. et al. (2007). Transforming growth factor-beta1 induces an epithelial-to-mesenchymal transition state in mouse hepatocytes in vitro. *J. Biol. Chem.* 282 (30): 22089–22101.
66 Kang, M., Choi, S., Jeong, S.J. et al. (2012). Cross-talk between TGFB1 and EGFR signalling pathways induces TM4SF5 expression and epithelial-mesenchymal transition. *Biochem. J.* 443 (3): 691–700.
67 Zeisberg, M., Yang, C., Martino, M. et al. (2007). Fibroblasts derive from hepatocytes in liver fibrosis via epithelial to mesenchymal transition. *J Biol. Chem.* 282 (32): 23337–23347.
68 Taura, K., Miura, K., Iwaisako, K. et al. (2010). Hepatocytes do not undergo epithelial-mesenchymal transition in liver fibrosis in mice. *Hepatology* 51 (3): 1027–1036.
69 Chu, A.S., Diaz, R., Hui, J.J. et al. (2011). Lineage tracing demonstrates no evidence of cholangiocyte epithelial-to-mesenchymal transition in murine models of hepatic fibrosis. *Hepatology* 53 (5): 1685–1695.
70 Greuter, T. and Shah, V.H. (2016). Hepatic sinusoids in liver injury, inflammation, and fibrosis: new pathophysiological insights. *J. Gastroenterol.* 51 (6): 511–519.
71 Ramachandran, P., Pellicoro, A., Vernon, M.A. et al. (2012). Differential Ly-6C expression identifies the recruited macrophage phenotype, which orchestrates the regression of murine liver fibrosis. *Proc. Natl. Acad. Sci. U.S.A.* 109 (46): 3186–3195.
72 Mochizuki, A., Pace, A., Rockwell, C.E. et al. (2014). Hepatic stellate cells orchestrate clearance of necrotic cells in a hypoxia-inducible factor-1a-dependent manner by modulating macrophage phenotype in mice. *J. Immunol.* 192 (8): 3847–3857.
73 Langer, D.A., Das, A., Semela, D. et al. (2008). Nitric oxide promotes caspase-independent hepatic stellate cell apoptosis through the generation of reactive oxygen species. *Hepatology* 47 (6): 1983–1993.
74 DeLeve, L.D. (2015). Liver sinusoidal endothelial cells in hepatic fibrosis. *Hepatology* 61 (5): 1740–1746.
75 Xie, G., Wang, X., Wang, L. et al. (2012). Role of differentiation of liver sinusoidal endothelial cells in progression and regression of hepatic fibrosis in rats. *Gastroenterology* 142 (4): 918–927.
76 Knorr, A., Hirth-Dietrich, C., Alonso-Alija, C. et al. (2008). Nitric oxide-independent activation of soluble guanylate cyclase by BAY 60-2770 in experimental liver fibrosis. *Arzneimittelforschung* 58 (2): 71–80.
77 Kramer, B., Korner, C., Kebschull, M. et al. (2012). Natural killer p46High expression defines a natural killer cell subset that is potentially involved in control of hepatitis C virus replication and modulation of liver fibrosis. *Hepatology* 56 (4): 1201–1213.
78 Glassner, A., Eisenhardt, M., Kramer, B. et al. (2012). NK cells from HCV-infected patients effectively induce apoptosis of activated primary human

hepatic stellate cells in a TRAIL-, FasL- and NKG2D-dependent manner. *Lab. Invest.* 92 (7): 967–977.

79 Tian, Z., Chen, Y., and Gao, B. (2013). Natural killer cells in liver disease. *Hepatology* 57 (4): 1654–1662.

80 Melhem, A., Muhanna, N., Bishara, A. et al. (2006). Anti-fibrotic activity of NK cells in experimental liver injury through killing of activated HSC. *J. Hepatol.* 45 (1): 60–71.

81 Radaeva, S., Sun, R., Jaruga, B. et al. (2006). Natural killer cells ameliorate liver fibrosis by killing activated stellate cells in NKG2D-dependent and tumor necrosis factor-related apoptosis-inducing ligand-dependent manners. *Gastroenterology* 130 (2): 435–452.

82 Miyake, T., Kumagai, Y., Kato, H. et al. (2009). Poly I:C-induced activation of NK cells by CD8 alpha+ dendritic cells via the IPS-1 and TRIF-dependent pathways. *J. Immunol.* 183 (4): 2522–2528.

83 Ngoi, S.M., Tovey, M.G., and Vella, A.T. (2008). Targeting poly (I:C) to the TLR3-independent pathway boosts effector CD8 T cell differentiation through IFN-alpha/beta. *J. Immunol.* 181 (11): 7670–7680.

84 Lafyatis, R. (2014). Transforming growth factor beta – at the centre of systemic sclerosis. *Nat. Rev. Rheumatol.* 10 (12): 706–719.

85 Popov, Y., Patsenker, E., Stickel, F. et al. (2008). Integrin alphavbeta6 is a marker of the progression of biliary and portal liver fibrosis and a novel target for antifibrotic therapies. *J. Hepatol.* 48 (3): 453–464.

86 Wipff, P.J., Rifkin, D.B., Meister, J.J., and Hinz, B. (2007). Myofibroblast contraction activates latent TGF-beta1 from the extracellular matrix. *J. Cell. Biol.* 179 (6): 1311–1323.

87 Wang, X.M., Yu, D.M., McCaughan, G.W., and Gorrell, M.D. (2005). Fibroblast activation protein increases apoptosis, cell adhesion, and migration by the LX-2 human stellate cell line. *Hepatology* 42 (4): 935–945.

88 Wrighton, K.H., Lin, X., and Feng, X.H. (2009). Phospho-control of TGF-beta superfamily signaling. *Cell Res.* 19 (1): 8–20.

89 Massague, J. (2012). TGFB signalling in context. *Nat. Rev. Mol. Cell Biol.* 13 (10): 616–630.

90 Xie, W., Aisner, S., Baredes, S. et al. (2013). Alterations of Smad expression and activation in defining 2 subtypes of human head and neck squamous cell carcinoma. *Head Neck* 35 (1): 76–85.

91 Meng, X.M., Chung, A.C., and Lan, H.Y. (2013). Role of the TGF-B/BMP-7/Smad pathways in renal diseases. *Clin. Sci. (Lond.)* 124 (4): 243–254.

92 Hill, C.S. (2009). Nucleocytoplasmic shuttling of Smad proteins. *Cell Res.* 19 (1): 36–46.

93 Dai, F., Lin, X., Chang, C., and Feng, X.H. (2009). Nuclear export of Smad2 and Smad3 by RanBP3 facilitates termination of TGF-B signaling. *Dev. Cell* 16 (3): 345–357.

94 Derynck, R. and Zhang, Y.E. (2003). Smad-dependent and Smad-independent pathways in TGF-beta family signalling. *Nature* 425 (6958): 577–584.

95 Schnabl, B., Kweon, Y.O., Frederick, J.P. et al. (2001). The role of Smad3 in mediating mouse hepatic stellate cell activation. *Hepatology* 34 (1): 89–100.

96 Roberts, A.B., Russo, A., Felici, A., and Flanders, K.C. (2003). Smad3: a key player in pathogenetic mechanisms dependent on TGF-beta. *Ann. N.Y. Acad. Sci.* 995: 1–10.

97 Meng, X.M., Huang, X.R., Chung, A.C. et al. (2010). Smad2 protects against TGF-beta/Smad3-mediated renal fibrosis. *J. Am. Soc. Nephrol.* 21 (9): 1477–1487.

98 Tsuchida, K., Zhu, Y., Siva, S. et al. (2003). Role of Smad4 on TGF-beta-induced extracellular matrix stimulation in mesangial cells. *Kidney Int.* 63 (6): 2000–2009.

99 Tan, R., He, W., Lin, X. et al. (2008). Smad ubiquitination regulatory factor-2 in the fibrotic kidney: regulation, target specificity, and functional implication. *Am. J. Physiol. Renal Physiol.* 294 (5): 1076–1083.

100 Lin, X., Liang, M., and Feng, X.H. (2000). Smurf2 is a ubiquitin E3 ligase mediating proteasome-dependent degradation of Smad2 in transforming growth factor-beta signaling. *J. Biol. Chem.* 275 (47): 36818–36822.

101 Dooley, S., Hamzavi, J., Ciuclan, L. et al. (2008). Hepatocyte-specific Smad7 expression attenuates TGF-beta-mediated fibrogenesis and protects against liver damage. *Gastroenterology* 135 (2): 642–659.

102 Bian, E.B., Huang, C., Wang, H. et al. (2014). Repression of Smad7 mediated by DNMT1 determines hepatic stellate cell activation and liver fibrosis in rats. *Toxicol. Lett.* 224 (2): 175–185.

103 Davies, M., Robinson, M., Smith, E. et al. (2005). Induction of an epithelial to mesenchymal transition in human immortal and malignant keratinocytes by TGF-beta1 involves MAPK, Smad and AP-1 signalling pathways. *J. Cell. Biochem.* 95 (5): 918–931.

104 Hartsough, M.T. and Mulder, K.M. (1995). Transforming growth factor beta activation of p44mapk in proliferating cultures of epithelial cells. *J. Biol. Chem.* 270 (13): 7117–7124.

105 Yu, L., Hebert, M.C., and Zhang, Y.E. (2002). TGF-beta receptor-activated p38 MAP kinase mediates Smad-independent TGF-beta responses. *EMBO J.* 21 (14): 3749–3759.

106 Bourbonnais, E., Raymond, V.A., Ethier, C. et al. (2012). Liver fibrosis protects mice from acute hepatocellular injury. *Gastroenterology* 142 (1): 130–139.

107 Mucsi, I., Skorecki, K.L., and Goldberg, H.J. (1996). Extracellular signal-regulated kinase and the small GTP-binding protein, Rac, contribute to the effects of transforming growth factor-beta1 on gene expression. *J. Biol. Chem.* 271 (28): 16567–16572.

108 Zavadil, J., Bitzer, M., Liang, D. et al. (2001). Genetic programs of epithelial cell plasticity directed by transforming growth factor-beta. *Proc. Natl. Acad. Sci. U.S.A.* 98 (12): 6686–6691.

109 Kretzschmar, M., Doody, J., Timokhina, I., and Massague, J. (1999). A mechanism of repression of TGFbeta/ Smad signaling by oncogenic Ras. *Genes Dev.* 13 (7): 804–816.

110 Gonzalez, D.M. and Medici, D. (2014). Signaling mechanisms of the epithelial-mesenchymal transition. *Sci. Signal.* 7 (344): 1–16.

111 Medici, D., Hay, E.D., and Olsen, B.R. (2008). Snail and slug promote epithelial-mesenchymal transition through beta-catenin-T-cell factor-4-dependent expression of transforming growth factor-beta3. *Mol. Biol. Cell* 19 (11): 4875–4887.

112 Cano, A., Perez-Moreno, M.A., Rodrigo, I. et al. (2000). The transcription factor snail controls epithelial-mesenchymal transitions by repressing E-cadherin expression. *Nat. Cell Biol.* 2 (2): 76–83.

113 Julien, S., Puig, I., Caretti, E. et al. (2007). Activation of NF-kappaB by Akt upregulates snail expression and induces epithelium mesenchyme transition. *Oncogene* 26 (53): 7445–7456.

114 Zhou, B.P., Deng, J., Xia, W. et al. (2004). Dual regulation of snail by GSK-3beta-mediated phosphorylation in control of epithelial-mesenchymal transition. *Nat. Cell Biol.* 6 (10): 931–940.

115 Chaudhury, A., Hussey, G.S., Ray, P.S. et al. (2010). TGF-beta-mediated phosphorylation of hnRNP E1 induces EMT via transcript-selective translational induction of Dab2 and ILEI. *Nat. Cell Biol.* 12 (3): 286–293.

116 Lamouille, S. and Derynck, R. (2007). Cell size and invasion in TGF-beta-induced epithelial to mesenchymal transition is regulated by activation of the mTOR pathway. *J. Cell Biol.* 178 (3): 437–451.

117 Lamouille, S., Connolly, E., Smyth, J.W. et al. (2012). TGF-B-induced activation of mTOR complex 2 drives epithelial-mesenchymal transition and cell invasion. *J. Cell Sci.* 125 (Pt. 5): 1259–1273.

118 Rosmorduc, O., Wendum, D., Corpechot, C. et al. (1999). Hepatocellular hypoxia-induced vascular endothelial growth factor expression and angiogenesis in experimental biliary cirrhosis. *Am. J. Pathol.* 155 (4): 1065–1073.

119 Ji, S., Lemasters, J.J., Christenson, V., and Thurman, R.G. (1982). Periportal and pericentral pyridine nucleotide fluorescence from the surface of the perfused liver: evaluation of the hypothesis that chronic treatment with ethanol produces pericentral hypoxia. *Proc. Natl. Acad. Sci. U.S.A.* 79 (17): 5415–5419.

120 Moon, J.O., Welch, T.P., Gonzalez, F.J., and Copple, B.L. (2009). Reduced liver fibrosis in hypoxia-inducible factor-1alpha-deficient mice. *Am. J. Physiol. Gastrointest. Liver Physiol.* 296 (3): 582–592.

121 Copple, B.L., Kaska, S., and Wentling, C. (2012). Hypoxia-inducible factor activation in myeloid cells contributes to the development of liver fibrosis in cholestatic mice. *J. Pharmacol. Exp. Ther.* 341 (2): 307–316.

122 Copple, B.L., Bai, S., and Moon, J.O. (2010). Hypoxia-inducible factor-dependent production of profibrotic mediators by hypoxic Kupffer cells. *Hepatol. Res.* 40 (5): 530–539.

123 Murakami, Y., Toyoda, H., Tanaka, M. et al. (2011). The progression of liver fibrosis is related with overexpression of the miR-199 and 200 families. *PloS One* 6 (1): e16081.

124 Wei, J., Feng, L., Li, Z. et al. (2013). MicroRNA-21 activates hepatic stellate cells via PTEN/Akt signaling. *Biomed. Pharmacother.* 67 (5): 387–392.

125 Zhao, J., Tang, N., Wu, K. et al. (2014). MiR-21 simultaneously regulates ERK1 signaling in HSC activation and hepatocyte EMT in hepatic fibrosis. *PloS One* 9 (10): e108005.

126 Roderburg, C., Urban, G.W., Bettermann, K. et al. (2011). Micro-RNA profiling reveals a role for miR-29 in human and murine liver fibrosis. *Hepatology* 53 (1): 209–218.

127 Zeng, C., Wang, Y.L., Xie, C. et al. (2015). Identification of a novel TGF-B-miR-122-fibronectin 1/serum response factor signaling cascade and its implication in hepatic fibrogenesis. *Oncotarget* 6 (14): 12224–12233.

128 Seki, E., De Minicis, S., Osterreicher, C.H. et al. (2007). TLR4 enhances TGF-beta signaling and hepatic fibrosis. *Nat. Med.* 13 (11): 1324–1332.

129 Ji, L., Xue, R., Tang, W. et al. (2014). Toll like receptor 2 knock-out attenuates carbon tetrachloride (CCl4)-induced liver fibrosis by downregulating MAPK and NF-kB signaling pathways. *FEBS Lett.* 588 (12): 2095–2100.

130 Shu, M., Huang, D.D., Hung, Z.A. et al. (2016). Inhibition of MAPK and NF-kB signaling pathways alleviate carbon tetrachloride (CCl4)-induced liver fibrosis in toll-like receptor 5 (TLR5) deficiency mice. *Biochem. Biophys. Res. Commun.* 471 (1): 233–239.

131 Pao, W., Miller, V., Zakowski, M. et al. (2004). EGF receptor gene mutations are common in lung cancers from "never smokers" and are associated with sensitivity of tumors to gefitinib and erlotinib. *Proc. Natl. Acad. Sci. U.S.A.* 101 (36): 13306–13311.

132 Sharma, S.V., Bell, D.W., Settleman, J., and Haber, D.A. (2007). Epidermal growth factor receptor mutations in lung cancer. *Nat. Rev. Cancer* 7 (3): 169–181.

133 Haber, D.A., Bell, D.W., Sordella, R. et al. (2005). Molecular targeted therapy of lung cancer: EGFR mutations and response to EGFR inhibitors. *Cold Spring Harb. Symp. Quant. Biol.* 70: 419–426.

134 Flaherty, K.T., Puzanov, I., Kim, K.B. et al. (2010). Inhibition of mutated, activated BRAF in metastatic melanoma. *N. Engl. J. Med.* 363 (9): 809–819.

135 Pao, W., Miller, V.A., Politi, K.A. et al. (2005). Acquired resistance of lung adenocarcinomas to gefitinib or erlotinib is associated with a second mutation in the EGFR kinase domain. *PLoS Med.* 2 (3): e73.

136 Kobayashi, S., Boggon, T.J., Dayaram, T. et al. (2005). EGFR mutation and resistance of non-small-cell lung cancer to gefitinib. *N. Engl. J. Med.* 352 (8): 786–792.

137 Kwak, E.L., Sordella, R., Bell, D.W. et al. (2005). Irreversible inhibitors of the EGF receptor may circumvent acquired resistance to gefitinib. *Proc. Natl. Acad. Sci. U.S.A.* 102 (21): 7665–7670.

138 Sos, M.L., Koker, M., Weir, B.A. et al. (2009). PTEN loss contributes to erlotinib resistance in EGFR-mutant lung cancer by activation of Akt and EGFR. *Cancer Res.* 69 (8): 3256–3261.

139 Steelman, L.S., Navolanic, P.M., Sokolosky, M.L. et al. (2008). Suppression of PTEN function increases breast cancer chemotherapeutic drug resistance while conferring sensitivity to mTOR inhibitors. *Oncogene* 27 (29): 4086–4095.

140 Paraiso, K.H., Xiang, Y., Rebecca, V.W. et al. (2011). PTEN loss confers BRAF inhibitor resistance to melanoma cells through the suppression of BIM expression. *Cancer Res.* 71 (7): 2750–2760.

141 Toy, W., Weir, H., Razavi, P. et al. (2017). Activating ESR1 mutations differentially affect the efficacy of ER antagonists. *Cancer Discov.* 7 (3): 277–287.

142 Nazarian, R., Shi, H., Wang, Q. et al. (2010). Melanomas acquire resistance to B-RAF(V600E) inhibition by RTK or N-RAS upregulation. *Nature* 468 (7326): 973–977.

143 Emery, C.M., Vijayendran, K.G., Zipser, M.C. et al. (2009). MEK1 mutations confer resistance to MEK and B-RAF inhibition. *Proc. Natl. Acad. Sci. U.S.A.* 106 (48): 20411–20416.

144 Brown, R., Curry, E., Magnani, L. et al. (2014). Poised epigenetic states and acquired drug resistance in cancer. *Nat. Rev. Cancer* 14 (11): 747–753.

145 Van Boeckel, T.P., Glennon, E.E., Chen, D. et al. (2017). Reducing antimicrobial use in food animals. *Science* 357 (6358): 1350–1352.

146 Sommer, M.O.A., Dantas, G., and Church, G.M. (2009). Functional characterization of the antibiotic resistance reservoir in the human microflora. *Science* 325 (5944): 1128–1131.

147 Gulick, R.M., Mellors, J.W., Havlir, D. et al. (1997). Treatment with indinavir, zidovudine, and lamivudine in adults with human immunodeficiency virus infection and prior antiretroviral therapy. *N. Engl. J. Med.* 337 (11): 734–739.

148 Hammer, S.M., Squires, K.E., Hughes, M.D. et al. (1997). A controlled trial of two nucleoside analogues plus indinavir in persons with human immunodeficiency virus infection and CD4 cell counts of 200 per cubic millimeter or less. AIDS Clinical Trials Group 320 study team. *N. Engl. J. Med.* 337 (11): 725–733.

149 Tsai, J., Lee, J.T., Wang, W. et al. (2008). Discovery of a selective inhibitor of oncogenic B-Raf kinase with potent antimelanoma activity. *Proc. Natl. Acad. Sci. U.S.A.* 105 (8): 3041–3046.

150 Goldstein, L.J., Galski, H., Fojo, A. et al. (1989). Expression of a multidrug resistance gene in human cancers. *J. Natl. Cancer Inst.* 81 (2): 116–124.

151 Gottesman, M.M. and Pastan, I.H. (2015). The role of multidrug resistance efflux pumps in cancer: revisiting a JNCI publication exploring expression of the MDR1 (P-glycoprotein) gene. *J. Natl. Cancer Inst.* 107 (9): https://doi.org/10.1093/jnci/djv222.

152 Nowell, P.C. (1976). The clonal evolution of tumor cell populations. *Science* 194 (4260): 23–28.

153 Moncada, R., Chiodin, M., Devlin, J.C., et al. (2018). Building a tumor atlas: integrating single-cell RNA-Seq data with spatial transcriptomics in pancreatic ductal adenocarcinoma. *bioRxiv*. https://doi.org/10.1101/254375.

154 Filbin, M.G., Tirosh, I., Hovestadt, V. et al. (2018). Developmental and oncogenic programs in H3K27M gliomas dissected by single-cell RNA-seq. *Science* 360 (6386): 331–335.

155 Puram, S.V., Tirosh, I., Parikh, A.S. et al. (2017). Single-cell transcriptomic analysis of primary and metastatic tumor ecosystems in head and neck cancer. *Cell* 171 (7): 1611–1624.

156 Tirosh, I., Izar, B., Prakadan, S.M. et al. (2016). Dissecting the multicellular ecosystem of metastatic melanoma by single-cell RNA-seq. *Science* 352 (6282): 189–196.

157 Tirosh, I., Venteicher, A.S., Hebert, C. et al. (2016). Single-cell RNA-seq supports a developmental hierarchy in human oligodendroglioma. *Nature* 539 (7628): 309–313.

158 Patel, A.P., Tirosh, I., Trombetta, J.J. et al. (2014). Single-cell RNA-seq highlights intratumoral heterogeneity in primary glioblastoma. *Science* 344 (6190): 1396–1401.

159 Konieczkowski, D.J., Johannessen, C.M., Abudayyeh, O. et al. A melanoma cell state distinction influences sensitivity to MAPK pathway inhibitors. *Cancer Discov.* 4 (7): 816–827.

160 Kumar, V., Donthireddy, L., Marvel, D. et al. (2017). Cancer-associated fibroblasts neutralize the anti-tumor effect of CSF1 receptor blockade by inducing PMN-MDSC infiltration of tumors. *Cancer Cell* 32 (5): 654–668.

161 Kaur, A., Webster, M.R., Marchbank, K. et al. (2016). sFRP2 in the aged microenvironment drives melanoma metastasis and therapy resistance. *Nature* 532 (7598): 250–254.

162 Werb, Z. and Lu, P. (2016). The role of stroma in tumor development. *Cancer J.* 21 (4): 250–253.

163 Hirata, E., Girotti, M.R., Viros, A. et al. (2015). Intravital imaging reveals how BRAF inhibition generates drug-tolerant microenvironments with high integrin beta1/FAK signaling. *Cancer Cell* 27 (4): 574–588.

164 Nair, S. and Dhodapkar, M.V. (2017). Natural killer T cells in cancer immunotherapy. *Front. Immunol.* 8: 1178.

165 Mantovani, A., Marchesi, F., Malesci, A. et al. (2017). Tumour-associated macrophages as treatment targets in oncology. *Nat. Rev. Clin. Oncol.* 14 (7): 399–416.

166 Kalluri, R. (2016). The biology and function of fibroblasts in cancer. *Nat. Rev. Cancer* 16 (9): 582–598.

167 Noy, R. and Pollard, J.W. (2014). Tumor-associated macrophages: from mechanisms to therapy. *Immunity* 41 (1): 49–61.
168 Hadrup, S., Donia, M., and Thor Straten, P. (2013). Effector CD4 and CD8 T cells and their role in the tumor microenvironment. *Cancer Microenviron.* 6 (2): 123–133.
169 Song, C., Piva, M., Sun, L. et al. (2017). Recurrent tumor cell-intrinsic and -extrinsic alterations during MAPKi-induced melanoma regression and early adaptation. *Cancer Discov.* 7 (11): 1248–1265.
170 Wherry, E.J. and Kurachi, M. (2015). Molecular and cellular insights into T cell exhaustion. *Nat. Rev. Immunol.* 15 (8): 486–499.
171 Yun, C.H., Mengwasser, K.E., Toms, A.V. et al. (2008). The T790M mutation in EGFR kinase causes drug resistance by increasing the affinity for ATP. *Proc. Natl. Acad. Sci. U.S.A.* 105 (6): 2070–2075.
172 Liu, Y., Sun, L., Xiong, Z.C. et al. (2017). Meta-analysis of the impact of de novo and acquired EGFR T790M mutations on the prognosis of patients with non-small cell lung cancer receiving EGFR-TKIs. *Onco. Targets Ther.* 10: 2267–2279.
173 Sequist, L.V., Soria, J.C., Goldman, J.W. et al. (2015). Rociletinib in EGFR-mutated non-small-cell lung cancer. *N. Engl. J. Med.* 372 (18): 1700–1709.
174 Janne, P.A., Yang, J.C., Kim, D.W. et al. (2015). AZD9291 in EGFR inhibitor-resistant non-small-cell lung cancer. *N. Engl. J. Med.* 372 (18): 1689–1699.
175 Johnson, D.B., Menzies, A.M., Zimmer, L. et al. (2015). Acquired BRAF inhibitor resistance: a multicenter meta-analysis of the spectrum and frequencies, clinical behaviour, and phenotypic associations of resistance mechanisms. *Eur. J. Cancer* 51 (18): 2792–2799.
176 Robert, C., Karaszewska, B., Schachter, J. et al. (2015). Improved overall survival in melanoma with combined dabrafenib and trametinib. *N. Engl. J. Med.* 372 (1): 30–39.
177 Larkin, J., Ascierto, P.A., Dreno, B. et al. (2014). Combined vemurafenib and cobimetinib in BRAF-mutated melanoma. *N. Engl. J. Med.* 371 (20): 1867–1876.
178 Knudsen, E.S. and Witkiewicz, A.K. (2017). The strange case of CDK4/6 inhibitors: mechanisms, resistance, and combination strategies. *Trends Cancer* 3 (1): 39–55.
179 Prahallad, A., Sun, C., Huang, S. et al. (2012). Unresponsiveness of colon cancer to BRAF(V600E) inhibition through feedback activation of EGFR. *Nature* 483 (7387): 100–103.
180 Villanueva, J., Vultur, A., and Herlyn, M. (2011). Resistance to BRAF inhibitors: unraveling mechanisms and future treatment options. *Cancer Res.* 71 (23): 7137–7140.
181 Alexander, P.B. and Wang, X.F. (2015). Resistance to receptor tyrosine kinase inhibition in cancer: molecular mechanisms and therapeutic strategies. *Front. Med.* 9 (2): 134–138.

182 Zhang, Z., Lee, J.C., Lin, L. et al. (2012). Activation of the AXL kinase causes resistance to EGFR-targeted therapy in lung cancer. *Nat. Genet.* 44 (8): 852–860.

183 Terai, H., Soejima, K., Yasuda, H. et al. (2013). Activation of the FGF2-FGFR1 autocrine pathway: a novel mechanism of acquired resistance to gefitinib in NSCLC. *Mol. Cancer Res.* 11 (7): 759–767.

184 Turke, A.B., Zejnullahu, K., Wu, Y.L. et al. (2010). Preexistence and clonal selection of MET amplification in EGFR mutant NSCLC. *Cancer Cell* 17 (1): 77–88.

185 Courtois-Cox, S., Jones, S.L., and Cichowski, K. (2008). Many roads lead to oncogene-induced senescence. *Oncogene* 27 (20): 2801–2809.

186 Liu, X.L., Ding, J., and Meng, L.H. (2018). Oncogene-induced senescence: a double edged sword in cancer. *Acta. Pharmacol. Sin* 39: 1553–1558.

187 Lund, A.H. and van Lohuizen, M. (2004). Epigenetics and cancer. *Genes Dev.* 18 (19): 2315–2335.

188 Baylin, S.B. and Jones, P.A. (2016). Epigenetic determinants of cancer. *Cold Spring Harb. Perspect. Biol.* 8 (9): a019505. https://doi.org/10.1101/cshperspect.a019505.

189 Hugo, W., Shi, H., Sun, L. et al. (2015). Non-genomic and immune evolution of melanoma acquiring MAPKi resistance. *Cell* 162 (6): 1271–1285.

190 Wang, J., Huang, S.K., Marzese, D.M. et al. (2015). Epigenetic changes of EGFR have an important role in BRAF inhibitor-resistant cutaneous melanomas. *J. Invest. Dermatol.* 135 (2): 532–541.

191 Jones, P.A., Issa, J.P., and Baylin, S. (2016). Targeting the cancer epigenome for therapy. *Nat. Rev. Genet.* 17 (10): 630–641.

192 Shaffer, S.M., Dunagin, M.C., Torborg, S.R. et al. (2017). Rare cell variability and drug-induced reprogramming as a mode of cancer drug resistance. *Nature* 546 (7658): 431–435.

193 Sun, C., Wang, L., Huang, S. et al. (2014). Reversible and adaptive resistance to BRAF(V600E) inhibition in melanoma. *Nature* 508 (7494): 118–122.

194 Sharma, S.V., Lee, D.Y., Li, B. et al. (2010). A chromatin-mediated reversible drug-tolerant state in cancer cell subpopulations. *Cell* 141 (1): 69–80.

195 Knoechel, B., Roderick, J.E., Williamson, K.E. et al. (2014). An epigenetic mechanism of resistance to targeted therapy in T cell acute lymphoblastic leukemia. *Nat. Genet.* 46 (4): 364–370.

5

Experimental Design in Signal Transduction

Weimin Gao[1], Meghan Cromie[2], Qian Wang[3], Zhongwei Liu[1], Song Tang[4], and Julie Vrana Miller[5]

[1] *Department of Occupational and Environmental Health Sciences, West Virginia University School of Public Health, Morgantown, WV, USA*
[2] *National Jewish Health, Denver, CO, USA*
[3] *Department of Respiratory Medicine, Jiangsu Province Hospital of Chinese Medicine, Affiliated Hospital of Nanjing University of Chinese Medicine, Nanjing, Jiangsu, China*
[4] *National Institute of Environmental Health, Chinese Center for Disease Control and Prevention, Beijing, China*
[5] *Cardno ChemRisk, Pittsburgh, PA, USA*

This chapter provides a comprehensive explanation for designing experiments with signal transduction pathways as the researcher's expected endpoint. Sound statistical testing is paramount for any study and should be evaluated prior to any experimentation; therefore, this chapter begins with a review of different statistical tests. Following a discussion of statistical testing, aseptic technique for both *in vitro* and *in vivo* studies is briefly discussed and should be stringently practiced throughout any well-designed experiment. Proper sample collection, processing, and storage are discussed sequentially, followed by brief descriptions of relevant and useful *in vitro, in vivo,* and epidemiological experiments and study designs. Finally, real applications of signal transduction experiments are provided. Overall, this chapter provides insight into the appropriate practices needed to design scientifically sound signal transduction experiments.

5.1 Overview of Basic Experimental Design

As with all scientific fields, experimental design in toxicology and pharmacology is necessary to test a hypothesis, to determine relationships and causality, and to draw educated scientific conclusions from generated data. It is important to understand the statistical tests to analyze the generated data prior to

Cellular Signal Transduction in Toxicology and Pharmacology: Data Collection, Analysis, and Interpretation, First Edition. Edited by Jonathan W. Boyd and Richard R. Neubig.

experimentation. Calculating the sample size needed based on statistical power and identifying possible study constraints results in more time- and cost-effective experiments. Thorough literature review and pilot studies can provide a wealth of information that can contribute significantly to one's study. The following section provides an explanation of statistical techniques commonly employed in the design and subsequent data analysis in toxicology and pharmacology studies, including *t* test for independent sample design, completely randomized design, *t* test for dependent sample design, randomized block design, and completely randomized factorial design.

5.1.1 Independent Sample *t* Test

The most common statistical test available to researchers is the independent sample *t* test, which is used to determine whether two different groups or populations have statistically different means. Both sets of data must have equal variance for this test to be used. The null hypothesis for an independent sample *t* test would be that the two sample groups have equal means. When there are two datasets and the data have been paired or one set has been tested twice, a dependent *t* test for paired samples is used, as described below.

As an example, the *t* test is applied to test the efficacy of two cancer therapeutics (a_1 or a_2), such as cisplatin or docetaxel in nude mice implanted with cancer cells. The subjects that receive treatment dose or concentration a_1 are termed Group 1; and those that receive treatment level a_2 are termed Group 2. The dependent variable is the tumor size after therapy. Assume that a total of 12 nude mice are to be implanted with cancer cells in this experiment and 6 nude mice are assigned to each of the 2 treatment groups. To maintain the same probability in assigning each mouse to the study groups, the nude mice can be numbered from 1 to 12 and randomly drawn in order to avoid selective bias inherent in the study and from the experimenter in order to ensure statistical independence.

5.1.2 Completely Randomized Analysis of Variance (ANOVA)

Unlike a student's *t* test, a one-way analysis of variance (ANOVA) is implemented in order to test whether there are significant differences among the means of two or more groups. A post hoc test can be used to identify the groups that are significantly different from one another. The null hypothesis in this statistical test determines whether two or more groups are derived from the same population, and an *F*-statistic is produced by taking the ratio of variance between the individual samples and total groups into consideration. If the individual samples are similar to the group or population means, then the *F*-statistic is small, and the samples follow the central limit theorem, which indicates the samples follow a normal distribution. In order to correctly test a one-way

ANOVA, the samples must adhere to the following assumptions: (i) variables are normally distributed, (ii) homogeneity of sample variance, and (iii) the samples are independent of one another. If the data fail any of these assumptions, then a nonparametric test such as a Kruskal–Wallis one-way ANOVA must be used. Assuming that a total of 18 nude mice are implanted with cancer cells, the subjects are randomly assigned to the treatment levels (n = 6/group). The three levels of treatment are denoted as a_1 for gefitinib, a_2 for docetaxel, and a_3 for cisplatin. The tumor sizes of the subjects after therapy are denoted as $Y_1...Y_{18}$. The subjects that receive treatments a_1, a_2, and a_3 are termed Group 1, Group 2, and Group 3, respectively. The three sample means, $\bar{Y}_1$, $\bar{Y}_2$, and $\bar{Y}_3$, reflect the tumor sizes as a result of the treatment. A completely randomized design differs from a t test for independent sample design in that an additional treatment level is accounted for in the study.

5.1.3 *t* Test for Dependent Sample Design

The aforementioned statistical tests require the use of independent samples. However, dependent samples can be obtained during the study design process, such as repeated measures and subject matching/pairing. A t test for dependent samples can be applied to this scenario. Consider again the previously discussed tumor effect experiments utilizing the t test for independent sample design. For example, if the mouse body weight influences the outcome of the anticancer treatment, then a dependent t test would be used to avoid the effects of body weight on the study. Therefore, instead of randomly assigning 12 nude mice implanted with cancer cells to the treatment levels, pairs of subjects can be formed in a block in which the subjects in each pair have a similar body weight. As a result, six blocks of dependent samples can be formed. The two mice in each block are randomly assigned to the two types of therapy. The t test for dependent sample design increases the power of the results and reduces the errors by isolating the nuisance variable.

5.1.4 Randomized Block Design

Similar to t tests for independent samples and one-way ANOVAs, comparable observations can be applied to a t test for dependent samples and a randomized block ANOVA. Following the same example as described in completely randomized design, randomized block design is used instead to evaluate the effectiveness of three kinds of therapy in reducing tumor size. As mentioned in the t test for dependent samples, the body weight of the mouse can influence the treatment outcome. If a total of 18 nude mice are used for the study and implanted with cancer cells, 6 blocks will be modeled to contain three subjects with similar body weights prior to treatment. The dependent variables for the study are denoted as $Y_1...Y_{18}$. A randomized block design enables a researcher to test two null hypotheses. The first hypothesis states that the means for the three treatments are equal. The second hypothesis, which is usually of little interest, is that the means for the six levels of the nuisance variable (the weight of mice prior to the experiment) are equal.

5.1.5 Completely Randomized Factorial Design

A completely randomized factorial design is developed based on a completely randomized ANOVA in which two or more independent variables can be tested simultaneously in a single experiment. To evaluate the anticancer effect of gefitinib (a_1: used, a_2: not used) and cisplatin (b_1: used, b_2: not used), the following 2×2 treatment conditions are formed: a_1b_1, a_1b_2, a_2b_1, and a_2b_2 to account for each combination. Assume that a total of 24 nude mice are implanted with cancer cells in the experiment. The nude mice are randomly assigned to the four treatment combinations, with the restriction that six nude mice receive each combination. Not only does completely randomized factorial analysis test for differences between individual independent variables, but it also accounts for possible interactions.

5.1.6 Summary

Performing statistical tests prior to experimentation is necessary for proper experimental designs and is important for both ethical and fiscal reasons. It is imperative to consider what statistical testing fits best with the experiment's hypothesis and endpoint as well as ensuring all assumptions of the statistical test of interest are considered. While proper data analysis is what allows scientists to form proper conclusions of the study, another key factor of proper experimental design is sample collection and processing. Therefore, an overview of lab techniques required for proper sample collection and processing is described in the following sections.

5.2 Aseptic Technique

Having proper aseptic technique in any laboratory is essential; not only is this to create a safe environment for the researcher, but it also serves to prevent possible contamination that can negatively impact the study at hand. The unintentional introduction of pathogens such as bacteria and fungus within *in vitro* and *in vivo* studies, via the researcher or contaminated workspaces and equipment, can prove detrimental to the results of the experiment. The implementation of an established aseptic technique is necessary to ensure reliable and reproducible experimental results. In cell culture studies, the contamination can be localized but quickly become widespread and compromise the entire study. In order to prevent and even catch early contamination, it is best to observe cultures every day while following proper aseptic technique. Environmental, workspace, and researcher areas of concern with regard to aseptic technique will be discussed in this section.

For animal studies, facilities and animal housing must be approved for the species and study. Facilities, equipment, and handlers must be properly sterilized upon working with the animals. It is best to store autoclaved or sterile materials in the room where they are needed in enclosed cabinets to prevent dust collection. When working with immunocompromised animals (such as athymic nude mouse), researchers must store lab coats in the same space as the study or in a sterile anteroom, and sterile booties must be worn over the researcher's shoes. It is required that animals be monitored daily to ensure their health. Floors should be swept daily, and floors and walls should be mopped weekly. Water changes should occur when needed, but best practice is to change them biweekly when a single animal is housed in a cage. Cages, bottles, bedding, and animal enrichment materials must be autoclaved prior to entering the animal facility. Irradiated animal feed must be purchased prior to the experiment and can be stored in a −80 ° C freezer no longer than six months beyond the manufacture date.

5.2.1 Sterile Work Environment and Laminar-Flow Hood

In order to maintain a sterile working environment, it is best to ensure that the dedicated culture area is cleaned regularly and contains only essential items. The best means of ensuring a sterile environment is to localize the culture area to a small room with a door or even a small enclave from the rest of the lab space. Laminar-flow hoods, incubators, and culture microscopes must remain clean, so spraying all work materials with at least a 70% ethanol solution and wiping them dry daily are important. Spraying and wiping down the laminar-flow hood with ethanol, followed by at least 15 minutes of ultraviolet (UV) light, are essential before starting any experiment. Wiping down the hood with a dilute bleach solution (1 : 10 in H_2O) is a good practice to follow. Furthermore, required maintenance and quality control of the hood must be strictly adhered and certified annually.

5.2.2 Good Personal Hygiene Practices

In order to prevent contamination of the workspace, good personal hygiene practices are of utmost importance. Regular washing of hands and forearms (up to the elbow) prior to any experiment can be useful. Additionally, it is best to wear new sterile gloves at the start of each experiment. If a glove must be removed during an active experiment, the soiled glove should be properly disposed of and replaced immediately. Researchers should spray gloves and lab coat sleeves with 70% alcohol before touching anything in the culture area. It is important to continually spray these items to prevent any possible sources of contamination to the experiment. Regular washing and/or bleaching of lab coats will also serve to protect both the experiment and the researcher. There are many routes for pathogens to compromise an experiment, but personal hygiene is one of the easiest and preventable aseptic techniques to implement.

5.2.3 Sterile Reagents and Materials

Commercial reagents and media undergo stringent quality control testing to safeguard sterility. If the researcher prepares their own reagents or media, it is best to autoclave or filter when appropriate and follow proper storage and handling protocols. Purchase of presterilized manufactured cell culture materials is a common means to avoid contamination, but it is important to remember to spray (with 70% ethanol) and tape any bags or boxes upon opening. If possible, storage of all cell culture materials in clean cabinets is ideal to prevent contamination and collection of dust. Materials such as pipette tips, incubator water, glass pipette tips, etc. should be properly autoclaved and either stored in the culture hood or a clean cabinet. The use of autoclave tape is necessary to determine if proper sterilization temperature of the item has been reached.

5.2.4 Sterile Handling

When removing items such as culture media from the water bath for use, the researcher should wipe the bottle dry and spray it with 70% ethanol twice to avoid contamination. All items that are dried and sprayed should immediately be placed in the culture hood to minimize potential pathogen contact in the workspace. Prior to any experiment, cleaning pipettors with ethanol will help to prevent cross-contamination from previous experiments. All materials should be opened in the sterile laminar-flow culture hood to avoid contamination. When removing a cap from a bottle, place the top part down on the hood surface to promote minimal contact with the contents of the bottle. Also, it is best to never directly remove or introduce pipettes into the original bottle. Carefully pouring the liquids to a secondary storage container can prevent widespread contamination. It is important to immediately open paper and/or plastic sheaths covering items such as pipette tips as soon as they are needed and be sure not to reuse these items. Finally, it is beneficial to complete experiments in a timely manner and properly clean the workspace to prevent contamination in future experiments.

5.3 Biological Sample Collection, Processing, and Pretreatment Technology

5.3.1 Sample Collection

5.3.1.1 Sample Collection *In Vivo*

Prior to starting any *in vivo* work, it is imperative that all personnel are properly trained in animal handling and that all protocols are approved by the Institutional Animal Care and Use Committee (IACUC). It is best to anticipate a minimum of three months for IACUC submissions and approval when planning *in vivo* studies. Facilities and animal housing must be approved for the species and study. Sample collection involving humans requires Institutional Review Board (IRB) approval, and researchers should expect the entire process to take a minimum of three to six months. Additionally, IRB-related experiments may take years for sample collection due to subject recruitment. All of these issues are contingent on the nature of the study, and these issues should be taken into consideration during the experimental design process.

5.3.1.1.1 Organs/Tissues

Sterile instruments must be used in the removal of organs and tissues from the animals. Samples needed for future RNA analysis can be cut to less than 0.5 cm, washed in 1× phosphate buffered saline (PBS), and then immersed in five volumes of RNAlater®, or they can be flash frozen in liquid nitrogen and stored in properly marked cryovials in a −80 °C freezer. Tissues or organs needed for histopathology can be dissected and stored in glass vials containing 10% formalin at least four times the volume of the organ or tissue.

5.3.1.1.2 Blood

Blood may be removed from the animal or human for a variety of reasons to test different molecular endpoints. The volume of blood collected by the researcher is contingent on the method of venipuncture, animal species, size of the animal, testing endpoints, multiple versus terminal blood withdrawals, and the percentage of blood desired for removal. The collection and storage of whole blood or blood fractions are also determined by future experimental endpoints to be tested. Sampling sites for venipuncture include the lateral tarsal vein, marginal ear vein, sublingual vein, lateral tail vein, cranial vena cava, cardiac puncture, and retrobulbar plexus. The blood is typically collected by the researchers using a syringe with a needle and a vacuum tube. In order to minimize stress to the test animal, it is best to handle the animal with care and follow the appropriate blood draw method generally accepted or recommended by facility veterinarians. In order to prevent coagulation, ethylenediaminetetraacetic acid (EDTA) or heparin can be added to collection vials before or immediately after sample collection, and both are useful for future DNA-based assays. If serum is to be collected from the blood sample, centrifugation upon collection is necessary prior to freezing [1]. Centrifugation of blood for 10 minutes at 14 000 rpm at 4 °C will separate the plasma/serum and allow for removal and storage. Serum and plasma both can be collected for analyses regarding antibodies, lipoproteins, lipids, and nutrients [2].

5.3.1.2 Cell Culture *In Vitro*

Cell culture is the continued growth of cells from animals, plants, or pathogens in an unnatural environment. The cells can be derived from primary or immortalized cell lines. In order to maintain the cells, they are maintained in an incubator with 5% CO_2 and at 37 °C for warm-blooded animals, 15–26 °C for cold water fish, and 38 °C for birds. Sterile water must be used in the incubator. Growth medium provides the cells with the appropriate nutrients, growth factors, and hormones in a properly pH-buffered and balanced solution that are necessary for survival. Most cell lines propagate at an optimum pH of 7.4, and the pH can be monitored by examining the color change from phenol red, which can be added to the growth medium. If the pH is either too high or too low, the cells can be damaged, and this will confound the results. Additionally, mammalian serum can be used as a growth factor and antibiotics can be added. For example, cancer cell lines such as the lung adenocarcinoma cell line (A549) are maintained in Roswell Park Memorial Institute (RPMI)-1640 growth medium supplemented with 5–10% fetal bovine serum, 50 U/ml penicillin, and 50 mg/ml streptomycin, but normal lung epithelial cells (human bronchial epithelial cells [BEAS-2B]) are cultured in LHC-9 completed growth medium (serum-free) containing 50 U penicillin/ml and 50 μg/ml streptomycin.

5.3.1.2.1 Primary Culture

Cells that are isolated directly from the test subject or patients are defined as primary cell cultures, and they are propagated as either explant cultures or the single cells can be mechanically or enzymatically isolated. Primary cultures maintain the original genotype of the test subject, and therefore, they have a limited lifespan.

5.3.1.2.2 Cell Line

The single cells isolated from the primary explant culture can continue to grow in an adherent monolayer, or they can grow suspended in the culture medium. Once those cells grow to confluency, they are passaged and considered as a cell line. Following the Hayflick limit, normal cells divide a finite number of times, and they eventually undergo senescence, which is the irreversible growth arrest of mitotic cells. In order to maintain a cell line indefinitely or over a larger number of passages, normal cells can be immortalized through the synthetic expression of telomerase genes. For reliable results, purchasing cell lines from commercial vendors is recommended.

5.3.1.2.3 New Cell Culture Techniques

Induced pluripotent stem cells (iPSCs) were created in response to ethical concerns surrounding embryonic stem cells and are often implemented in studies involving regenerative medicine and cell therapy [3]. A multidisciplinary approach to cell culture is 3D culture systems, which can provide a more informed approach to studying drug screening for clinical purposes. Incorporation of multiple cell types or even the inclusion of iPSCs is currently being investigated.

5.3.2 Sample Processing

5.3.2.1 DNA Isolation

Commercial kits are available for DNA isolation, but the phenol–chloroform method is commonly used and includes the following steps: cell lysis, proteinase K treatment, phenol–chloroform treatment, phase separation, and DNA precipitation. DNA that has been isolated and purified from a sample can be used for complementary DNA (cDNA) microarrays, Southern blots, and DNA damage assays. Cell lysis is achieved through the administration of Tris-chloride, which is a buffer that acts to maintain the pH. A chelator, EDTA, is added to inhibit DNase in the sample, and ionic and nonionic detergents can be used to disrupt cellular membranes and protein interactions. Proteinase K is active in the presence of the detergents, and it digests nucleases while simultaneously protecting the DNA. In the phenol–chloroform treatment, the phenol denatures the secondary structure of the proteins in the sample, allowing for them to precipitate out of the solution. The chloroform solubilizes lipids in order to eliminate them from the DNA. When the solution is at a pH of 7–8, the DNA and RNA are maintained in the aqueous phase. If the pH is 6 or lower, the DNA is not as soluble, and it remains in the organic phase, while the RNA is in the aqueous phase. The proteins and DNA are located in the interphase, and the RNA is in the aqueous phase at a pH of 4.5. Isopropanol or ethanol can be added to precipitate the DNA from solution, and diethylpyrocarbonate (DEPC)-treated water can be added to solubilize the DNA. UV spectrophotometry or electrophoretic techniques are used to determine the DNA quality and quantity prior to further experimentation. DNA with an A260/280 ratio around 1.8 is indicative of pure DNA, and the A260/230 value is a measure of nucleic acid purity and should be between 2.0 and 2.2.

5.3.2.2 RNA Extraction

The extraction of RNA is useful for studies involving reverse transcription polymerase chain reaction (RT-PCR), microarrays, Northern blots, promoter deletion analyses, and RNase protein assays. RNA extraction is subject to contamination through a variety of routes including sample degradation, the presence of RNases, and human contamination. It is important that the lab space used to collect RNA is clean and free of RNase. It is best to conduct RNA studies in an enclosed space similar to a cell culture hood and to use autoclaved or RNase-free pipette tips and microtubes. The bench or workspace and pipettes should be sprayed with RNaseZAP™ prior to the experiment. There are many commercially available kits for researchers for RNA extraction, but a common method is the guanidinium thiocyanate–phenol–chloroform extraction method. For this method, the cells should be washed three times with 1× PBS before the guanidinium thiocyanate is added to the cells to denature nucleases that could potentially degrade the RNA. The cells are then broken up manually with a pipette to extract the RNA, and the lysate is then transferred to a sterile microtube. Chloroform is added to the lysate, and the aqueous phase contains the RNA due to the acidic conditions. The RNA is carefully removed so that the interphase and organic phases are undisturbed since they contain DNA, proteins, and lipids that can cause RNA degradation. Isopropanol is added to precipitate out the RNA from the aqueous phase, and the pellet is dried with ethanol. The RNA is solubilized with DEPC-treated water that contains no RNase. The RNA quality and quantity are determined using UV spectroscopy or electrophoretic techniques. RNA with an absorbance ratio value of A260/280 near 2.0 is generally regarded as pure RNA with little protein contamination, and like DNA, the A260/230 value should be between 2.0 and 2.2.

5.3.2.3 Protein Extraction

In order to characterize proteins for Western blot, enzyme-linked immunosorbent assay (ELISA), and immunoprecipitation studies, protein must be extracted from cells or tissues. For cultured cells, radioimmunoprecipitation assay buffer (RIPA buffer) with protease inhibitor is mostly used as a lysis buffer for protein extraction. The RIPA buffer is added to the cells and then scraped on ice and collected. Sonication of the sample on ice further breaks the cells, and centrifugation at 4 °C separates the cell debris into the pellet and the protein is contained in the supernatant where it is collected and stored. Phosphatase inhibitors can be added to the lysis buffer for the extraction of phosphorylated proteins. The addition of DNase for DNA digestion is not recommended as this introduces protein contamination from the enzyme.

5.4 Sample Storage

To preserve cells for future use, cryopreservation of cell lines is necessary. When freezing cells, it is important to follow the manufacturer's instructions on preparing the appropriate freeze medium with cryoprotectants such as dimethyl sulfoxide (DMSO) and allow for slow freezing of the cells. In order to do this, the cells must be aliquoted to cryovials and stored at cooling temperatures of −1 to −3 °C every minute, using alcohol-free freezing containers or programmable freezers, and then temporarily stored in −80 °C and finally liquid nitrogen (~ −140 °C). When resuscitating a cell line, the cells must be gently thawed in a water bath at 37 °C for several minutes and then immediately spun so that the freeze medium can be removed. The freeze medium contains toxic levels of cryoprotectant that can damage the cells. Due to the greater stability of DNA and protein, they can be stored at −20 to −80 °C, and RNA is best stored at −80 °C for long-term storage. It is best to prepare aliquots of the sample to reduce the number of freeze–thaw cycles and maintain the integrity of the sample.

5.5 Common *In Vitro* Studies in Toxicology/Pharmacology

In vitro studies can provide a plethora of useful information that can serve as a foundation for *in vivo* studies. *In vitro* studies can provide cytotoxicity information and insight into possible cellular mechanisms affected upon exposure. The results can often times be translated into useful information to support future studies. It is important to include the necessary controls during any study in order to exclude the potential factors interfering with experiment results. Solvent or vehicle controls are needed to ensure that the solvent used in the experiment does not cause measurable effects to the experiment. For example, solvent controls can include culture medium containing a negligible concentration of ethanol, DMSO, or saline, to name a few, and they are often compared with a control containing medium alone.

5.5.1 Cytotoxicity Studies

Cytotoxicity studies are usually necessary to establish the boundaries of cellular response to treatments. Measurable endpoints to consider include changes in cell permeability, mitochondrial function, cell morphology changes, and cellular proliferation [4].

5.5.2 Viability Assays

Cell viability studies can be used to test membrane permeability by measuring the uptake of dyes such as trypan blue, erythrosin, or naphthalene black. Conversely, these studies can measure the release of dyes such as diacetyl fluorescein or neutral red from the damaged cell membranes [4, 5].

5.5.2.1 Trypan Blue

A simple method to identify live versus dead cells is the trypan blue exclusion assay. This technique is based on the principle that live cells have intact membranes that will prohibit the dye from entering the cell. On the other hand, dead cells have leaky membranes and will take up the dye. By examining the ratio of cells with dye excluded (live) and dye present (dead), it is possible to determine cell viability in your sample. To perform this assay, a 0.1% solution of trypan blue dye and cell suspension are incubated together at room temperature and centrifuged. Following centrifugation, the supernatant must be removed, and the pellet is resuspended with clear media, PBS, or Hank's balanced salt solution. The cell suspension is transferred to a hemocytometer for counting using a standard inverted microscope. It is important to know the dimensions and associated calculations with your lab's hemocytometer, since that information is subject to change depending on the type. The following equation can be used to calculate the percentage of dead cells in the sample of interest [6]: % of stained cells in sample = number of stained cells/total number of cells counted × 100.

5.5.2.2 Erythrosin

The use of erythrosin B dye and the fluorophore fluorescein diacetate (FDA) to analyze the integrity of cell membranes after treatment can be used to quantify the number of viable cells. Erythrosin B, like trypan blue, is an exclusion dye, and it will not stain viable cells with an intact membrane. FDA is used in conjunction (as a control) because it will stain viable cells that retain their membrane integrity. The primary difference between this assay and the trypan blue assay is the detection method. Trypan blue is an absorbance-based dye, while erythrosin B and FDA are fluorescent dyes. While there are other differences between the two assays, the choice for their use ultimately comes down to the ability to effectively stain cells in a monolayer. Krause, Carley, and Webb [7] found that the erythrosin B exclusion dye assay had more reliable results when compared with trypan blue in mammalian monolayer cell cultures. Therefore, the erythrosin B/FDA staining method is generally regarded as more effective due to its ability to stain a higher percentage of the cell population.

5.5.2.3 Crystal Violet Staining

Crystal violet can be implemented in monolayer cell culture to identify the number of viable cells after treatment. The crystal violet stain acts by staining DNA in living cells, as it preferentially binds to helical DNA. The color of the dye can be influenced by pH, which must be accounted for during experimentation. A solution of 0.25 mg/ml of crystal violet in water can be made, and the cells are incubated in this mixture at room temperature. The solubilizing solution is a 33% acetic acid solution, and the optimal absorbance wavelength is 570 nm [8], which can be read using a microplate reader. This method can be preferable to more toxic intercalating DNA binding dyes.

5.5.2.4 Neutral Red Staining

The neutral red absorption assay is a spectrophotometric test that can be used in *in vitro* toxicity screening. It is based on the ability of viable cells to incorporate the dye into lysosomes, which allows for determination of viable/nonviable cells. Cells can be seeded into a 96-well flat-bottom plate 24 hours prior to treatment. Neutral red dye can be prepared as a 0.4% stock solution in the culture medium used during treatment to make a final concentration of 50 µg/ml and then stored at 37 °C overnight. The dye should be prepared the day before it is needed so the precipitated crystals can be removed prior to use. To remove the precipitate, the dye must be centrifuged at 1500 × *g* for 10 minutes. One hundred to 200 µl of the 50 µg/ml neutral red dye can be added to each well and then incubated at 37 °C for three hours. Viable cells will uptake the dye, and after the incubation period, the cells can be quickly washed with 40% formaldehyde and 10% $CaCl_2$ to remove unwanted dye. After the formaldehyde mixture is removed, the addition of 100–200 µl of 1% acetic acid and 50% ethanol will cause the extraction of the neutral red dye from the cells into the solution. After allowing the plate to incubate for 20 minutes, the plate can be mixed and then read with a spectrophotometer at an absorbance of 540 nm. The treated replicates are then averaged and then compared with the control average to form a ratio [9]. A decrease in uptake of neutral red dye following exposure to toxicants is related to a decrease in viability, which can be used to determine relative toxicity.

5.5.3 Survival Assays

Survival assays in toxicology or pharmacology are often long-term assays when compared with other types of cytotoxicity studies. The most common assays in this category are clonogenic assays and cell counts. These assays can examine the ability of the cells to proliferate and, in some cases, form colonies. Treated and non-treated cells can be compared in order to quantify changes upon treatment.

5.5.3.1 Clonogenic or Colony Formation Assay

The use of clonogenic assays as a cellular sensitivity assay is common in experiments that aim to determine the ability of individual cells to form colonies after treatment. Cells are seeded into petri dishes or six-well plates at the desired cell density. Cells are treated with the drug/chemical of choice for an amount of time and maintained in an incubator at 37 °C. After treatment, the treatment medium is removed, and the cells are gently rinsed with 1× PBS. The PBS is aspirated, and 2–3 ml of the fixation solution is added at room temperature for five minutes. This solution is removed after five minutes, and 0.5% crystal violet stain is added and incubated at room temperature for two hours. The plate or dish is gently washed with tap water to remove the crystal violet stain, and they are allowed to dry at room temperature for several days. The number of colonies can be counted using a standard microscope [10]. The plating efficiency (PE) and surviving fraction (SF) can be measured using the following equations: PE = (# of colonies formed/# of cells seeded) × 100%; SF = (# of colonies formed after treatment/# of cells seeded) × PE.

5.5.3.2 Cell Cycle Analysis: Flow Cytometry

Many compounds can elicit cell cycle arrest or induce damage at different points during the cell cycle. The following cell cycle checkpoints, G1/S and G2/M, can give researchers important information in regard to the progression or arrest of the cell cycle. One means to more deeply understand cell cycle is to measure apoptotic cells. Endonucleases degrade DNA during apoptosis, and this is an effective means to estimate the level of apoptosis. Alternatively, researchers can study growth kinetics or selective growth by comparing cells cultured in serum-free medium and supplemented medium. Observing cycle-related changes can also be achieved by growing cells in serum-free medium to ensure that all cells are in the same cell cycle followed by exposure to a compound of interest.

The Nicoletti assay measures apoptosis in cells by sorting and analyzing cells with diploid or haploid DNA content. The fluorophore propidium iodide (PI) can permeate through damaged cells and intercalate with double-stranded DNA, preferentially staining the damaged cells. In a typical Nicoletti assay, approximately 5×10^3 cells (based on the cell type) are in each well of a 96-well plate and then treated with a compound of interest. Control cells are incubated with 25 µl of 50 µg/ml PI for 10 minutes. The fluorescence is read initially before the cells are frozen at −20 °C for 24 hours, which allows PI to invade all the cells. The final fluorescence reading is made after the 24 hour freezing. The cell viability is then determined by finding the difference between the first and second measurements [11]. When using a Millipore system, 2×10^5 cells from all treatment and control groups need to be collected and fixed in 70% ethanol for more than 24 hours at 4 °C. Cells are then stained (using Guava Cell Cycle Reagent, for example), sorted, and analyzed using a Guava easyCyte Flow Cytometer. The percentage of cells in pre-G1, G0/G1, S, and G2/M phases of the cell cycle are determined using the GuavaSoft software. It is best to run all of the samples in triplicate when doing flow cytometry [12].

5.5.4 DNA Damage Assays

There are many assays that have been developed to measure endpoints such as DNA damage, changes in gene expression, copy number changes, DNA–protein interactions, gene silencing/knockdown, and transfection. These data provide a great amount of useful information to further characterize the state of a cell upon treatment.

5.5.4.1 Comet Assay

The comet assay that is used today was created in the 1980s, and unlike previous methods, it allows for examination of DNA damage in a single cell, which gave rise to its name, "the single cell gel electrophoresis assay." Put simply, the head of the comet contains heavy DNA, and the tail is composed of damaged DNA fragments. Tail moment, tail length, and tail DNA percentage are common measurable endpoints. In order to measure single- and double-strand breaks, two different comet assays can be implemented. Single-strand break assays have a longer lysis time, while the double-strand assay is performed in neutral conditions, and hydrogen peroxide (H_2O_2) is used to help measure anywhere from 50 to 10 000 breaks per cell. Abasic sites (AP) and excision repair sites can even be detected under alkaline conditions. Modifications of the comet assay have included the addition of enzymes to detect specific lesions in the DNA. Imaging software is used to measure total intensity, tail length, percent DNA in tail, and tail moment of each cell. Each treatment or control should be done in triplicate [13].

5.5.4.2 Sister Chromatid Exchange Assay

Sister chromatid exchange is considered as a sensitive indicator of chromosomal damage in cells exposed to mutagens, and this assay can serve to discern the mutagenicity of a compound. In the sister chromatid exchange assay, treated and untreated cells are cultured with 5-bromodeoxyuridine (BUdr) for 24 hours so that two DNA divisions can occur. During the last two hours of culture, colcemid is added. After this step, the mitotic cells are shaken and then treated with potassium chloride for six minutes in order to separate the chromosomes, and then they are fixed in a 3 : 1 methanol/acetic acid mixture. The cells are pipetted onto a slide and allowed to air dry. After the slides are dry, the cells are stained with Hoechst 33258 and washed with distilled water. The coverslips are mounted with rubber cement in the same buffer, and the slides are exposed to fluorescent light in an X-ray viewing box for two hours. The coverslips can be removed at this time, and the slides are incubated at 62 °C in saline sodium citrate (SSC) buffer for 20 minutes. Finally, the slides are stained using 3% Giemsa stain for 15–30 minutes. The chromosomes have now undergone irreversible staining [14], which allows for visualization of chromosomal damage.

5.5.5 Southern Blot and DNA Sequencing

5.5.5.1 Southern Blot

Southern blot can be used as a method to detect a specific sequence of DNA in a sample, identify restriction fragment length polymorphisms, and determine copy number changes. In Southern blot, DNA is cut into smaller fragments by restriction endonucleases, and those smaller fragments are then separated by gel electrophoresis. The DNA fragments are transferred to a membrane and then permanently attached by heat or UV radiation. The membrane is incubated with a radiolabeled or fluorescently labeled probe where it hybridizes with the sequence of interest and it is visualized using X-ray film.

5.5.5.2 DNA Sequencing

DNA sequencing is a method used to identify the sequence of nucleotides in DNA, which can be used to better characterize the genome and associated proteins. DNA sequences can be used in the diagnosis and prognosis of diseases, identification of biomarkers, discovery of new genes and their function, etc. There are several newer sequencing methods, such as whole-genome sequencing (WGS) and next-generation sequencing (NGS).

5.5.5.3 Transfection and Gene Silencing

Transfection is the process of transferring foreign DNA or RNA into a eukaryotic cell. The uptake of foreign nucleic acids into a cell can be accomplished through chemical- or physical-based methods. Chemical-based transfections include calcium phosphate-mediated transfection, liposome-mediated transfection, and diethylethanolamine (DEAE)-dextran-mediated gene transfer. Physical-based transfections include electroporation, protoplast fusion, and microinjection [15]. In order to reduce the expression of a particular gene, gene knockdown can be used. RNA interference can be used in gene silencing by introducing exogenous small interfering RNAs (siRNA) that are complementary to the target mRNA sequence, and the siRNA is processed by the RNA-induced silencing complex (RISC). Ribonucleases act to cleave the target mRNA [16]. A newer gene manipulation/editing technique is clustered regularly interspaced short palindromic repeats (CRISPR)/Cas9 system. It has been discovered that bacteria have a CRISPR system that acts to protect the bacteria from foreign DNA through the implementation of RNA-guided DNA cleavage. This same concept can be applied to gene silencing through the co-delivery of a plasmid containing Cas9 endonuclease activity and the appropriate short CRISPR RNAs [17]. Gene transfections or knockdowns can be useful in testing the impact of the gene of interest under certain conditions. For example, many oncogenes are overexpressed in cancer, so the knockout or knockin of an oncogene of interest is often applied with the CRISPR/Cas9 method. The knockout of genes can provide insight into the overall function of an unknown gene.

5.5.6 RNA Quantification and Identification

5.5.6.1 Northern Blot

Gene expression can be studied by detecting RNA in samples undergoing various stages of differentiation in normal and diseased cells. Specifically, changes to gene expression can be monitored in the cells of interest and compared to normal cells. Novel genes can be discovered using Northern blot and gene sequencing; mRNA size, expression levels, and splice variants of the gene of interest can be detected or quantified using Northern blot. When using the Northern blot method, total RNA is first extracted from a sample, and the RNA types are separated by denaturing gel electrophoresis. This step is followed by transfer and cross-linking onto a membrane. Either heat or UV light is applied to the membrane to immobilize the RNA, and a labeled probe hybridizes to the RNA of interest on the membrane. Finally, the membrane is washed, and the signal is visualized using X-ray imaging [18]. Northern blots are considered limited, as this type of analysis can only test a small number of genes compared with higher-throughput methods like microarrays and RNA sequencing (RNA-Seq).

5.5.6.2 Promoter Deletion Analysis

Promoter deletion analysis is a method that identifies sequences in the promoter region responsible for controlling transcription. Reporter genes, such as luciferase or chloramphenicol acetyltransferase (which are oxidative enzymes), are fused to promoter deletion fragments and inserted into a vector, ultimately leading to insertion into the cell of interest. Both qualitative and quantitative means are used to assess the expression level of the reporter gene. Possible reporters to use include renilla and firefly luciferases, which have differing enzymatic activities, allowing for a dual-luciferase reporter assay. This method uses measurement of luciferase intensity to quantitatively or qualitatively indicate transcriptional activity [19].

5.5.6.3 RNase Protection Assay

In order to characterize changes in mRNA expression temporally, in various tissues, or at different developmental stages, the RNase protection assay can be employed. An antisense radiolabeled riboprobe that is 100–200 nucleotides long is positioned between an intron and an exon. The probe hybridizes with the mRNA and becomes protected, but unbound probes are digested by RNase. The amount of hybridized probes is detected using a high-resolution denaturing polyacrylamide gel [20]. Alternative splicing may also be studied with the assistance of exon-specific probes.

5.5.7 Gene Expression

Analyzing changes to gene expression upon treatment can provide valuable information with regard to the specific molecular mechanisms upregulated or downregulated in the cell. Gene expression data can provide insight into changes in cellular proliferation, metabolism, cell cycle regulation, tumorigenesis, oxidative stress, etc. There are many technologies that can be used to examine gene expression *in vitro*. Several are discussed in the following sections.

5.5.7.1 Quantitative Real-Time Polymerase Chain Reaction (qRT-PCR)

A common technique used in many molecular laboratories is PCR, which is the amplification of target cDNA to produce thousands to millions of copies that can subsequently be quantified. PCR is generally broken into three major steps. First, the denaturation of hydrogen bonds in the cDNA at high temperatures (94–98 °C for 20–30 seconds) yields two single strands of DNA. Next, annealing of the desired primers to the single-stranded DNA occurs at lower temperatures (50–65 °C for 20–40 seconds), and then the DNA polymerase can bind. Finally, the temperature at the elongation step is specific to the type of DNA polymerase in use. Taq polymerase is a common DNA polymerase used in PCR; its desired temperature range is 75–80 °C. Taq polymerase can then begin to add the deoxynucleotide triphosphates (dNTPs) to synthesize the new strand of DNA. These three steps combined are part of a cycle, and a PCR experiment will typically have 20–40 cycles. Housekeeping or constitutive genes are also measured to act as a control and help quantify the gene expression by comparing the gene of interest to an abundant gene found in all cell types under any condition. The cycle threshold (Ct) value is the number of cycles needed for the fluorescent signal to reach the threshold value. There is an inverse relationship between the Ct value and the amount of target cDNA. The ΔΔCt method calculates the fold change in gene expression by normalizing the amount of target product to the housekeeping gene relative to the control. Quantitative real-time polymerase chain reaction (qRT-PCR) is frequently used, and it involves the quantitative measurement of a target sequence in real time. Reverse transcription-PCR (RT^2-PCR) differs from PCR in that RNA can be added, and the reverse transcription into cDNA and the PCR all happen in the same sample, thereby eliminating a step, which is why it is often referred to as one-step PCR. If greater specificity is needed, TaqMan probe PCR can be used. In this variation, a fluorophore is attached to the 5′ end of an oligonucleotide probe, and a quencher is attached at the 3′ end. Before TaqMan probes anneal with the target cDNA, the quencher reduces the fluorescence of the reporter fluorophore. After the probe anneals with the DNA, the quencher and reporter are no longer in close proximity to one another, so the fluorescent signal can be measured. Common applications of PCR technology include disease diagnosis and prognosis, viral detection, and forensic analyses. However, it is important for all applications to be aware of some common problems that can occur when conducting a PCR experiment. These common problems include primer dimers, DNA polymerase error rates, and large target lengths, to name a few.

5.5.7.2 Microarray

Traditional PCR methods are often limited in the number of genes that can be simultaneously tested, but one alternative to avoid this issue is to use DNA microarray/chip technologies. DNA microarrays are small chips with thousands of bound probes covalently attached, and, like PCR, microarrays can detect changes in gene expression in samples. Similar to PCR, tissues or cells undergo RNA extraction, followed by reverse transcription to cDNA. The cDNA is coupled with cyanine (Cy) dyes and is measured onto the array, allowing for hybridization with complementary sequences. After washing away the nonspecific binding, the fluorescent signal can be read. The fluorescence data must be normalized with proper controls before analysis.

5.5.8 Protein-Related Assays

5.5.8.1 Bradford Assay

In order to determine total protein concentrations, the colorimetric Bradford assay can be implemented. Protein is added to Coomassie Brilliant Blue G250 dye, which produces a color change from red to blue that is dependent on protein concentration; as more dye binds to protein, the solution becomes more blue. The dye is cationic alone, but once the protein is added and conditions become acidic, the dye becomes anionic, which corresponds with an absorbance maximum shift from 470 to 595 nm [21]. The Bradford assay is a simple and quick method to measure total protein concentration, and the researcher can determine how much protein is present in the sample for subsequent experiments.

5.5.8.2 Enzyme-Linked Immunosorbent Assay (ELISA)

One of the most common methods found in a biochemical laboratory, especially for signal transduction research, is the enzyme-linked immunosorbent assay, commonly referred to as ELISA. The introduction of enzyme-antigen immunoassays was first described in 1971 by Eva Engvall and Peter Perlmann to elongate shelf life of reagents and to simplify instrumentation used to measure the assay over the previously used radioimmunoassay, which used an isotope-labeled antigen. This test can measure specific antigens, antibodies, hormones, and even viruses in a sample. Direct and competitive ELISA assays are the two major types, both of which will be discussed in further detail in Chapter 6.

5.5.8.3 Western Blot and 2D Gel Electrophoresis

For over 30 years, the workhorse of protein research has been the Western blot, also referred to as immunoblotting. This technique not only has the ability to identify a protein or phosphoprotein of interest in a given protein sample but can also quantify the amount of a desired analyte compared with standard samples. A successful Western relies on many factors – good separation via gel electrophoresis, which will separate based on charge, size, isoelectric point, or any combination thereof for native or denatured proteins; successful transfer via electroblotting to a membrane that has a high affinity for protein (nitrocellulose or PVDF); effective blocking to prevent nonspecific binding of the antibody to the membrane; and, most importantly, a specific mono- or polyclonal primary antibody. A successful Western truly lives and dies by the primary antibody, which is why preliminary testing is useful before valuable samples are tested. An issue that may arise during an experiment is the problem of the protein expression not reflecting the gene expression, which can be a result of posttranscriptional mechanisms.

5.5.8.4 Immunolocalization

Immunolocalization is a technique that is used to identify the location of an antigen of interest in a cell or tissue through the use of antibodies as probes. Direct immunolocalization is a single step, and the probe and marker are coupled prior to the localization. Indirect immunolocalization is similar to Western blot; a primary antibody binds to the target, and the secondary antibody that is conjugated with a marker binds to the primary antibody. Finally, a tertiary probe can bind to the secondary antibody in order to amplify the signal. Possible uses of immunolocalization include identifying diagnostic markers, locating protein distributions in cells and membranes, and even determining the origin of cell types, to name a few [22].

5.5.8.5 Immunoprecipitation Assays

Immunoprecipitation assays use protein–protein interactions, in which an antibody acts to precipitate the antigen of interest from a solution. This same technique can be applied to a protein of interest within a complex of known and unknown proteins; this method is referred to as protein complex immunoprecipitation.

5.5.8.6 Chromatin Immunoprecipitation (ChIP)

Interactions between protein and DNA in a cell are investigated using chromatin immunoprecipitation (ChIP). ChIP can be used to identify the DNA binding sites on the genome for the protein of interest to better understand the mechanisms involved in gene regulation pathways. The cells are first cross-linked and fixed using formaldehyde; micrococcal nucleases are used to digest the chromatin into 150–900 base pair (bp) DNA/protein fragments. Coprecipitation occurs when antibodies selective for histones of nonhistone proteins are added and seized by protein G agarose or protein G magnetic beads. Finally, the cross-links are reversed, and the DNA is purified, which allows for analysis. The protein of interest couples with the cross-linked fragments during the immunoprecipitation process and can be sorted and identified using PCR, microarrays, or sequencing. This technique can be used to elucidate the protein and DNA interactions involved in cell signaling, DNA replication and repair, cell cycle, epigenetic events, and chromosomal stability pathways [23].

5.5.9 Epigenetics

In addition to the more classical *in vitro* studies, epigenetic changes can be detected and quantified. Epigenetics is the study of the effects of the external environment on genes. Epigenetic events of interest include DNA methylation, histone modifications, and RNA-associated silencing. Some diseases, including some cancers, are influenced by epigenetic changes, with many researchers identifying new ways to monitor these alterations. Commonly employed downstream methylation analyses include the following: endpoint and real-time PCR, primer extension, microarray analysis, blotting, ion-pair reversed-phase high-performance liquid chromatography (HPLC), and sequencing. No single technique can be applied to the entirety of the epigenome, and it is best to use them in combination [24]. Bisulfite pyrosequencing for DNA methylation is the method discussed in the following section. The classical methods used to measure histone modifications include specific gel systems and even the addition of a radioactive precursor followed by protein hydrolysis and amino acid analysis. Both mass spectrometry (MS) and "ChIP-on-chip" are commonly used today, with the focus being on the latter [25]. Finally, RNA silencing can be measured by qPCR, miRNA arrays, RNA-Seq, and multiplex miRNA profiling, but multiplex miRNA profiling will be discussed in greater detail.

5.5.9.1 Bisulfite Pyrosequencing

In order to quantify the amount of DNA methylation in a sample, bisulfite conversion of DNA followed by pyrosequencing is an appropriate method to use. First, primers must be designed to amplify 200–300 bp in the promoter region. One primer should be biotinylated at the 5′ end to help with the attachment of the PCR products to the streptavidin-coated magnetic beads. Bisulfite conversion of the DNA under harsh conditions converts unmethylated cytosine to uracil so that methylated and unmethylated cytosine can be differentiated. The DNA is purified, and PCR amplification of the pyrosequenced template follows. The PCR product is immobilized once it couples with the streptavidin beads. The single-stranded pyrosequenced template of DNA is prepared, and DNA polymerase causes the release of pyrophosphate. The pyrophosphate and adenosine phosphosulfate are converted to adenosine triphosphate (ATP) and sulfate by ATP sulfurylase. ATP, luciferin, and oxygen are converted to oxyluciferin and light by luciferase. The resulting signal is based on the number of bases present and is represented on the pyrogram, which can then be translated into the appropriate sequence. The remaining cytosine is degraded by apyrase.

5.5.9.2 ChIP-on-Chip

ChIP assay has been described above. ChIP-on-chip allows for the mapping of genome histone modifications and analysis of chromatin-associated proteins. First, the array and targets of interest (usually for specific histone modifications) are selected. Chromatin is precipitated using the probes, and the DNA undergoes purification. The chromatin is fluorescently labeled with different dyes, such as Cy5 or Alexa 647, using ligation-mediated PCR. The labeled DNA is poured on the microarray, allowed to hybridize, and then read. The fluorescent signals are normalized to controls located on the chip or from a different array to account for nonspecific binding.

5.5.9.3 Multiplex miRNA Profiling

This technique utilizes a firefly luciferase-type reporter in order to profile multiple miRNAs in a sample. In this technique, probes are embedded with particles that have an miRNA-specific binding site. When these probes are mixed with a sample, the target miRNAs bind to their specific probe, and the miRNA-reporter complex is hybridized to allow subsequent removal of unbound materials (which is especially useful when using crude biological samples). A labeling mix that contains universal adaptors and ligation enzymes allows ligation of adaptors on the target. Finally, these ligated miRNAs and adaptors are eluted from the probes, and PCR amplification occurs. During PCR amplification, biotin is incorporated at the target miRNAs to facilitate labeling with the fluorescent firefly reporter. This is what allows for data collection via flow cytometry and subsequent analysis and miRNA profiling.

5.6 Common *In Vivo* Studies in Toxicology

In vivo studies are useful in providing a comprehensive analysis of all biological systems involved, whereas *in vitro* assays simply provide a snapshot. Animal testing, clinical trials, and epidemiology studies are the primary means of gathering *in vivo* data. These data can be used to determine the toxicity of a compound and its elicited effects. *In vivo* data can be extrapolated and utilized in a risk assessment. Animal *in vivo* models are often utilized and provide insightful information that is often extrapolated to human populations.

There are many endpoints that can be used to examine the varying levels of toxicity to a system *in vivo*. Some common toxicity endpoints include maximum tolerated doses (MTDs), acute toxicity, subchronic toxicity, chronic toxicity, reproductive toxicity, developmental toxicity, genotoxicity, and carcinogenicity, and these are discussed in the following sections.

5.6.1 Toxicological Endpoints

5.6.1.1 Maximum Tolerated Dose (MTD)

MTD is the amount of a compound necessary to produce the preferred effect, without unwanted toxicity to the individual. In clinical trials, the MTD can help identify negative health effects due to prolonged exposure and can help eliminate the testing of unnecessary concentrations that are too toxic or biologically irrelevant.

5.6.1.2 Acute, Subchronic, and Chronic Toxicity

Acute toxicity is characterized as the brief time after exposure, typically 24 hours or less, that results in adverse changes. It can be continuous exposure to a toxicant over a short time period. The median lethal dose (LD_{50}) is a measure of acute toxicity, and it is the dose necessary to cause death in 50% of the test population [26]. LD_{50} calculations are important values in toxicology, as they can indicate the acute toxicity of the compound of interest. Acute toxicity studies typically investigate the impact of one large dose to the individual and are often needed to characterize emergency exposure situations. In order to understand the effects of long-term exposure similar to pharmaceutical use or prolonged exposure at low concentrations, subchronic and chronic toxicity tests are used. A subchronic test usually lasts for 90 days, while a chronic toxicity test can be used to study the adverse effects of a compound over months and even years. These results are useful in determining the side effects associated with long-term exposure to a compound. Important values that can be calculated from these tests include no-observed-adverse-effect level (NOAEL), lowest-observed-adverse-effect level (LOAEL), ceiling values, and threshold limit values. The half maximal effective concentration (EC_{50}) is often used to quantify the effectiveness of the drug in question *in vivo*. The EC_{50} value is the median response between the starting point and the maximum effect. The *in vitro* equivalent of this value is the IC_{50} (half maximal inhibitory concentration). Data from acute, subchronic, and chronic toxicity tests provide invaluable information for researchers in academia, regulation, and industry.

5.6.1.3 Reproductive and Developmental Toxicity

Many compounds elicit effects that can have reproductive and developmental toxicological effects in the exposed individual, as well as the offspring. Reproductive toxicity concerns the unfavorable health effects on male and female internal and external sex organs upon exposure to a compound. Common reproductive toxicity endpoints that are indicative of reproductive dysfunction include the following: sperm abnormalities, decreased libido, and subfecundity. Developmental toxicology concerns the adverse effects observed in a maturing organism, whether it be effects in a fertilized egg to effects in an adult as a result of early exposure. Within developmental toxicology, teratology, or the study of birth defects, attributed to developmental exposure prior to birth. During gestation, there are many different toxic windows that will result in varying effects based on the time of exposure. These important developmental stages include preimplantation, embryonic, fetal, and neonatal stages. In animal *in vivo* developmental studies, multigenerational studies can be conducted in order to characterize the effect of exposure in subsequent generations. This information can be useful in the identification of unforeseeable health effects attributed to the compound of interest in future generations.

5.6.1.4 Genotoxicity and Carcinogenicity Studies

Genotoxicity is the study of DNA damage, gene mutations, or chromosomal aberrations incurred after exposure to a compound. These genotoxic events have the ability to contribute to the formation of cancer in the organism. Genotoxicity studies aim to identify any indirect or direct changes to the DNA or chromosomes that can have transmissible effects. The Ames test is a useful study to determine whether a compound has genotoxicity that results in DNA mutations. Briefly, *Salmonella typhimurium* containing mutations preventing endogenous histidine synthesis are incubated with rat liver lysates containing enzymes and the test compound. If the compound is mutagenic, the histidine-mutated *Salmonella* will undergo further mutations that will allow it to synthesize histidine and grow colonies [27]. Unlike genotoxicity studies, carcinogenicity studies deal with compounds that directly cause cancer. Some compounds like benzo[a]pyrene (BaP) are not directly carcinogenic but become so once they are metabolized; therefore, they are categorized as procarcinogens. Long-term tests using animals can help to determine the carcinogenicity of a compound, while short-term assays such as cell transformation and gene mutation assays can be done *in vitro*.

5.6.2 Routes of Exposure

It is important that *in vivo* tests take the most relevant, real-life route of exposure into consideration when studying a particular compound. The absorption, distribution, metabolism, and excretion (abbreviated as ADME) of a compound is an important concept in toxicology, and the route of exposure can greatly influence the ADME of the compound of interest. The main routes of exposure include oral, dermal, inhalation, and injection; further, in animal *in vivo* studies, intravenous, subcutaneous, intraperitoneal, and intramuscular are methods of injection that can be used.

5.6.2.1 Oral, Dermal, and Inhalation

Oral routes of exposure provide a simple, efficient, and cost-effective means of delivering certain compounds to the test subject. Metabolism and absorption of the compound in the gastrointestinal tract can greatly influence whether the test compound can be delivered orally. Ingestion through diet is the most common route of oral exposure, but oral gavage is an alternative, more invasive method that is often used in rodent studies to get the compound directly to the stomach. Dermal exposure is often avoided in animal studies due to minimal uptake; furthermore, many of the compounds can promote skin irritation. The chemistry of the compound in question is important when examining the dermal absorption. Finally, inhalation of some compounds can be taken into consideration to simulate environmental exposure. In some cases, both dermal and inhalation exposure routes are used to reduce systemic exposure and directly target the organ system that is being investigated through the study.

5.6.2.2 Exposure via Injection

First-pass elimination is a phenomenon in which the concentration (or dose) is greatly reduced (via gastrointestinal and/or hepatic metabolism) prior to reaching the bloodstream. Injection is a method used to quickly administer the test compound to the subject and avoid the first-pass elimination that occurs in the gastrointestinal tract. Intravenous injections allow for the compound to be introduced through the vein in an efficient manner. Subcutaneous injections are administered underneath the dermis and epidermis. Unlike intravenous injections, the release of the compound occurs over a longer period of time due to the lack of vasculature below the skin. This method is often employed in mouse xenograft tumor models. Intramuscular injections are faster acting than subcutaneous injections due to the presence of blood vessels in the muscle. Injection volumes are generally much smaller when using this method. Finally, intraperitoneal injections are in the body cavity of the test subject and are done when venous injection is too difficult or if there is a large volume to be injected.

5.6.3 Animal Models

Clinical studies in humans are often preceded by *in vivo* animal studies. Animal studies provide an informative risk assessment of a compound, sometimes over the comparatively short lifetime of the animal, which differs from the long testing process required in humans. Based on preliminary *in vitro* data, the researchers can model the subsequent animal studies according to the characteristics of the test compound. Additionally, there are standard protocols and experiments that are conducted in clinical studies that utilize numerous animal models ranging from rodents to primates. Unlike clinical trials conducted by private companies, academic institutions rely heavily on murine studies for *in vivo* tests. When designing animal protocols, researchers need to plan the animal studies by taking into consideration the time/process of IACUC approval.

5.6.3.1 Rodent Studies

Rodent *in vivo* studies are extremely common, and they are often conducted in mice, rats, guinea pigs, hamsters, and gerbils. A lifetime murine study can last approximately two years with these types of animals, and these studies provide data that may be translated to humans. Many rodent studies use nude athymic mice, a strain that is advantageous for use in xenograft studies, since the animal is immunocompromised and therefore unable to reject xenograft injections. Transgenic/knockdown/knockout mice are also used when studies are interested in adding a gene or removing a gene. Cancer bioassays in rodents are especially common for testing a dose/response.

5.6.3.2 Other Studies

During the *in vivo* studies associated with pharmaceutical testing, trials use animals such as nonhuman primates, dogs, cats, and rabbits prior to human clinical trials. Depending on the compound of interest, there are standard testing procedures in place that must receive regulatory approval for continued testing. The well-being of the test subject is always a priority, and these regulatory processes are in place to ensure ethical practices are strictly followed.

5.7 Basic Advantages and Disadvantages Associated with Sample Types

There are many advantages and disadvantages to using *in vitro* and *in vivo* studies in toxicology. *In vitro* studies are useful because they are considerably cheaper and require less oversight from regulatory agencies. *In vitro* experimentation often allows for a higher-throughput approach; cells are less expensive and require less maintenance than animal test subjects. When choosing an *in vivo* approach, scientists must always keep the three R's in mind – replacement, reduction, and refinement – especially for ethical reasons. Another concern in both *in vitro* and *in vivo* studies is the lack of evidence supporting the extrapolation of results to humans, with some studies finding little evidence that cell or animal models even provide any predictive measures for humans [28]. A lack of correlation in these studies highlights the importance of careful selection of the model, as it can greatly affect the ability for researchers to translate the findings to humans. For example, mice have vastly different innate and adaptive immunity compared with humans, and that must be taken into consideration during studies relating to the immune system [29]. All in all, there are many advantages and disadvantages when choosing an *in vitro* or *in vivo* model for an experiment, and it is critical for researchers to be acutely aware of these during the experimental design process all the way through to interpretation and analysis of results.

5.8 Human Epidemiology Studies

Epidemiology is the study of the distribution of a disease in a population and the variables that impact that distribution. It is important that epidemiological studies identify the etiology and risk factors of the disease, but they must also establish the magnitude of the disease in the population tested. History of the disease, prognosis, and assessment of current and new preventative/therapeutic measures are other important factors for consideration. Finally, it is desirable to use the epidemiological findings to inform public policy and regulations in the affected community. Epidemiology studies can be broadly characterized into nonexperimental studies and experimental studies. Common nonexperimental studies that are practiced include cross-sectional, case–control, cohort, and ecological studies; examples of experimental studies include clinical, field, community intervention, cluster randomized, and chemoprevention trials. Similar to IACUC approval for animal studies, IRB approval is required, and this process can be time consuming.

5.8.1 Nonexperimental Studies

In nonexperimental epidemiology studies, cases of the disease are detected, and the exposure status is ascertained. Cross-sectional studies examine the association between the disease and the possible risk factors. This type of study is a simple means of gathering data and making conclusions, but it only examines the prevalence of the disease, not the incidence. Based on this type of study, it is difficult to deduce the temporal sequence of events, as measurements are made prior to the formation of the disease in the population. Unlike a cross-sectional study, a case–control study provides a retrospective outlook that takes measurements after the cases have developed. Case–control studies can consist of regular-, hospital-, population-, and nested-based studies. They are inexpensive and efficient, they are appropriate when studying diseases with long dormancy or latency periods, and they can evaluate the effects of several exposures on the disease. Despite the positive attributes of case–control studies, the possibility of bias when measuring exposure and ascribing comparison groups can be problematic. Cohort studies are prospective studies in which the exposed and unexposed cohorts are compared and incidence rates/relative risk can be calculated based on the data. Cohort studies require a large number of subjects and are therefore expensive to conduct. Follow-up often presents an issue, which is one of the reasons why large groups are necessary. Ecological studies involve groups of people rather than individuals. These groups are based on factors such as geographical location, socioeconomic status, and health. These studies compare the frequency of a disease in a population and correlate the frequency to exposure values. Previously collected data can be used, making it a cost-effective approach.

5.8.2 Experimental Studies

Clinical trials are a well-known type of experimental study utilized in toxicology and pharmacology. After appropriate *in vitro* and *in vivo* studies are conducted, a test compound can undergo clinical trial testing in humans to determine the safety and effectiveness of the drug. Phase I trials involve a small number of patients, and they often seek to determine the optimal route of administration for the drug, the frequency of doses, and, most importantly, the safest dose. Phase II trials investigate the safety and effectiveness of the drug, and they involve more patients (over 100). New drugs and combination therapies can be tested in phase III trials, and the patients are separated into new therapy or old treatment groups to allow for comparison. Phase III trials examine thousands of human subjects, and successful drugs or treatments are able to be implemented after passing this phase. After the drug or treatment regimen has been approved, phase IV trials study prolonged use to identify side effects and benefits [30]. Field trials are conducted in subjects that have not acquired the disease being tested. Community intervention trials are similar to field trials, but instead they study the entire community. Cluster randomized trials are field trials in which the treatment is allotted to the subjects at random. Chemoprevention trials use drugs or supplements to inhibit or reverse the onset of disease.

5.8.3 Molecular Epidemiology

Unlike traditional epidemiology, molecular epidemiology assesses the involvement of genetic and environmental risk factors at the molecular level. Biomarker targets in these studies can be any factors that can aid in the measurement of disease. Biomarkers of exposure provide evidence of previous or current exposure. Biomarkers of effect are changes that occur within the subject that are a result of a disease. Finally, biomarkers of susceptibility indicate the presence of acquired or intrinsic factors that are specific to the compound of interest. Molecular epidemiology aims to prevent diseases by utilizing risk assessment and uses biomarkers of susceptibility in conjunction with genetic testing to identify high-risk individuals and populations. This type of experimental design also uses analytical and descriptive statistics to increase monitoring in an attempt to fully characterize the different states of disease. Molecular epidemiology is an exciting field in that it incorporates epidemiology, biostatistics, healthcare professionals, genetics, and environmental scientists; this extensive collaboration between experts creates a comprehensive assessment of the disease at a molecular level.

5.9 Examples of Tox- and Pharm-Based Experiments Relevant to Signal Transduction Endpoints

Cell-based assays are currently considered essential to pharmacology and toxicity testing, and they provide useful screening bridges between *in vitro* assessments and *in vivo* applications. Most *in vitro* assays have been designed using various types of target cells with different endpoints including cytotoxicity, genotoxicity, and cellular signaling changes. For example, assays of live versus dead cells to determine cellular metabolism are the focus of many cytotoxic assays; these provide preliminary estimates of the cellular response to exposure to a toxicant or pharmaceutical agent and are frequently used as a criterion of overall biosafety evaluation. Gene mutations, oxidized guanine bases, chromosome damage, and DNA strand breaks manifest the genotoxicity of environmental contaminants that could be examined by corresponding assays. Furthermore, alterations in cell signaling and function result from drug treatments may lead to cell cycle arrest and changes in selected gene/protein expression patterns as well as reactive oxygen species (ROS) overproduction, which could be detected by different types of *in vitro* assays.

Treatments that induce or inhibit signal transduction can ultimately result in alterations in cellular behaviors at various physiological levels, including cellular viability, gene expression, and protein expression. This section will select some examples to explain how to choose the appropriate targets and treatments using appropriate cell-based *in vitro* assays in toxicology and pharmacology as described above. Epidermal growth factor receptor (EGFR) pathway, cell cycle, cell apoptosis and necrosis, and ROS are selected as examples to be discussed below.

5.9.1 Cytotoxicity

Cytotoxicity can be measured using several different methods, some of which were explained in Section 5.5.2; the choice of cytotoxicity assay is dependent on the assay sensitivity and incubation time with chemical treatments. Assays for proliferation and viability range from the direct measurement of the release of cytosolic enzymes such as lactate dehydrogenase (LDH) after cellular membrane damage to the use of reagents such as 3-(4,5-dimethylthiazol-2-yl)-2,5-diphenyltetrazolium bromide (MTT) or 3-(4,5-dimethylthiazol-2-yl)-5-(3-carboxymethoxyphenyl)-2-(4-sulfophenyl)-2H-tetrazolium (MTS) to determine the overall reducing capacity of the culture (the specifics of MTT and MTS assays are discussed in more detail in Chapter 6). Cytotoxicity assessment is one of the most important steps for selecting appropriate dose and time for further signaling characterization. An example of each chemical, drug, natural compound, and combination treatment are provided in the following sections.

5.9.1.1 Nicotine-Derived Nitrosamine Ketone (NNK)

Normal human lung epithelial cells, BEAS-2B, were treated with the following doses of the cigarette carcinogen nicotine-derived nitrosamine ketone (NNK): vehicle control (0.1% DMSO), 1, 10, 100, 150, and 500 μM for 4, 24, 48, and 72 hours (Figure 5.1). MTS assay was used to determine the number of viable cells, and a significant dose- and time-dependent response was observed ($P < 0.01$).

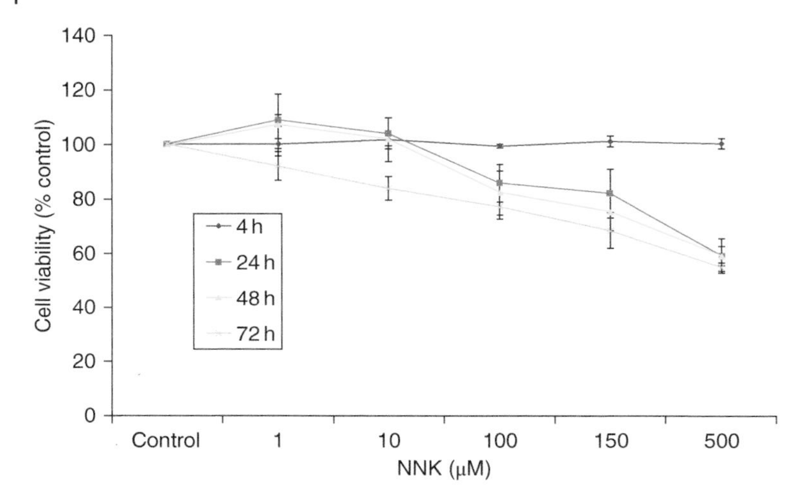

Figure 5.1 The cytotoxic effects of 4-(methylnitrosamino)-1-(3-pyridyl)-1-butanone (NNK) on BEAS-2B cells. BEAS-2B cells were exposed to 0–500 μM NNK for 4, 24, 48, and 72 hours. The MTS assay was conducted to test cell viability. Data were expressed as the percentage of the value as compared with the vehicle control ($n = 6$/data point).

A significant inhibition in cell viability was observed at 100–500 μM NNK at 24 hours and from 1 to 500 μM for 72 hours ($P < 0.05$). The IC_{50} values could not be calculated at even the highest test concentration, so 10, 75, or 150 μM NNK test concentrations were utilized for further studies of signaling transduction pathways in response to DNA damage, such as p53 signaling pathways in this study [31].

5.9.1.2 Doxorubicin (DOX)

Doxorubicin (DOX) is a commonly used chemotherapeutic drug that has cytotoxic effects in cancer cells due to its intercalation with DNA and activation of p53-related pathways. Cancer cells become resistant to DOX treatment, in addition to the strong side effects from DOX treatment alone. MTT was conducted to determine the cytotoxicity of DOX to lung adenocarcinoma A549 cells after 24 and 48 hours of treatment (Figure 5.2). A dose-dependent inhibition of cell growth was observed in 0.1–5 μM DOX-treated cells at 24 and 48 hours ($P < 0.001$), and based on these results, the IC_{50} was calculated to be 2.2 μM DOX at 48 hours [32].

5.9.1.3 Curcumin

In the study by Chen et al. [33], an MTT assay was performed to determine the cytotoxicity of curcumin (5–240 μM) in A549 cells at each time point (12, 24, 48, and 96 hours). The cytotoxic effects of curcumin on A549 cells occurred in a time- and dose-dependent manner ($P < 0.001$) with the maximum inhibitory

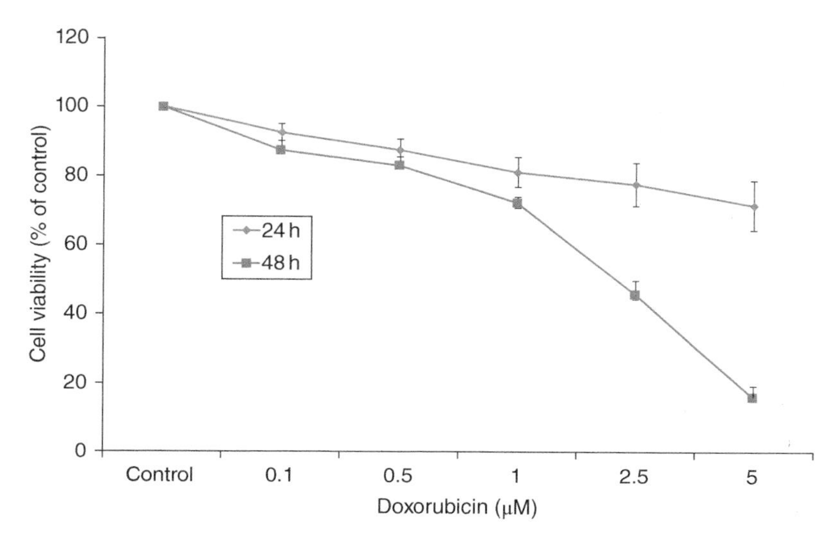

Figure 5.2 Cytotoxic effects of doxorubicin (DOX) on A549 cells. The cell viability after DOX treatment for 24 and 48 hours was determined by the MTT assay. Data are expressed as a percentage compared to the vehicle control (n = 6/data point).

effect after 96 hours of exposure to 240 μM curcumin (Figure 5.3). The IC_{50}s of curcumin in A549 cells decreased with the increased treatment time (93, 65, and 24 μM at 24, 48, and 96 hours, respectively).

5.9.1.4 Combination Effects of Cisplatin and/or Leptomycin B (LMB)

The cytotoxic effects of cisplatin alone or cisplatin + 0.5 nM leptomycin B (LMB) on A549 cells were evaluated (Figure 5.4) [34]. Cisplatin alone significantly suppressed the proliferation of A549 in a dose- and time-dependent manner ($P < 0.001$). In addition, cisplatin and LMB co-treatment significantly enhanced the cytotoxic effects of cisplatin on A549 cells ($P < 0.05$). These data suggest that targeting chromosomal region maintenance 1 (CRM1) using LMB provides a novel and effective adjunct therapy for cisplatin in lung cancer treatment.

5.9.2 DNA Damage

Based on the cell viability assay, BEAS-2B cells were treated with either 5 or 20 μM BaP for 24 hours. Cells were collected, and DNA was extracted to measure the DNA adducts with HPLC after acid hydrolysis. Significant amounts of benzo[a]pyrene-r-7,t-8,t-9,c-10-tetrahydrotetrol(+/−) (RTTC) were detected in BaP-treated cells.

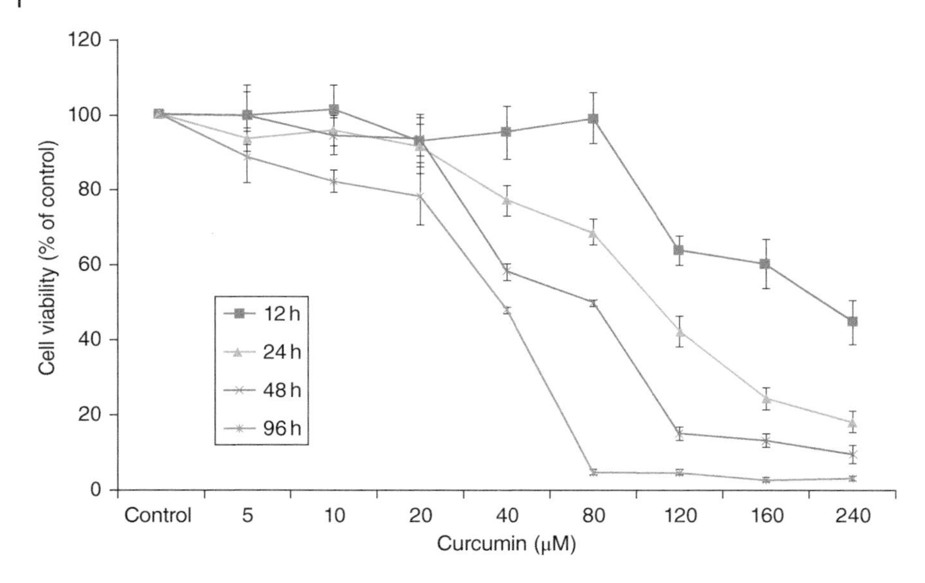

Figure 5.3 Effects of curcumin on the cytotoxicity of A549 cells. A549 cells were treated with 5–240 μM curcumin for 12–96 hours. The number of viable cells was determined by the MTT assay (*n* = 6/data point).

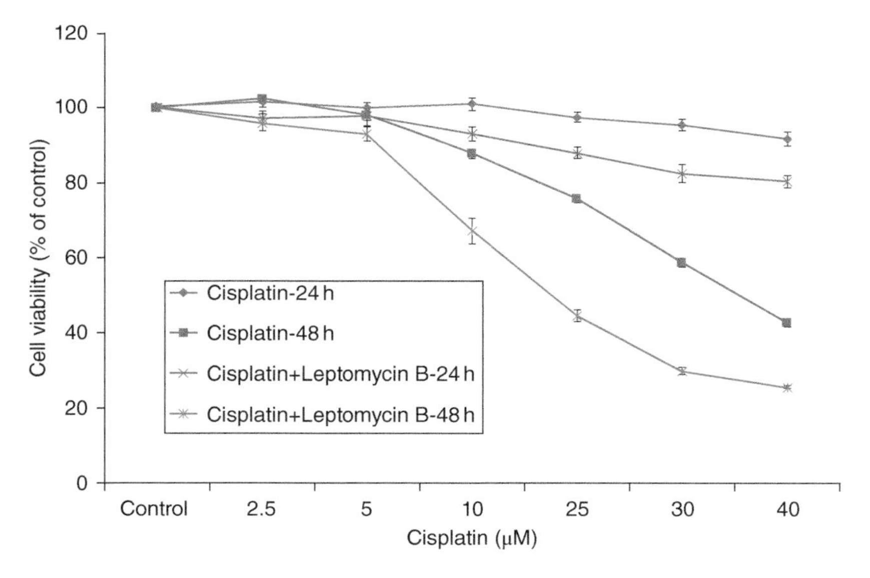

Figure 5.4 Cytotoxic effects of leptomycin B (LMB) and/or cisplatin on A549 cells. The cell viability was determined by MTT assay after LMB and/or cisplatin treatment for 24 and 48 hours. Data are expressed as a percentage compared to the vehicle control for cisplatin and LMB for LMB cisplatin (n = 6/data point).

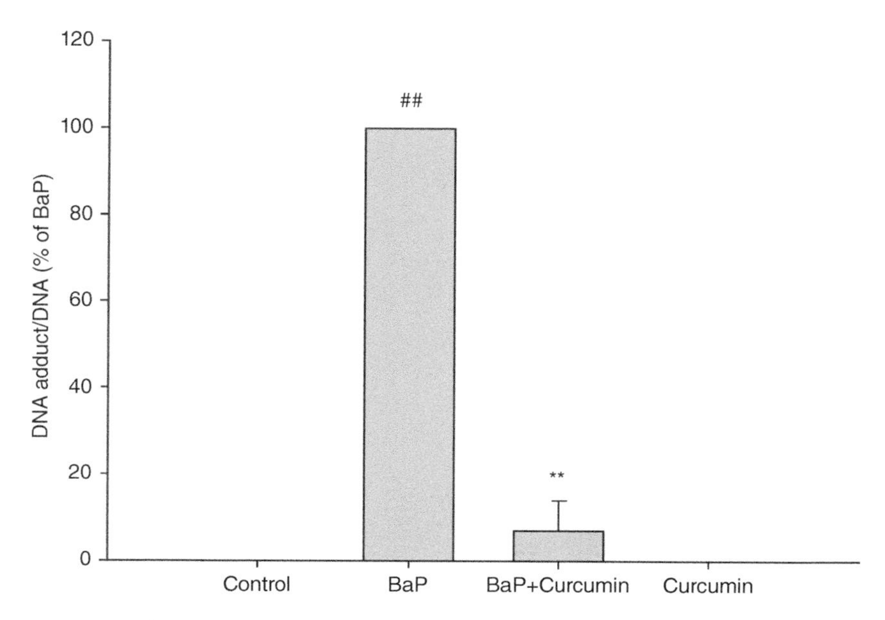

Figure 5.5 Effects of curcumin on benzo[a]pyrene (BaP)-induced DNA adducts in BEAS-2B cells. Relative BPDE tetrol adducts after BaP and/or curcumin treatment for 24 hours. The adduct level in each group was quantified by comparing to BaP group. $^{\#\#}P < 0.01$, BaP alone compared with control; $^{*}P < 0.05$, $^{**}P < 0.01$, BaP + curcumin compared with BaP alone.

Thus, BEAS-2B cells are capable of directly metabolizing BaP. BaP co-treatment with a low concentration of curcumin (5 μM) was found to significantly increase the number of viable cells and reduce the number of benzo(a)pyrene diol epoxide (BPDE)-DNA adducts ($P < 0.01$, Figure 5.5) [12]. Therefore, a biologically relevant concentration of curcumin protected against damage induced by exposure of BEAS-2B cells to BaP for 24 hours. Additional study has found that co-treatment significantly reduced BaP-induced phospho-p53 (Ser15) and PARP-1, which helps to explain the protective role or curcumin observed in the study.

5.9.3 Cell Cycle and Apoptosis

Cell proliferation inhibition could result from either cell cycle arrest or apoptosis; thus both aspects were examined by flow cytometry analysis of A549 cells after LMB treatment [32]. The cell cycle analysis revealed that the percentages of cells in pre-G1 + G2/M and apoptosis were significantly and dose-dependently increased in A549 cells after LMB treatment (Figure 5.6).

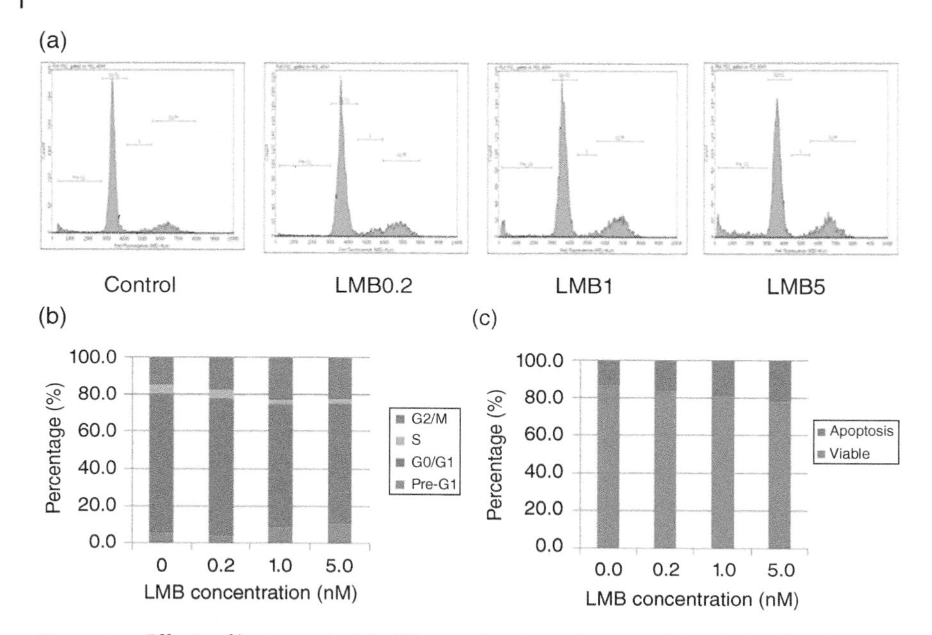

Figure 5.6 Effects of leptomycin B (LMB) on cell cycle and apoptosis in A549 cells. (a) Representative histograms of flow cytometry cell cycle analyses in LMB-treated A549 cells. (b) Cell cycle phases after LMB treatment. (c) The proportion of apoptotic cells after LMB treatment. LMB0.2, 0.2 nM LMB; LMB1, 1 nM LMB; and LMB5, 5 nM LMB.

5.9.4 ROS Induction in A549 Cells After LMB and Epigallocatechin Gallate (EGCG) Treatment

In a study conducted by Cromie and Gao [35], LMB and epigallocatechin gallate (EGCG) co-treatment was found to induce a significant increase in ROS after two hours of treatment when compared with LMB treatment alone in A549 cells (Figure 5.7). In order to confirm these findings, the ROS quencher N-acetyl-L-cysteine (NAC) was commonly used in the pretreatment of cells. NAC pre-treatment significantly decreased ROS in the treated A549 cells, thereby supporting the finding that EGCG contributes to LMB-induced oxidative stress in this particular model. Further experiments have shown that this phenomenon could be related to the inhibition of EGCG on the metabolism of LMB.

5.9.5 Signaling Pathways

5.9.5.1 Metabolizing Alterations After Chemical Exposure

qRT-PCR was implemented to determine the role of phase I and II enzymes in LMB and EGCG combination treatment in A549 cells for 48 hours in a study conducted by Cromie and Gao (Figure 5.8) [35]. The commonly tested phase I cytochrome P450 (CYP) enzymes, such as CYP1A1, CYP1B1, CYP1A2, and CYP3A4, were measured, and CYP3A4 was found to be the only phase I

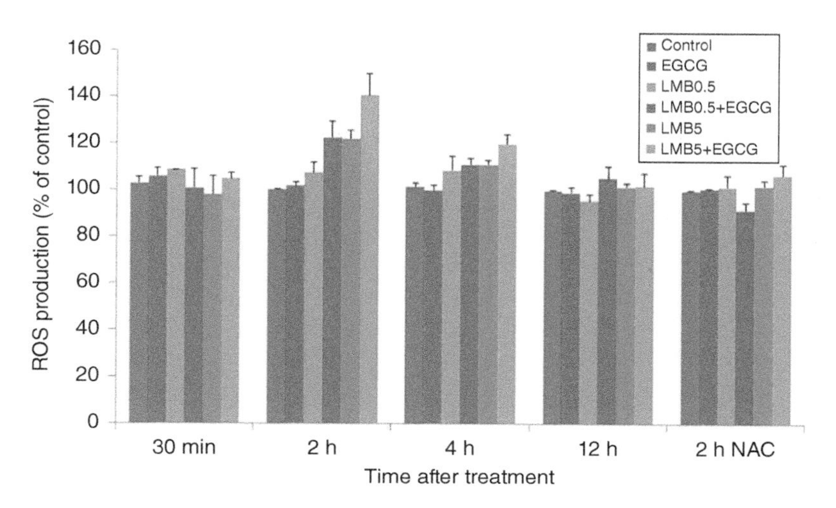

Figure 5.7 Effects of leptomycin B (LMB) and/or epigallocatechin gallate (EGCG) on reactive oxygen species (ROS) production in A549 cells. EGCG, 20 μM EGCG; LMB0.5, 0.5 nM LMB; and LMB5, 5 nM LMB. Cells were treated with LMB and/or EGCG for 30 minutes, 2, 4, and 12 hours. After treatment, ROS formation in cells was measured by the 2,7-dichlorofluorescein diacetate (DCFDA) assay. The fluorescence in treated cells at each time point was compared with the respective control. A549 cells were also pretreated for 2 hours with 10 mM *N*-acetyl-D-cysteine (NAC) and then treated with LMB and/or EGCG for 2 hours upon NAC removal.

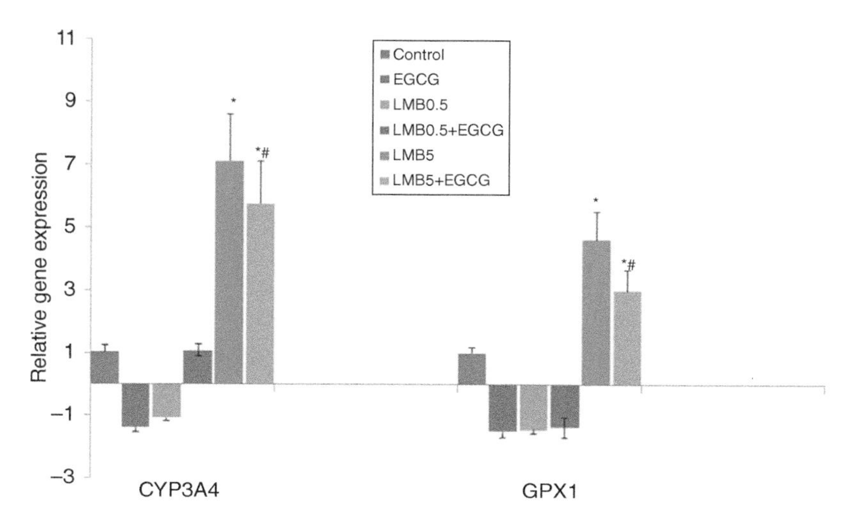

Figure 5.8 CYP3A4 and glutathione peroxidase 1 (GPX1) gene expression after 48 hours of leptomycin B (LMB) and/or epigallocatechin gallate (EGCG) treatments in A549 cells. Gene expression levels of CYP3A4 and GPX1 were measured by one-step quantitative real-time PCR (qRT-PCR). The Ct values of CYP3A4 and GPX1 were normalized to GAPDH, and the fold change in gene expression was calculated using the ΔΔCt method. EGCG, 20 μM EGCG; LMB0.5, 0.5 nM LMB; and LMB5, 5 nM LMB; $^*P < 0.05$ as compared with control; $\#P < 0.05$ as compared with 5 nM LMB.

enzyme significantly upregulated. There was a 7.1-fold ($P < 0.05$) and 5.8-fold ($P < 0.05$) increase in 5 nM LMB and 5 nM LMB + EGCG (20 μM), respectively, compared with control (0.1% DMSO). A significant reduction in CYP3A4 gene expression was observed when 5 nM LMB + EGCG-treated cells were compared with 5 nM LMB alone ($P < 0.05$). Glutathione S-transferase pi 1 (GSTP1) and glutathione peroxidase 1 (GPX1) were phase II enzymes tested in this study. GPX1 gene expression significantly increased 4.6-fold and 3.0-fold in cells treated alone and in combination when compared with control, respectively ($P < 0.05$). The 5 nM LMB + EGCG combination-treated cells had significantly lower GPX1 expression when compared with 5 nM LMB alone ($P < 0.05$).

5.9.5.2 p53 Signaling Pathways

5.9.5.2.1 p53 Posttranslational Modifications (PTMs) and Cellular Localization

A study by Chen et al. [31] found that NNK promoted the nuclear retention and accumulation of p53 in the nucleus through the downregulation of CRM1 protein expression in BEAS-2B cells. After 72 hours, phospho-p53 (Thr55), an important indicator of p53 cytoplasmic degradation, was decreased after the cells were exposed to NNK (Figure 5.9).

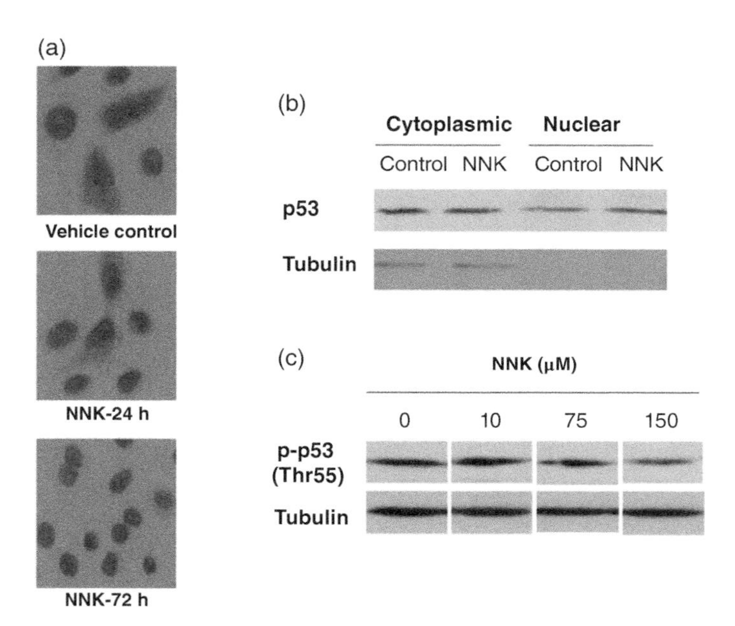

Figure 5.9 The nuclear accumulation of p53 in BEAS-2B cells after 72 hours of 4-(methylnitrosamino)-1-(3-pyridyl)-1-butanone (NNK) exposure. (a) Immunocytochemistry of p53 staining in BEAS-2B cells after vehicle control (0.1% DMSO) or 150 μM NNK exposure. (b) p53 protein in the cytoplasm and nucleus of BEAS-2B cells after 150 μM NNK exposure. (c) Phospho-p53 (Thr55) expression in BEAS-2B cells treated with NNK or vehicle control.

5.9.5.2.2 p53 Signaling Using RT-PCR Array

A study conducted by Shao et al. [36] implemented a gene microarray to test 84 genes involved in p53-mediated signaling pathways (apoptosis, cell cycle, cell proliferation, and DNA repair; Figure 5.10a) upon 24-hour 20 nM LMB

(a)

	A	B	C	D	E	F	G	H
1	APAF1	CASP2	CHEK1	GADD45A	MDM2	PRKCA	TNF	B2M
2	ATM	CASP9	CHEK2	GML	MDM4	PTEN	TNFRSF10B	HPRT1
3	ATR	CCNB2	CRADD	HDAC1	MLH1	PTTG1	TNFRSF10D	RPL13A
4	BAI1	CCNE2	DNMT1	HK2	MSH2	RB1	TP53	GAPDH
5	BAX	CCNG2	E2F1	IFNB1	MYC	RELA	TP53AIP1	ACTB
6	BCL2	CCNH	E2F3	IGF1R	MYOD1	RPRM	TP53BP2	HGDC
7	BCL2A1	CDC2	EGR1	IL6	NF1	SESN1	TP63	RTC
8	BID	CDC25A	EI24	JUN	NFKB1	SESN2	TP73	RTC
9	BIRC5	CDC25C	ESR1	KAT2B	PCBP4	SIAH1	TRAF2	RTC
10	BRCA1	CDK4	FADD	KRAS	PCNA	SIRT1	TSC1	PPC
11	BRCA2	CDKN1A	FASLG	LRDD	PPM1D	STAT1	WT1	PPC
12	BTG2	CDKN2A	FOXO3	MCL1	PRC1	TADA3L	XRCC5	PPC

apoptosis
cell cycle
cell proliferation
DNA repair

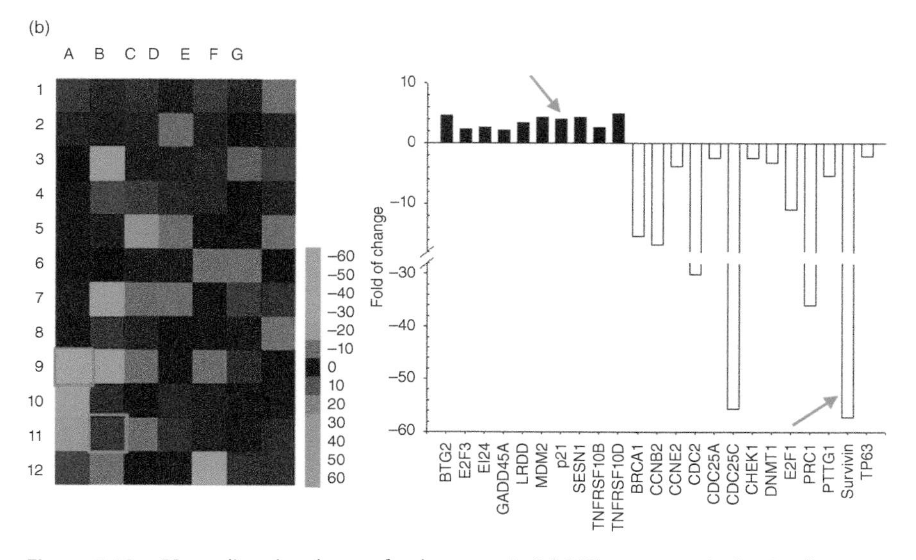

Figure 5.10 p53-mediated pathway after leptomycin B (LMB) treatment in A549 cells. (a) Functional gene grouping of the RT-PCR array for p53 pathways. (b) The heat map and histogram demonstrating fold changes of the gene expression after 20 nM LMB treatment. Gray: genes were not measurable.

treatment in p53 wild-type A549 cells. Upon LMB treatment, 13 genes were downregulated more than twofold, and this included apoptotic genes such as survivin. A total of 10 genes were upregulated twofold, including cell cycle genes, such as p21 (Figure 5.10b).

5.9.6 Protein Kinase B (Akt/PKB)/Mechanistic Target of Rapamycin (mTOR) Pathway Analysis Using Multiblot

The antibody-based multiblot approach is another useful technique to identify proteins of interest, and quantification is already built into this approach without any additional effort. Multiblot was used to characterize the AKT/mechanistic target of rapamycin (mTOR) pathway in A549 cells after curcumin treatment [33]. Curcumin treatment (20 μM for 24 hours) of A549 cells was found to downregulate protein expression of elF2α and elF4E, decrease the phosphorylation of 4E-BP1, and increase the phosphorylation of elF2α (Figure 5.11), which may lead to the inhibition of A549 cell proliferation. These findings suggest that multiblot screening could be useful to gain insight into molecular pathways of interest.

5.9.7 Discovery of Unrecognized Pathways/Molecules Using Proteomics

Nuclear proteins from A549 cells treated with LMB (20 nM) or vehicle control were separated by 2D difference gel electrophoresis (2D-DIGE). Among approximately 1000 protein spots detected in nuclear extractions of A549 cells, the highest changes were identified by MS as sequestosome 1 (SQSTM1/p62) (Figure 5.12a). These data were further validated by Western blot analysis, in

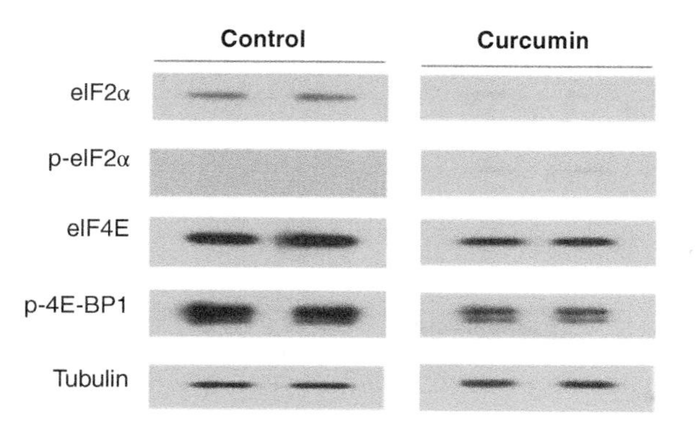

Figure 5.11 Multiblot screening of AKT/mTOR pathway including protein expressions of elF2α, p-elF2α, elF4E, and p-4E-BP1 after 20 μM curcumin treatment for 24 hours in A549 cells. Blots were also probed for tubulin to confirm equal protein loading.

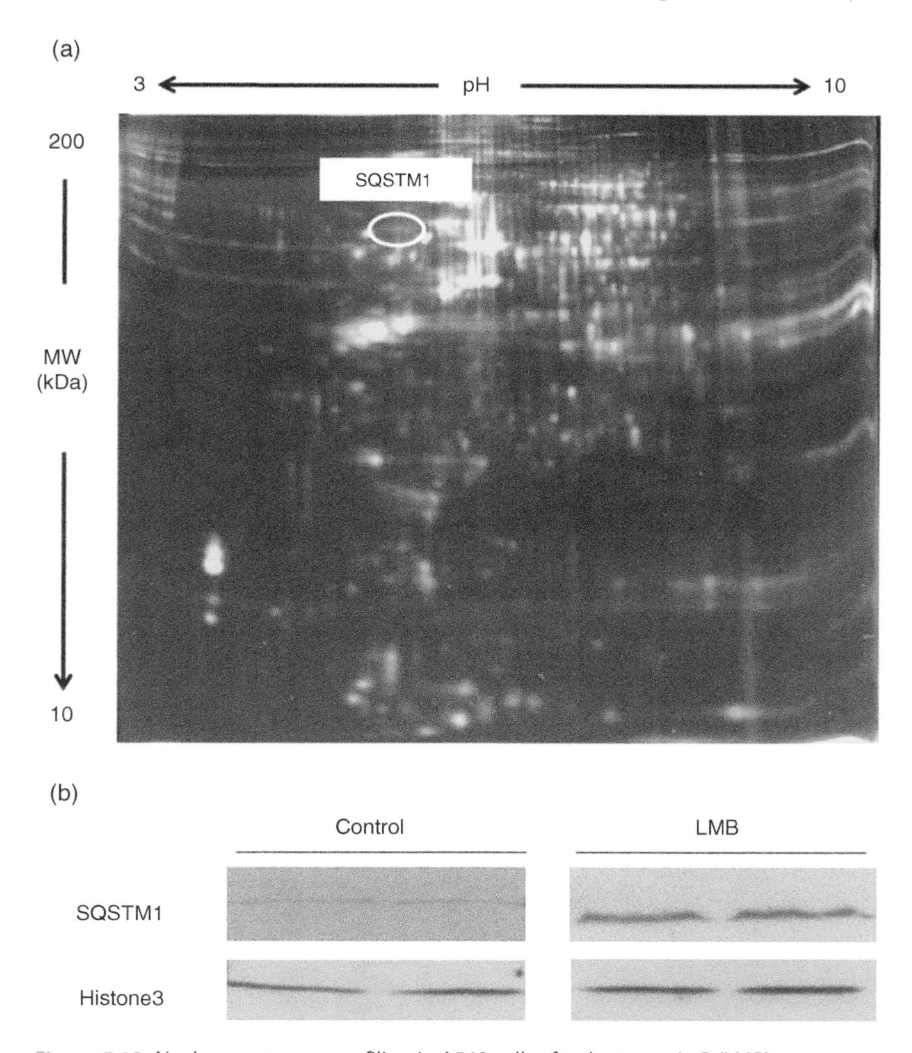

Figure 5.12 Nuclear proteome profiling in A549 cells after leptomycin B (LMB) treatment. (a) Two-dimensional difference gel electrophoresis analyses of nuclear proteins. A549 cells were treated with 20 nM LMB for 24 hours. Nuclear proteins treated with LMB or vehicle control were labeled with Cy3 (green channel) and Cy5 (red channel), respectively. Sequestosome 1 (SQSTM1/p62) was identified by LC/MS/MS. (b) Western blot analysis of SQSTM1 in the nucleus of A549 cells after 20 nM LMB treatment for 24 hours. Blots were also probed for histone 3 to confirm equal protein loading.

which the expression of SQSTM1 in the nucleus of LMB-treated A549 cells was significantly upregulated compared with control (Figure 5.12b). Additionally, the different isoelectric point of SQSTM1 suggests the possibility of posttranslational modifications (PTMs) of this same protein (Figure 5.12a).

5.10 Coupling Experimental Results Within the Larger Literature Framework to Generate Information

5.10.1 Cell Proliferation–EGFR Pathway

EGFR is a member of the human epidermal growth factor receptor (HER)/avian erythroblastosis oncogene B (erbB) family of receptor tyrosine kinases (RTKs) including four closely relevant RTKs: HER1 (EGFR/erbB1), HER2 (neu, erbB2), HER3 (erbB3), and HER4 (erbB4). This family is characterized as having an extracellular, cysteine-rich epidermal growth factor (EGF) binding domain, a single α-helix transmembrane domain, a cytoplasmic tyrosine kinase (TK) domain, and a carboxy-terminal signaling domain [37]. Homodimerization of EGFR in response to the binding of its specific ligands (EGF, transforming growth factor alpha [TGF-α], etc.) stimulates TK activity and elicits autophosphorylation of tyrosine residues in the C-terminal domain, resulting in activation of EGFR downstream signaling pathways, mainly through PI3K/AKT/mTOR, RAS/RAF/MEK/ERK, and the signal transducer and activator of transcription (STAT) pathways (Figure 5.13) [38]. These downstream EGFR signaling

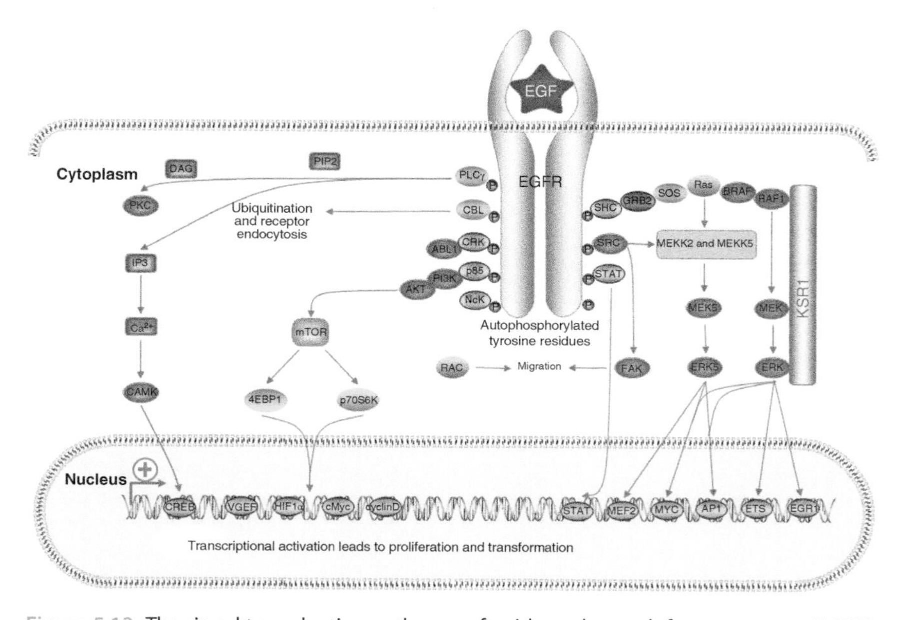

Figure 5.13 The signal transduction pathways of epidermal growth factor receptor (EGFR) provide a multitude of potential drug targets. The autophosphorylation of tyrosine residues in the EGFR cytoplasmic domain elicits its downstream pathway to activate specific biological responses.

pathways consequently lead to the increases of cell proliferation, angiogenesis, and metastasis as well as inhibition of apoptosis [39].

Besides EGFR, there are a variety of other factors that enhance the PI3K/AKT/mTOR pathway, including insulin-like growth factor 1 (IGF-1) [40], insulin [41], and cell adhesion molecule (CAM) [42]. The PI3K/AKT/mTOR pathway is an intracellular signaling pathway directly relevant to cellular quiescence, proliferation, cancer, and regulation of cell cycle. PI3K activation stimulates the phosphorylation and activation of AKT, which has a number of downstream effects such as the activation of cAMP response element-binding protein (CREB) [40], inhibition of p27 [41], the translocation of forkhead box (FOXO) into the cytoplasm [42], and the activation of mTOR that affects the transcription of p70 or eukaryotic translation initiation factor 4E-binding protein 1 (4E-BP1) [41]. Over activation of mTOR signaling significantly promotes the initiation and propagation of various types of cancers including lung, breast, bladder, prostate, brain, melanoma, and renal carcinomas [43]. The most common reason for constitutive activation of mTOR is the mutation of the tumor suppressor phosphatase and tensin homolog (PTEN) gene [44], which negatively modulates mTOR activity by interfering with PI3K. Additionally, increased mTOR activity was reported to drive cell cycle progression and enhance cell proliferation because of its effects on synthesis and inhibition of autophagy [45]. Constitutively activated mTOR also aids in the growth of cancerous cells by increasing the translation of HIF-1α, which further supports angiogenesis [46] and activates glycolytic metabolism via the upregulated expression of a glycolytic enzyme, pyruvate kinase isozyme M2 (PKM2) [47]. Overexpression of the mTOR downstream effectors 4E-BP1, S6K, and eukaryotic translation initiation factor 4E (elF4E) are indicative of a poor cancer prognosis [48]. Some mTOR inhibitors, such as temsirolimus and everolimus, are currently being applied in the therapy of cancer, and many natural compounds such as caffeine, curcumin, EGCG, and resveratrol have also been shown to inhibit mTOR activity in cancer cells *in vivo* [49, 50]. For example, in one of our previous studies [33], curcumin treatment of human lung adenocarcinoma epithelial cells, A549, was found to downregulate protein expression of elF2α and elF4E expression, decrease 4E-BP1 phosphorylation, and increase the phosphorylation of elF2α and elF4E, which could potentially lead to the inhibition of A549 cell proliferation (Figures 5.3 and 5.11). However, there is no report supporting that these compounds could inhibit mTOR when they are orally administrated as dietary supplements.

Non-small-cell lung carcinoma (NSCLC) occupies 75–80% of all lung cancer cases, and EGFR is highly expressed in over 60% of NSCLCs [51, 52]. The TK activity of EGFR may be upregulated by several oncogenic mechanisms, including increased gene copy number of EGFR, gene mutation of the TK domain, and the overexpression of EGFR protein [53]. TK mutation of EGFR destabilizes its domain conformation and constitutively activates TK activity and downstream signaling pathways, which inhibits apoptosis of tumor cells and contributes to tumor propagation [54]. TK mutations may occur in different sites on exons 18–21 of EGFR, but >85% of TK mutations are constituted by exon 19 deletions and an exon 21 L858R point mutation [55].

Cell line studies play important roles in demonstrating the relationship between EGFR-TK mutations and sensitivity to tyrosine kinase inhibitors (TKIs) [56], which have become a category of therapeutics based on their ability to directly impact cellular functions. However, acquired resistance to TKIs is common and has become one of the greatest hindrances to their use. These studies have made significant contributions to the biological characterization of EGFR-TK mutations in cases of both intrinsic and acquired resistance to TKIs [57–59]. In the late 1980s, the unsuccessful rodent model systems used to screen out effective anticancer compounds caused a shift in testing drugs from rodent model systems to *in vitro* cell line-based panels [59]. The National Cancer Institute (NCI) established a panel composed of 60 human cancer cell lines (assembled from about 1500 available lines) representing 9 human cancer types for anticancer drug screening [60]. The NCI panel includes 8 NSCLC cell lines such as NCI-H1975 and NCI-H820, both of which contain an EGFR-activating mutation (L858R point mutation) and the TKI resistance-associated T790 M mutation, which are preferentially applied to inform *in vitro* studies of EGFR-targeted therapy.

5.10.2 Cell Cycle

In a population of dividing cells, either *in vivo* or *in vitro*, each cell goes through a series of defined stages, which constitutes the cell cycle. Two of the major phases include M phase and interphase. M phase consists of mitosis and cytokinesis, while interphase includes successive G1, S, and G2 phases, with S phase being a period of DNA replication and chromosome duplication. The machinery of cell cycle control is frequently altered in human cancer, and cyclin-dependent kinases (CDKs) play a crucial role in regulating the initiation, progression, and completion of cell cycle events. In cancer cells, overexpression of cyclins and the loss of expression of CDK inhibitors deregulate the CDK activity that provides cancer cells with a selective survival advantage over normal cells. When cells are subjected to DNA damage treatment via DNA-altering toxicants or drugs, their progress through the cell cycle is halted due to surveillance checkpoints enacted to prevent cell cycle progression in response to the recognition of DNA damage and cellular abnormalities. The cell can then utilize the temporary delay to repair DNA damage or correct cellular abnormalities rather than entering the next stage. If the DNA damage is irreparable, the checkpoint mechanism can activate a signal that leads either to cell death or to a state of permanent cell cycle arrest senescence. This is especially important because normal cells dividing with DNA damage run the risk of being transformed into malignant cells.

The incompetent checkpoint control of malignant transformed cells has elicited remarkable interest in the development of anticancer drugs that would be applied to induce cell cycle arrest and block cell cycle progression. Particularly, combinations of different pharmaceuticals administrated in a specific sequence may be applied to convert tumor cells to a state in which they are more sensitive to the cytotoxic effects of chemotherapeutic agents. For example, flow cytometry analysis validated G2/M arrest of pre-DOX/LMB-treated A549 cells [32], which is consistent with previous reports on DOX-induced G2 arrest [61] and LMB-induced reversible G1 and G2 arrest [62]. DOX is a commonly used chemotherapeutic drug, and LMB is a CRM1 (exportin 1) inhibitor. DNA damage also leads to the synthesis of inhibitors of cyclin-CDK complexes that drive the progress of the cell cycle. For example, p21 is one of the few known CDK inhibitors that suppress kinase activity of the G1-CDK and prevent the cell from phosphorylating key signal molecules and progressing into S phase. Ataxia-telangiectasia mutated (ATM) gene encodes a protein kinase that is involved in this checkpoint mechanism and could be activated by DNA lesions. In response to DNA damage, DNA breaks caused by chemical treatment recruit the protein complex Mre11-Rad50-Nbs1, a sensor for DNA breaks that recruits ATM to phosphorylate another checkpoint kinase, Chk2. Chk2, in turn, phosphorylates the p53 transcription factor that plays a critical role in the control of cell proliferation. The transcription and translation of p21 is activated through p53 phosphorylation, which leads to subsequent inhibition of CDK. In the study of Lu et al. [32], after pre-DOX and LMB treatment, the p21 protein level was elevated, which could lead to hypo-phosphorylation of the Rb protein that, in turn binds with the E2F transcription factor and consequently impedes the cell cycle [34, 63, 64]. In addition, the repression of survivin, a member of the apoptosis inhibitor family, also contributed to the activation of p53-dependent G2/M checkpoint arrest in pre-DOX/LMB-treated A549 cells. In our recent study [34], LMB (0.5 nM) was found to be ineffective in altering the cell cycle distribution of A549 cells. However, when LMB (0.5 nM) was combined with cisplatin (25 μM) to treat A549 cells, the proportion of cells in G2/M phase, and apoptotic cells were significantly higher than that of cells treated by LMB (0.5 nM) or cisplatin (25 μM) alone (Figure 5.4). Toxic agents could also induce cell cycle arrest and alteration of the phase distribution in normal cells. For example, in our study of BaP, a common environmental carcinogen, we found that BaP-induced G2/M phase accumulation of human normal lung epithelial cells, BEAS-2B, which was revealed by flow cytometry analysis, but this cell cycle arrest was suppressed by co-treatment with the antioxidant vitamin E (VE) or curcumin [12].

5.10.3 Signal-Mediated Cell Death

Besides cell proliferation, external stimuli, such as certain cytokines, granzymes (granule enzymes), or pharmaceutical agents, and internal stimuli, such as DNA damage, could activate the cell's suicide program and trigger cell apoptosis. Studies have shown that external stimuli activate apoptosis by the extrinsic pathway, which is different from the intrinsic pathway, activated by internal stimuli. The extrinsic pathway of apoptosis is stimulated by an extracellular messenger ligand called tumor necrosis factor (TNF) that binds to the transmembrane receptor TNFR1, which belongs to a family of related death receptors that ultimately turn "on" the apoptotic process. The apoptosis induced by granzymes, secretory granules reserved in the natural killer (NK) cells, and cytotoxic T lymphocytes, is responsible for the elimination of malignant transformed or virus-infected cells; this process is controlled by immune regulation and inflammation [65]. Granzymes are a family of highly homologous serine proteases packaged with perforin in cytotoxic granules of T cells and NK cells, which trigger cell death via two principal pathways: granule-mediated and death ligand-mediated cytotoxicity [66]. Both of these apoptosis pathways depend on cell-to-cell contact and result in activation of executioner caspases [67, 68]. The death receptor pathway requires expression of ligands (e.g. Fas ligand) on the immune cells, which will then bind to their receptors located on the target cell (e.g. Fas). As for the granule exocytosis pathway, the pore-forming perforin mediates the delivery of granzymes into the target cell cytosol, causing cleavage and activation of effector molecules within the target cell.

Internal stimuli, including irreparable genetic damage, lack of oxygen (hypoxia), and severe oxidative stress (overproduction of ROS), can trigger apoptosis through the intrinsic pathway. This pathway activation is modulated by the Bcl-2 family of proteins. The major apoptotic pathways can be categorized into caspase-dependent pathway and caspase-independent pathway; in both cases, mitochondria work as the crosstalk organelles to integrate diverse apoptotic pathways.

The caspase-dependent pathway is the classic apoptosis pathway, and it involves initiator caspases (2, 8, 9, 10, 11, and 12) and executioner caspases (3, 6, and 7). Initiator caspases are activated by binding to specific oligomeric activator proteins including FasL receptor, the TNF-α receptor, Toll-like receptor (TLR), death receptor, and so on. Executioner caspases are then proteolytically cleaved by the initiator caspases, and these active executioner caspases degrade diverse intracellular proteins to carry out the apoptotic sentence. Mitochondria and other nuclear organelles play key roles in caspase-dependent apoptosis, and these organelles can incorporate different signals to induce caspase activation, ROS production, cytochrome C release, and mitochondrial membrane potential alteration. As of now, the apoptotic mechanisms of the caspase-independent pathway have not been fully characterized. However, researchers have reported that there are many other ligands and signaling molecules that can act as apoptotic factors involved in the caspase-independent pathway, including ROS, drugs, and p53 tumor suppression factors. Since caspase, the apoptotic marker, may be transient and undetected by traditional Western blot, high-throughput luminescent caspase 8/9 assays are needed to monitor early apoptotic induction, and caspase 3/7 assays have been developed to monitor later events [69].

Granzyme A and granzyme B (GzmA, GzmB) are the most abundant granzymes. They can induce cell apoptosis by different pathways [65, 66]. GzmB is the most studied granzyme, and it activates apoptosis via various mechanisms, including direct activation of caspases 3/7 (triggered by ROS production) and cytochrome C release (due to mitochondrial damage caused by cleavage and activation of Bid, Bak, and Bax) [70]. GzmA-induced cell death is different from GzmB-induced cell death based on several distinctive features: GzmA causes cell death independently of caspase activation, and GzmA functions in synergy with perforin to induce single-stranded DNA nicks instead of DNA fragmentation [71–73]. Unlike GzmB-induced outer mitochondrial damage, GzmA-induced inner mitochondrial damage avoids the release of cytochrome C but leads to ROS production, which causes DNase activation and subsequent DNA damage [66].

Some cytotoxic drugs such as DOX, LMB, and cisplatin can trigger apoptosis of cancer cells via the tumor suppressor p53. Stress-induced p53 activation depends on PTM of p53 on at least 18 sites by phosphorylation, acetylation, and SUMOylation [74–76]. Among p53 modifications, phosphorylation has been mostly studied, and this modification plays a key role in the stabilization and activation of p53 [77]. Since antibodies recognizing p53-specific phosphoserine or phosphothreonine residues have been developed, multiple p53 serine residues (6, 9, 15, 20, 33, 37, 46, 315, 371, 376, 378, and 392) and three p53 threonine residues (18, 55, and 81) were reported to undergo phosphorylation induced by various stresses and treatments [77]. For example, our study [32] showed significant inhibitory effects of both DOX and LMB on A549 cells, and Western blot results demonstrated that the combination treatment of pre-DOX (0.5 μM) and LMB (5 nM) effectively induced both increased total p53 expression and p53 phosphorylation at Ser15 compared with control. Although the explicit mechanism of p53 phosphorylation and its exact regulation pathway are not clarified, a variety of serine/threonine kinases including ATM, ataxia telangiectasia related kinase (ATR), checkpoint kinase 1 (Chk1), Chk2, CK1, cyclin-dependent kinase 2 (CDK2), DNA-dependent protein kinase (DNA-PK), c-Jun N-terminal kinases (JNK), and p38 mitogen-activated protein kinases (p38MAPK) have been implicated in upstreaming signaling, leading to phosphorylated p53 activation [78, 79]. Besides phosphorylation, the MDM2 protein plays a role of paramount importance in controlling p53 activity, as it binds p53 via N-terminal transactivation domain and suppresses p53 function by negative transcriptional regulation [77]. Moreover, MDM2 is an E3 ubiquitin ligase and controls p53 half-life by ubiquitin-induced degradation, assuring that p53 protein is at very low levels in proliferating cell [80]. The stress-induced p53 phosphorylation significantly reduces the affinity between p53 and MDM2 and results in separation of MDM2 from p53, which is an essential step for p53 activation [80].

Since p53 has three nuclear localization signals (NLS) at its C-terminus and two nuclear export signals (NES) [80], p53 should be able to be transported between the nucleus and cytoplasm, mediated by nuclear import and nuclear export mechanisms. For example, our study found both p53 and phospho-p53 were accumulated in the nucleus when the CRM1-dependent nuclear export pathway needed for p53 export into the cytoplasm was blocked by LMB, which correlated with more significant cytotoxic effects of pre-DOX + LMB treatment [32]. Another study conducted in our lab [31] revealed that 4-(methylnitrosamino)-1-(3-pyridyl)-1-butanone (NNK), a known human lung carcinogen from tobacco, could inhibit proliferation of BEAS-2B cells and increases in p53 phosphorylation and p53 nuclear accumulation were observed in NNK-treated cells with DNA damage as compared with the controls. p53 is ubiquitinated and degraded quite efficiently within the nucleus by nuclear proteasomes, and most p53 proteins in the activated state are nuclear [80–82]. MDM2 may be a key modulator in controlling the subcellular compartmentation of p53. It was also found that nuclear p53, upon MDM-mediated ubiquitination, can still be exported and degraded in cytoplasm [80], suggesting this mechanism is the fastest way to dampen p53 activity due to removal from nucleus, where p53 works as a transcription factor. Therefore, more research will be needed in order to clarify the definitive role of p53 translocation and its relationship with p53 activation or degradation in different compartments under diverse stress conditions.

The tumor-suppressing function of p53 can be attributed to its ability to promote apoptosis or inhibit cell proliferation in response to DNA damage and oncogenic activation. However, recent studies reported that p53 can also promote cell survival by regulating various metabolic pathways, modulating the balance of glycolysis and oxidative phosphorylation, decreasing ROS formation, and enhancing the ability of cells to adapt and moderate metabolic stresses [83, 84]. Most normal cells use mitochondrial oxidative phosphorylation for their energy needs, but tumor cells primarily depend on glycolysis to provide energy even with abundant oxygen supply; this phenomenon is commonly referred to as the Warburg effect [85]. Since glycolysis produces energy (two ATPs per molecule of glucose) much less efficiently than oxidative phosphorylation (36 ATPs per molecule of glucose) in mitochondria, cancer cells compensate by increasing glucose uptake, consuming it at a higher rate than normal cells [83]. This shift in metabolic activity has been termed as a hallmark of cancer, which is used to diagnose cancer by pathologists and clinicians [86]. A previous study showed that loss of p53 leads to deficiency of mitochondrial oxidative phosphorylation and increased glycolysis in both cell culture and mouse models [87]. These results indicate that p53 may function to inhibit glycolysis and promote oxidative phosphorylation, which acts as a mechanism of suppressing tumorigenesis and the Warburg effect [84]. However, the study also found that p53 can both promote glycolysis and oxidative phosphorylation [84]. This complexity might be explained by p53's ability to interject at many points in both pathways of glycolysis and oxidative phosphorylation, and thus p53 loss causes alterations in metabolism as well as transcription of different sets of genes that impact almost every aspect of cell behavior in tissue-specific contexts [84, 87]. Some studies demonstrated that p53 could influence the expression levels of many gene products, affecting metabolic fates and metabolic products. For example, p53 represses glucose uptake by decreasing the expression of glucose transporters 1 and 4 (GLUT1 and GLUT4) directly [88] and decreasing the expression of glucose transporter 3 (GLUT 3) indirectly through downregulation of nuclear factor kappa-light-chain-enhancer of activated B cells (NF-κB) signaling [89]. p53 triggers the transcription of TP53-induced glycolysis and apoptosis regulator (TIGAR) gene and reduces the expression of phosphoglycerate mutase (PGM), which decreases glycolysis and limits ROS-induced cell death [84]. In addition to downregulation of glycolysis, p53 promotes oxidative phosphorylation by inducing the expression of several target genes including apoptosis-inducing factor (AIF), glutaminase 2 (GLS2), Parkin, p53R2, and synthesis of cytochrome c oxidase 2 (SCO2) [83]. Since mitochondrial oxidative phosphorylation is a primary source of ROS within cells, p53 also plays a crucial role in decreasing intracellular ROS produced from mitochondrial oxidative phosphorylation to maintain the cellular redox balance [83]. In order to serve its function of antioxidant defense, p53 induces a variety of antioxidant genes, such as aldehyde dehydrogenase family 4 (ALDH4), glutaminase 2 (GLS2), glutathione peroxidase 1 (GPX1), Parkin, sestrins 1/2, and TP53-induced glycolysis regulatory phosphatase (TIGAR) [83]. In summary, apoptosis and cell cycle arrest are typically accepted as the primary functions of p53, while more findings have provided strong evidence that p53 could contribute significantly to the redox balance and homeostasis of cellular energy metabolism. This may provide a new perspective for comprehensively understanding functions of p53 beyond tumor suppression.

Anoikis is a special form of apoptosis, and it occurs when anchorage-dependent cells detach from the surrounding extracellular matrix (ECM) [90], which prevents dysplastic growth by eliminating the cells detached form their original sites [91]. Thus, anoikis is a pivotal mechanism in maintaining the homeostasis of the whole organism and tissue development [92]. On the other hand, anoikis resistance is a hallmark of cancer cells [93] and contributes to the tumorigenic development and metastasis of different cancers including colon, lung, and prostrate [92]. Autophagy, a lysosome-dependent process involving degradation and recycling of cytosolic components in stressful conditions [94], is suggested to promote the development of anoikis resistance in cancer cells due to the intimate interaction of autophagy and anoikis [92]. Autophagy plays essential but controversial roles in the cellular homeostasis of pathophysiological processes, as it can promote cell survival or cell death, contingent upon the biological context and the environmental stimuli [95, 96]. After normal cells detach from the ECM, they go through autophagy, leading to anoikis [97]. However, in malignant transformed cells, autophagy may cooperate with antiapoptotic signals triggered by ECM detachment to allow the cells to gain anoikis resistance and reattach to the ECM in a timely manner [95]. As accumulating evidence has suggested, manipulation of autophagy-dependent anoikis pathways may inhibit metastasis of cancer cells with loss of anchorage dependence [92]. The delicate balance between autophagy and anoikis is highly exploitable for tumor therapy.

Compared with apoptosis, necrosis has been regarded as an uncontrolled and non-programmed form of cell death, in which significant changes of cell metabolism and structure take place. Necrosis caused by physical or chemical injury typically does not involve activation of caspases, but it may result from caspase inhibition [98]. Necrotic cell death is regarded as a pathologic form of cell death and is characterized by swelling of cytoplasm, endoplasmic reticulum, and mitochondria with subsequent disintegration of the dynamic plasma membrane in the dying cell [99]. The spillage of intracellular proteins from necrotic cells activates an immediate inflammatory response and immune activation of a damage signal, which is quite different from apoptotic cells that are silently engulfed by macrophages without an inflammatory response. The difference between apoptosis and necrosis is shown in Table 5.1.

TNF plays an important role in inducing mitochondrial control of necrosis by its effect on cellular ATP levels. Mitochondrial dysfunction propagates necrosis through excessive ROS production, mitochondrial permeability dysfunction, and ATP depletion [100]. TNF also activate poly(ADP-ribose) polymerase 1 (PARP-1) in response to DNA damage caused by overproduction of mitochondrial ROS [98, 101]. Besides mitochondria, the other two primary modulators in the process of necrosis are serine/threonine kinase receptor-interacting protein kinase 1 (RIPK1) and Ca^{2+}. The translocation of RIPK1 to the inner mitochondrial membrane disrupts the connection between the cyclophilin D and adenine nucleotide translocator, leading to ROS accumulation and ATP depletion [102], while Ca^{2+} controls the activation of calpains, polylactic acid (PLA), and nitric oxide synthase (NOS), which play important roles in a series of events resulting in necrosis [103].

Table 5.1 The difference between apoptosis and necrosis.

	Apoptosis	Necrosis
Cause	Physiological and pathological	Pathological
Stimuli	Intrinsic and extrinsic	Extrinsic
Energy requirement	Energy (ATP) dependent	No energy requirement
Tissue damage	No inflammation No secondary tissue damage	Inflammation and surrounding normal tissue damage
DNA breakdown	Random digestion of DNA	Nonrandom mono- and oligonucleosomal length fragmentation of DNA
Histology	Cell shrinkage and formation of apoptotic bodies Membrane blebbing without loss of integrity Increased mitochondrial membrane permeability due to pore formation, release of proapoptotic proteins from bcl-2 family Apoptotic bodies ingested by neighboring cells	Cell swelling and lysis Loss of membrane integrity Organelle disintegration and lysosomal leakage Lysed cells ingested by macrophages

5.10.4 Reactive Oxygen Species (ROS)

ROS, as a product of normal metabolism and xenobiotic exposure, may be the main cause of caspase-independent apoptosis. However, ROS is also involved in the caspase-dependent pathway, in which ROS play important roles in bridging two types of apoptosis. ROS at physiologically low levels functions as "redox messengers" in intracellular signal transduction and regulation, while excessive ROS causes oxidative modification of cellular macromolecules, inhibition of protein function, and promotion of cell apoptosis. For example, activation of the TNF receptor or Fas receptor leads to nicotinamide adenine dinucleotide phosphate (NADPH) oxidase-mediated ROS generation and extracellular release of ROS, most likely H_2O_2 and HOCl, which might diffuse intracellularly back into cells to deactivate the PI3K/AKT survival signal [104, 105] and activate caspase cascades. Therefore, the quantity, chemical nature, and intra- and extracellular production of NADPH oxidase-derived ROS are potentially important indicators of the proapoptosis potential of ROS.

Cellular redox potential is primarily determined by glutathione (GSH), which makes up over 90% of cellular nonprotein thiols [106] and functions as a principal antioxidant protecting against oxidative damage to important cellular molecules caused by various sources of ROS including free radicals, heavy metals, lipid peroxides, and peroxides [107]. GSH is a ubiquitous tripeptide thiol and can be oxidized to glutathione disulfide (GSSG), which is reduced back to GSH by GSH reductase utilizing NADPH as an electron donor [107]. GSH is the only intracellular molecule attaining millimolar concentrations, while extracellular concentrations of GSH are at low micromolar levels due to its rapid catabolism [108]. As a reducing agent and a major antioxidant, GSH maintains a tight control of the intracellular redox status, and the ratio of cellular GSH to GSSH is often used as an indicator of cellular oxidative stress [109, 110]. The redox potential for GSH/GSSG generally ranges from −260 to −150 mv in most cells and tissues.

Although the majority of GSH is distributed in the cytosol, a small but significant percentage (10–15%) of total cellular GSH is found in the organelles including mitochondria, endoplasmic reticulum, nuclear matrix, and peroxisomes [108]. Mitochondrial GSH plays an ultra-important role in protecting the organelle from oxidative stress caused by the coupled mitochondrial electron transport and oxidative phosphorylation [111]. Depletion of GSH allowing for overproduction of ROS has been suggested to function as an important mediator of apoptosis [112, 113]. Besides participating in cellular antioxidant defenses, GSH has reportedly several additional functions in cells, such as regulation of gene transcription and metabolism and detoxification of many endogenous molecules and xenobiotics [113].

References

1 Diehl, K.H., Hull, R., Morton, D. et al. (2001). A good practice guide to the administration of substances and removal of blood, including routes and volumes. *J. Appl. Toxicol.* 21 (1): 15–23.

2 Landi, M.T. and Caporaso, N. (1997). Sample collection, processing, and storage. *IARC Sci. Publ.* 142: 223–236.

3 Jiang, Z., Han, Y., and Cao, X. (2014). Induced pluripotent stem cell (iPSCs) and their application in immunotherapy. *Cell. Mol. Immunol.* 11 (1): 17–24.

4 Eisenbrand, G., Pool-Zobel, B., Baker, V. et al. (2002). Methods of in vitro toxicology. *Food Chem. Toxicol.* 40 (2–3): 193–236.

5 Freshney, R.I. (2005). *Culture of Animal Cells*, 5e. Hoboken: Wiley.

6 Tolnai, S. (1975). A method for viable cell count. *Tissue Cult. Assoc. Man.* 1: 37–38.

7 Krause, A.W., Carley, W.W., and Webb, W.W. (1984). Fluorescent erythrosin B is preferable to trypan blue as a vital exclusion dye for mammalian cells in monolayer culture. *J. Histochem. Cytochem.* 32 (10): 1084–1090.

8 Vega-Avila, E. and Pugsley, M.K. (2011). An overview of colorimetric assay methods used to assess survival or proliferation of mammalian cells. *Proc. West. Pharmacol. Soc.* 54: 10–14.

9 Borenfreund, E. and Puerner, J.A. (1985). Toxicity determined in vitro by morphological alterations and neutral red absorption. *Toxicol. Lett.* 24 (2–3): 119–124.

10 Yang, X. (2012). Clonogenic assay. *Cancer Biol. Ther.* 2: 1–3.

11 Dengler, W.A., Schulte, J., Berger, D.P. et al. (1995). Development of a propidium iodide fluorescence assay for proliferation and cytotoxicity assays. *Anti-Cancer Drugs* 6 (4): 522–532.

12 Zhu, W., Cromie, M.M., Cai, Q. et al. (2014). Curcumin and vitamin E protect against adverse effects of benzo[a]pyrene in lung epithelial cells. *PLoS One* 29 (3): e92992.

13 Olive, P.L. and Banath, J.P. (2006). The comet assay: a method to measure DNA damage in individual cells. *Nat. Protoc.* 1 (1): 23–29.

14 Stetka, D.G. and Wolff, S. (1976). Sister chromatid exchange as an assay for genetic damage induced by mutagen-carcinogens. I. In vivo test for compounds requiring metabolic activation. *Mutat. Res.* 41 (2–3): 333–342.

15 Kingston, R.E., Chen, C.A., Okayama, H. et al. (1996). Transfection of DNA into eukaryotic cells. *Current Protocols in Molecular Biology* 9 (1): 1–11.

16 Elbashir, S.M., Harborth, J., Weber, K., and Tuschl, T. (2002). Analysis of gene function in somatic mammalian cells using small interfering RNAs. *Methods* 26 (2): 199–213.

17 Gaj, T., Gersbach, C.A., and Barbas, C.F. (2013). ZFN, TALEN, and CRISPR/Cas-based methods for genome engineering. *Trends Biotechnol.* 31 (7): 397–405.

18 Pall, G.S. and Hamilton, A.J. (2008). Improved northern blot method for enhanced detection of small RNA. *Nat. Protoc.* 3 (6): 1077–1084.

19 Minou, B. (2013). Gene regulation: methods and protocols. In: *Methods in Molecular Biology* (ed. J.M. Walker), 79–93. Totowa, NJ: Humana Press Inc.

20 Emery, P. (2007). RNase protection assay. *Methods Mol. Biol.* 362: 343–348.

21 Kruger, N.J. (1994). The Bradford method for protein quantitation. *Methods Mol. Biol.* 32: 9–15.

22 Hoppert, M. and Wrede, C. (2011). Immunolocalization. In: *Encyclopedia of Geobiology* (ed. J. Reitner and V. Thiel), 482–486. Netherlands: Springer.

23 Das, P.M., Ramachandran, K., vanWert, J., and Singal, R. (2004). Chromatin immunoprecipitation assay. *BioTechniques* 37 (6): 961–969.

24 Zuo, T., Tycko, B., Liu, T.M. et al. (2009). Methods in DNA methylation profiling. *Epigenomics* 1 (2): 331–345.

25 Villar-Garea, A. and Imhof, A. (2006). The analysis of histone modifications. *Biochim. Biophys. Acta* 1764 (12): 1932–1939.

26 Walum, E. (1998). Acute oral toxicity. *Environ. Health Perspect.* 106 (Suppl. 2): 497–503.

27 Ames, B.N., McCann, J., and Yamasaki, E. (1975). Proceedings: carcinogens are mutagens: a simple test system. *Mutat. Res.* 33 (1 Spec. No.): 27–28.

28 Greek, R. and Menache, A. (2013). Systematic reviews of animal models: methodology versus epistemology. *Int. J. Med. Sci.* 10 (3): 206–221.

29 Mestas, J. and Hughes, C.C. (2004). Of mice and not men: differences between mouse and human immunology. *J. Immunol.* 172 (5): 2731–2738.

30 American Cancer Society (2016) Cancer Facts & Figures 2016, Atlanta, GA.

31 Chen, L., Shao, C., Cobos, E. et al. (2010). 4-(methylnitrosamino)-1-(3-pyridyl)-1-butanone [corrected] induces CRM1-dependent p53 nuclear accumulation in human bronchial epithelial cells. *Toxicol. Sci.* 116 (1): 206–215.

32 Lu, C., Shao, C., Cobos, E. et al. (2012). Chemotherapeutic sensitization of leptomycin B resistant lung cancer cells by pretreatment with doxorubicin. *PLoS One* 7 (3): e32895.

33 Chen, L., Tian, G., Shao, C. et al. (2010). Curcumin modulates eukaryotic initiation factors in human lung adenocarcinoma epithelial cells. *Mol. Biol. Rep.* 37 (7): 3105–3110.

34 Gao, W., Lu, C., Chen, L., and Keohavong, P. (2015). Overexpression of CRM1: a characteristic feature in a transformed phenotype of lung carcinogenesis and a molecular target for lung cancer adjuvant therapy. *J. Thorac. Oncol.* 10 (5): 815–825.

35 Cromie, M.M. and Gao, W. (2015). Epigallocatechin-3-gallate enhances the therapeutic effects of leptomycin B on human lung cancer a549 cells. *Oxidative Med. Cell. Longev.* 2015 (217304): 1–10.

36 Shao, C., Lu, C., Chen, L. et al. (2011). p53-Dependent anticancer effects of leptomycin B on lung adenocarcinoma. *Cancer Chemother. Pharmacol.* 67 (6): 1369–1380.

37 Wells, A. (1999). EGF receptor. *Int. J. Biochem. Cell Biol.* 31 (6): 637–643.

38 Jimeno, A. and Hidalgo, M. (2006). Pharmacogenomics of epidermal growth factor receptor (EGFR) tyrosine kinase inhibitors. *Biochim. Biophys. Acta* 1766 (2): 217–229.

39 Mosesson, Y. and Yarden, Y. (2004). Oncogenic growth factor receptors: implications for signal transduction therapy. *Semin. Cancer Biol.* 14 (4): 262–270.

40 Peltier, J., O'Neill, A., and Schaffer, D.V. (2007). PI3K/Akt and CREB regulate adult neural hippocampal progenitor proliferation and differentiation. *Dev. Neurobiol.* 67 (10): 1348–1361.

41 Rafalski, V.A. and Brunet, A. (2011). Energy metabolism in adult neural stem cell fate. *Prog. Neurobiol.* 93 (2): 182–203.

42 Man, H.Y., Wang, Q., Lu, W.Y. et al. (2003). Activation of PI3-kinase is required for AMPA receptor insertion during LTP of mEPSCs in cultured hippocampal neurons. *Neuron* 38 (4): 611–624.

43 Xu, K., Liu, P., and Wei, W. (2014). mTOR signaling in tumorigenesis. *Biochim. Biophys. Acta* 1846 (2): 638–654.

44 Wyatt, L.A., Filbin, M.T., and Keirstead, H.S. (2014). PTEN inhibition enhances neurite outgrowth in human embryonic stem cell-derived neuronal progenitor cells. *J. Comp. Neurol.* 522 (12): 2741–2755.
45 Hay, N. and Sonenberg, N. (2004). Upstream and downstream of mTOR. *Genes Dev.* 18 (16): 1926–1945.
46 Thomas, G.V., Tran, C., Mellinghoff, I.K. et al. (2006). Hypoxia-inducible factor determines sensitivity to inhibitors of mTOR in kidney cancer. *Nat. Med.* 12 (1): 122–127.
47 Nemazanyy, I., Espeillac, C., Pende, M., and Panasyuk, G. (2013). Role of PI3K, mTOR and Akt2 signalling in hepatic tumorigenesis via the control of PKM2 expression. *Biochem. Soc. Trans.* 41 (4): 917–922.
48 Populo, H., Lopes, J.M., and Soares, P. (2012). The mTOR signalling pathway in human cancer. *Int. J. Mol. Sci.* 13 (2): 1886–1918.
49 Beevers, C.S., Li, F., Liu, L., and Huang, S. (2006). Curcumin inhibits the mammalian target of rapamycin-mediated signaling pathways in cancer cells. *Int. J. Cancer* 119 (4): 757–764.
50 Zhou, H., Luo, Y., and Huang, S. (2010). Updates of mTOR inhibitors. *Anticancer Agents Med. Chem.* 10 (7): 571–581.
51 Visbal, A.L., Williams, B.A., Nichols, F.C. et al. (2004). Gender differences in non-small-cell lung cancer survival: an analysis of 4,618 patients diagnosed between 1997 and 2002. *Ann. Thorac. Surg.* 78 (1): 209–215.
52 da Cunha Santos, G., Shepherd, F.A., and Tsao, M.S. (2011). EGFR mutations and lung cancer. *Annu. Rev. Pathol.* 6: 49–69.
53 Ciardiello, F. and Tortora, G. (2008). EGFR antagonists in cancer treatment. *N. Engl. J. Med.* 358 (11): 1160–1174.
54 Woodburn, J.R. (1999). The epidermal growth factor receptor and its inhibition in cancer therapy. *Pharmacol. Ther.* 82 (2–3): 241–250.
55 Sakurada, A., Shepherd, F.A., and Tsao, M.S. (2006). Epidermal growth factor receptor tyrosine kinase inhibitors in lung cancer: impact of primary or secondary mutations. *Clin. Lung Cancer* 7 (Suppl 4): S138–S144.
56 Paez, J.G., Janne, P.A., Lee, J.C. et al. (2004). EGFR mutations in lung cancer: correlation with clinical response to gefitinib therapy. *Science* 304 (5676): 1497–1500.
57 Engelman, J.A. and Janne, P.A. (2008). Mechanisms of acquired resistance to epidermal growth factor receptor tyrosine kinase inhibitors in non-small cell lung cancer. *Clin. Cancer Res.* 14 (10): 2895–2899.
58 Gazdar, A.F. (2009). Activating and resistance mutations of EGFR in non-small-cell lung cancer: role in clinical response to EGFR tyrosine kinase inhibitors. *Oncogene* 28 (Suppl 1): S24–S31.
59 Gazdar, A.F., Girard, L., Lockwood, W.W. et al. (2010). Lung cancer cell lines as tools for biomedical discovery and research. *J. Natl. Cancer Inst.* 102 (17): 1310–1321.

60 Shoemaker, R.H. (2006). The NCI60 human tumour cell line anticancer drug screen. *Nat. Rev. Cancer* 6 (10): 813–823.
61 Blagosklonny, M.V. (2002). Sequential activation and inactivation of G2 checkpoints for selective killing of p53-deficient cells by microtubule-active drugs. *Oncogene* 21 (41): 6249–6254.
62 Yoshida, M., Nishikawa, M., Nishi, K. et al. (1990). Effects of leptomycin B on the cell cycle of fibroblasts and fission yeast cells. *Exp. Cell Res.* 187 (1): 150–156.
63 el- Deiry, W.S., Tokino, T., Velculescu, V.E. et al. (1993). WAF1, a potential mediator of p53 tumor suppression. *Cell* 75 (4): 817–825.
64 Lohr, K., Moritz, C., Contente, A., and Dobblestein, M. (2003). p21/CDKN1A mediates negative regulation of transcription by p53. *J. Biol. Chem.* 278 (35): 32507–32516.
65 Chowdhury, D. and Lieberman, J. (2008). Death by a thousand cuts: granzyme pathways of programmed cell death. *Annu. Rev. Immunol.* 26: 389–420.
66 Anthony, D.A., Andrews, D.M., Watt, S.V. et al. (2010). Functional dissection of the granzyme family: cell death and inflammation. *Immunol. Rev.* 235 (1): 73–92.
67 Degli-Esposti, M.A. and Smyth, M.J. (2005). Close encounters of different kinds: dendritic cells and NK cells take centre stage. *Nat. Rev. Immunol.* 5 (2): 112–124.
68 Trapani, J.A. and Smyth, M.J. (2002). Functional significance of the perforin/ granzyme cell death pathway. *Nat. Rev. Immunol.* 2 (10): 735–747.
69 Deyes, R., Arduengo, M., and Allard, S.T.M. (2008). Choosing the right assay to monitor your signal transduction pathway. *Cell Notes* 21: 13–16.
70 Pardo, J., Wallich, R., Martin, P. et al. (2008). Granzyme B-induced cell death exerted by ex vivo CTL: discriminating requirements for cell death and some of its signs. *Cell Death Differ.* 15 (3): 567–579.
71 Beresford, P.J., Xia, Z., Greenberg, A.H., and Lieberman, J. (1999). Granzyme A loading induces rapid cytolysis and a novel form of DNA damage independently of caspase activation. *Immunity* 10 (5): 585–595.
72 Fan, Z., Beresford, P.J., Oh, D.Y. et al. (2003). Tumor suppressor NM23-H1 is a granzyme A-activated DNase during CTL-mediated apoptosis, and the nucleosome assembly protein SET is its inhibitor. *Cell* 112 (5): 659–672.
73 Martinvalet, D., Zhu, P., and Lieberman, J. (2005). Granzyme A induces caspase-independent mitochondrial damage, a required first step for apoptosis. *Immunity* 22 (3): 355–370.
74 Appella, E. and Anderson, C.W. (2001). Post-translational modifications and activation of p53 by genotoxic stresses. *Eur. J. Biochem.* 268 (10): 2764–2772.
75 Brooks, C.L. and Gu, W. (2003). Ubiquitination, phosphorylation and acetylation: the molecular basis for p53 regulation. *Curr. Opin. Cell Biol.* 15 (2): 164–171.
76 Vogelstein, B., Lane, D., and Levine, A.J. (2000). Surfing the p53 network. *Nature* 408 (6810): 307–310.

77 Thompson, T., Tovar, C., Yang, H. et al. (2004). Phosphorylation of p53 on key serines is dispensable for transcriptional activation and apoptosis. *J. Biol. Chem.* 279 (51): 53015–53022.
78 Jimenez, G.S., Khan, S.H., Stommel, J.M., and Wahl, G.M. (1999). p53 regulation by post-translational modification and nuclear retention in response to diverse stresses. *Oncogene* 18 (53): 7656–7665.
79 Ljungman, M. (2000). Dial 9-1-1 for p53: mechanisms of p53 activation by cellular stress. *Neoplasia* 2 (3): 208–225.
80 Michael, D. and Oren, M. (2003). The p53-Mdm2 module and the ubiquitin system. *Semin. Cancer Biol.* 13 (1): 49–58.
81 Xirodimas, D.P., Stephen, C.W., and Lane, D.P. (2001). Cocompartmentalization of p53 and Mdm2 is a major determinant for Mdm2-mediated degradation of p53. *Exp. Cell Res.* 270 (1): 66–77.
82 Yu, Z.K., Geyer, R.K., and Maki, C.G. (2000). MDM2-dependent ubiquitination of nuclear and cytoplasmic P53. *Oncogene* 19: 5892–5897.
83 Liang, Y., Liu, J., and Feng, Z. The regulation of cellular metabolism by tumor suppressor p53. *Cell Biosci.* 3 (1): 9.
84 Puzio-Kuter, A.M. (2011). The role of p53 in metabolic regulation. *Genes Cancer* 2 (4): 385–391.
85 Warburg, O. (1956). On respiratory impairment in cancer cells. *Science* 124 (3215): 269–270.
86 Yeung, S., Pan, J., and Lee, M.H. (2008). Roles of p53, MYC and HIF-1 in regulating glycolysis—the seventh hallmark of cancer. *Cell. Mol. Life Sci.* 65 (24): 3981–3999.
87 Matoba, S., Kang, J.G., Patino, W.D. et al. (2006). p53 regulates mitochondrial respiration. *Science* 312 (5780): 1650–1653.
88 Schwartzenberg-Bar-Yoseph, F., Armoni, M., and Karnieli, E. (2004). The tumor suppressor p53 down-regulates glucose transporters GLUT1 and GLUT4 gene expression. *Cancer Res.* 64 (7): 2627–2633.
89 Kawauchi, K., Araki, K., Tobiume, K., and Tanaka, N. (2008). p53 regulates glucose metabolism through an IKK-NF-κB pathway and inhibits cell transformation. *Nat. Cell Biol.* 10 (5): 611–618.
90 Frisch, S.M. and Ruoslahti, E. (1997). Integrins and anoikis. *Curr. Opin. Cell Biol.* 9 (5): 701–706.
91 Guadamillas, M.C., Cerezo, A., and Del Pozo, M.A. (2011). Overcoming anoikis–pathways to anchorage-independent growth in cancer. *J. Cell Sci.* 124 (Pt 19): 3189–3197.
92 Yang, J., Zheng, Z., Yan, X. et al. (2013). Integration of autophagy and anoikis resistance in solid tumors. *Anat. Rec. (Hoboken)* 296 (10): 1501–1508.
93 Paoli, P., Giannoni, E., and Chiarugi, P. (2013). Anoikis molecular pathways and its role in cancer progression. *Biochim. Biophys. Acta* 1833 (12): 3481–3498.

94 Levine, B. (2007). Cell biology: autophagy and cancer. *Nature* 446 (7137): 745–747.
95 Fung, C., Lock, R., Gao, S. et al. (2008). Induction of autophagy during extracellular matrix detachment promotes cell survival. *Mol. Biol. Cell* 19 (3): 797–806.
96 Shang, L., Chen, S., Du, F. et al. (2011). Nutrient starvation elicits an acute autophagic response mediated by Ulk1 dephosphorylation and its subsequent dissociation from AMPK. *Proc. Natl. Acad. Sci. U. S. A.* 108 (12): 4788–4793.
97 Debnath, J. (2008). Detachment-induced autophagy during anoikis and lumen formation in epithelial acini. *Autophagy* 4 (3): 351–353.
98 Los, M., Mozoluk, M., Ferrari, D. et al. (2002). Activation and caspase-mediated inhibition of PARP: a molecular switch between fibroblast necrosis and apoptosis in death receptor signaling. *Mol. Biol. Cell* 13 (3): 978–988.
99 Schweichel, J.U. and Merker, H.J. (1973). The morphology of various types of cell death in prenatal tissues. *Teratology* 7 (3): 253–266.
100 Skulachev, V.P. (2006). Bioenergetic aspects of apoptosis, necrosis and mitoptosis. *Apoptosis* 11 (4): 473–485.
101 Leist, M. and Jaattela, M. (2001). Four deaths and a funeral: from caspases to alternative mechanisms. *Nat. Rev. Mol. Cell Biol.* 2 (8): 589–598.
102 Holler, N., Zaru, R., Micheau, O. et al. (2000). Fas triggers an alternative, caspase-8-independent cell death pathway using the kinase RIP as effector molecule. *Nat. Immunol.* 1 (6): 489–495.
103 Denecker, G., Vercammen, D., Steemans, M. et al. (2001). Death receptor-induced apoptotic and necrotic cell death: differential role of caspases and mitochondria. *Cell Death Differ.* 8 (8): 829–840.
104 Xu, Y., Loison, F., and Luo, H.R. (2010). Neutrophil spontaneous death is mediated by down-regulation of autocrine signaling through GPCR, PI3Kgamma, ROS, and actin. *Proc. Natl. Acad. Sci. U.S.A.* 107 (7): 2950–2955.
105 Zhu, D., Hattori, H., Jo, H. et al. (2006). Deactivation of phosphatidylinositol 3,4,5-trisphosphate/Akt signaling mediates neutrophil spontaneous death. *Proc. Natl. Acad. Sci. U.S.A.* 103 (40): 14836–14841.
106 Deneke, S.M. and Fanburg, B.L. (1989). Regulation of cellular glutathione. *Am. J. Phys. Lung Cell. Mol. Phys.* 257 (4): L163–L173.
107 Couto, N., Malys, N., Gaskell, S.J., and Barber, J. (2013). Partition and turnover of glutathione reductase from Saccharomyces cerevisiae: a proteomic approach. *J. Proteome Res.* 12 (6): 2885–2894.
108 Franco, R. and Cidlowski, J. (2009). Apoptosis and glutathione: beyond an antioxidant. *Cell Death Differ.* 16 (10): 1303–1314.
109 Lu, S.C. (2013). Glutathione synthesis. *Biochim. Biophys. Acta* 1830 (5): 3143–3153.

110 Pastore, A., Piemonte, F., Locatelli, M. et al. (2001). Determination of blood total, reduced, and oxidized glutathione in pediatric subjects. *Clin. Chem.* 47 (8): 1467–1469.

111 Fernandez-Checa, J.C., Kaplowitz, N., Garcia-Ruiz, C. et al. (1997). GSH transport in mitochondria: defense against TNF-induced oxidative stress and alcohol-induced defect. *Am. J. Phys.* 273 (Pt 1): G7–G17.

112 Langer, C., Jürgensmeier, J.M., and Bauer, G. (1996). Reactive oxygen species act at both TGF-β-dependent and-independent steps during induction of apoptosis of transformed cells by normal cells. *Exp. Cell Res.* 222 (1): 117–124.

113 Lushchak, V.I. (2012). Glutathione homeostasis and functions: potential targets for medical interventions. *J. Amino Acids* 2012 (736837): 1–26.

6

Techniques for Measuring Cellular Signal Transduction

Julie Vrana Miller[1], *Weimin Gao*[2], *Meghan Cromie*[3], *and Zhongwei Liu*[2]

[1] *Cardno ChemRisk, Pittsburgh, PA, USA*
[2] *Department of Occupational and Environmental Health Sciences, West Virginia University School of Public Health, Morgantown, WV, USA*
[3] *National Jewish Health, Denver, CO, USA*

6.1 Introduction

Understanding the cellular response to an extracellular cue, whether it be exogenous or endogenous stimuli, remains to be a significant challenge, especially due to the rapid and dynamic processes involved in cellular signal transduction. Recent advances in cellular signaling experimental techniques have significantly increased the amount of reliable data regarding signal transduction research for a wide range of scientific interests, whether that be toxicology, pharmacology, oncology and carcinogenesis, neurology, disease etiology, or many more. From these recent advances, especially high-throughput screening (HTS) assays and "big" dataset generating techniques, it is now possible to collect information regarding the response of diverse biomolecules, such as genes, proteins, RNA, metabolites, and small molecules. The dynamic interaction between these diverse biomolecules results in the transduction of information from an extracellular cue to the corresponding cellular response, which can range from a host of cellular functions, such as proliferation or apoptosis. While there have been numerous advances in cellular signaling techniques, there still exists a significant lack of data regarding temporal, mechanistic, and even cell-type specific information, which would be useful for constructing reliable models for signal transduction pathways. In this chapter, experimental techniques that are both useful and critical for conducting signal transduction research will be introduced and discussed. This chapter will cover not only target-specific or single-point assays but also kinetic

Cellular Signal Transduction in Toxicology and Pharmacology: Data Collection, Analysis, and Interpretation, First Edition. Edited by Jonathan W. Boyd and Richard R. Neubig.

assays and methods to generate "big" datasets (e.g. 'omics techniques, next-generation sequencing, and multiplexable enzyme-linked immunosorbent assays [ELISAs]).

6.2 High-Throughput Versus High-Content Data

The multitude of assays that can be employed for cellular signaling research can generally be divided into two camps: high throughput or high content. High-throughput assays or HTS enables the rapid interrogation of effects from an exposure (chemical, biological, environmental, etc.) in various cell-based or *in vitro* assays, whereas high-content analysis trades rapid delivery of large datasets for valuable single-cell-level information typically derived from microscopy techniques.

HTS are typically automated assays that have quick turnaround times, can analyze many samples quickly, and provide rapid readout for quantitation or relative response. Also, HTS utilizes population-based intracellular biological information (bioenergetics, specific protein or RNA information, posttranslational modifications [PTMs], or viability). The ability to rapidly acquire subcellular data on a large volume of samples can reveal a wealth of information regarding affected cellular targets or cellular perturbations after exposure in a time- and dose-dependent manner. The disadvantages of HTS assays are that they may yield data that has low quality since they rely on population-averaged outputs as opposed to the rich phenotypic assays of high-content analysis that can account for an individual's biological variation. However, the results from HTS assays allow for further and more in-depth investigation of identified targets for high-content assays, such as immunohistochemistry (IHC).

High-content analysis has traditionally been defined as techniques used to obtain single-cell-level information via microscopy, but has recently evolved to include an array of single-cell techniques such as flow cytometry, label-free assays, and even proteomic approaches to analyze intact cells. Additionally, the recent advances in microscopy and imaging automation have evolved high-content analysis into high-content screening (HCS), improving the throughput via automated acquisition of both multiplexed imaging and analysis output. HCS can validate targets identified via HTS. By combining fluorescence signals and sophisticated optics, epifluorescence microscopes have been widely used for HCS to obtain subcellular resolution of targets identified in HTS assays. This technology can yield both spatial and temporal information critical to cellular signaling studies. While HTS and HCS yield different information, for example, quantity (number of targets) and quality (subcellular spatiotemporal data), cellular signaling investigations must consider both types of analysis to get a true picture of cellular perturbations and response to an exposure or disease state.

6.2.1 Ergodic and Nonergodic Systems

Traditional HTS assays or population cell culture assays rely on the assumption of cell population homogeneity. Valuable information can be collected from the population of cells studied in an experiment, such as glucose uptake, oxygen consumption, adenosine triphosphate (ATP) production, and protein phosphorylation, to name a few. For this assumption of homogeneity to hold, we must assume that our population of cells are identical and at equilibrium, known as an ergodic system. However, each cell is not identical and cell subpopulations are at different steady states; thus cell culture populations can be ergodic and nonergodic in practice. For example, cancer cells may appear in equilibrium as a whole; however, cell subpopulations within cancerous tumors exist in several distinct yet dynamic states, yielding phenotypic diversity within cancer cell culture populations. The interface between ergodic and nonergodic cellular processes presents several interesting experimental questions: what type of experimental technique should I use to answer my hypothesis? Methods used to interrogate responses at a cell-by-cell basis are IHC and confocal microscopy, which will be discussed in this chapter. Population-based methods and assays useful for both *in vivo* and *in vitro* studies include many of the techniques discussed in this chapter, such as ELISA, array technologies, next-generation sequencing, proteomics, and viability assays.

6.3 Methods to Measure Signal Transduction Data

6.3.1 Microscopy

Microscopy has primarily been used for high-content analysis and, recently, HCS. Microscopy, such as traditional epifluorescence or confocal, can yield valuable spatial and temporal signaling information in a variety of samples from fixed cells in tissue slices to living cells *in vitro*. The type of microscopy used depends on the information that is desired, such as the presence of target proteins, protein–protein interaction, viability, cell number, or mitochondrial distribution. The various types of assays that can be used for signal transduction research include IHC, Förster resonance energy transfer (FRET), terminal deoxynucleotidyl transferase dUTP nick end labeling (TUNEL), live/dead (calcein-AM and ethidium homodimer-1 dyes), and many others.

6.3.1.1 Widefield Epifluorescence Microscopy

In traditional widefield microscopy, the entire sample is illuminated with light. The most recognizable widefield microscopes are the ones used in primary or secondary school biology courses, where a sample is illuminated from under the sample stage with white light and the sample is observed from above. This microscope is also referred to as a bright-field microscope.

While the simplicity of this microscope keeps costs low and makes it easy to use, they suffer from poor resolution due to the absorption of light by cellular specimen, yielding a low contrast image.

In the 1920s, fluorescence microscopy using reflected light, also known as episcopic, to illuminate the sample was first introduced. Their popularity soared with the introduction of mercury-vapor and xenon arc-discharge lamps in the 1930s, as well as different-colored glass and gelatin filters, which allowed for the use of various fluorophores that are excited by green or blue visible light from tungsten-halogen lamps. These first fluorescent microscopes used half-mirror beam splitters, resulting in significant light loss. However, in the 1940s, the development of dichromatic beam splitters, also known as dichroic mirrors, significantly reduced the amount of light loss. This type of episcopic fluorescence, commonly referred to as epifluorescence, is the most popular type of fluorescence microscopy for widefield, confocal, and multiphoton microscopic investigations.

A conventional modern epifluorescence microscope has several main components: an observation head equipped with a charge-coupled device (CCD) camera, two illumination sources for transmitted light (tungsten halogen) and episcopic (mercury arc-discharge) observations, excitation and emission filters for the wavelengths corresponding to the fluorophore(s) being used, and the dichromatic beam splitter, which reflects the incident (excitation) light that has already passed through the excitation filter and allows the emission light to pass through to the emission filter for the desired emission wavelength. Epifluorescence microscopes can be the traditional vertical setup, where the episcopic light comes from above the sample stage, or inverted (also known as tissue culture) epifluorescence microscope, where the excitation incident light from the mercury or xenon arc-discharge lamp comes from below the sample stage, shown in Figure 6.1. With the modern advances in filters, objective lenses, and phase contrast and the ability to be coupled to confocal or multiphoton microscopy, the inverted epifluorescence microscope is a versatile and critical tool for fluorescence-based techniques for single-point fixed-tissue (e.g. IHC) or kinetic/real-time *in vitro* (e.g. oxygen consumption, FRET, plasma membrane degradation) investigations.

6.3.1.2 Confocal Microscopy

Traditional widefield epifluorescence microscopy (vertical or inverted) collects a snapshot of a fluorophore's emission from the excitation incident light and yields a two-dimensional (2D) image of the sample being observed. Laser scanning confocal microscopy allows for three-dimensional observations, even deep into thick tissue samples, by controlling the depth of field and significantly decreasing the out-of-focus background signal away from the focal plane of interest. Confocal microscopes utilize lasers as the excitation light source that passes through a pinhole aperture in a conjugate focal plane

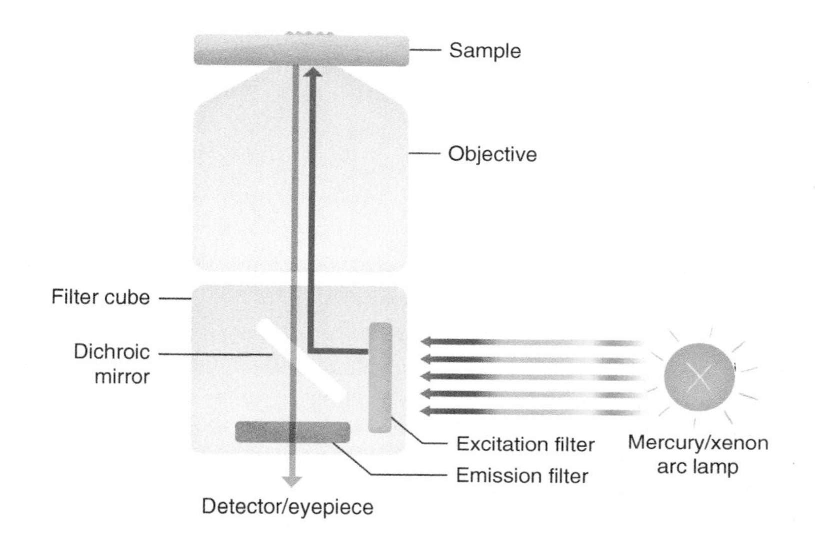

Figure 6.1 Typical light path in an epifluorescence microscope.

(confocal) to the laser and then through an excitation filter, dichromatic beam splitter, objective lens, and scanning point onto the sample. The fluorescence emission from the scanning point on the sample passes through the objective and dichromatic beam splitter and emission filter and then through the pinhole aperture in the confocal plane to the sample and detected by a photomultiplier tube. Software packages accompanying commercial confocal microscopes can construct complex three-dimensional renderings of fixed or live tissue as well as time-lapse imaging to observe cellular dynamics over time and even incorporate more than one fluorophore at a time. These applications offer significant advantages of confocal microscopy over traditional widefield epifluorescence. A disadvantage of confocal microscopy is not only cost but also the limited excitation wavelengths available due to the number of commercial lasers. Traditional xenon or mercury arc-discharge lamps used in widefield epifluorescence emit a wide spectrum of wavelengths, ultraviolet to visible and even near infrared, and rely on excitation filters to dial in the desired excitation wavelength for the fluorophore of interest. The microscope selected for a signal transduction study depends on the information desired and relative cost to benefit for a given study.

6.3.1.3 Immunohistochemistry

IHC has been a vital technique for many different fields, from pathology analyses in a hospital setting to cancer therapeutic development in the pharmaceutical industry. Therefore, it comes as no surprise that IHC is also an invaluable tool for signal transduction analyses *in vivo*. Spatial information for specific targets, such as intracellular or transmembrane proteins, in a tissue sample can be visualized under a microscope using a specific antibody selective for the biomarker of interest. Sample preparation is the most critical step of any histological technique. Samples can be obtained from biopsies or collection of whole organs that have the blood rinsed off and are rapidly immersed in a fixative best suited for your IHC assay, most often formaldehyde. This type of tissue extraction and fixation is best for small tissue samples or small organs. Alternatively, larger organs, such as the brain, benefit from whole-animal perfusion. This technique is carried out with a deeply anesthetized animal and perfused using a peristaltic pump. First, phosphate buffer is perfused via the animal's circulatory system, such as the ascending aorta, to remove blood. Following exsanguination, evidenced by a blanching of the liver, fixative, such as paraformaldehyde, is perfused for an optimized amount of time and flow rate for the species, tissue, and assay used in your study. Following adequate fixation time, the desired organ(s) can be carefully removed and placed in the post-fixation buffer and stored at 4 °C. The post-fixation time and composition of buffer vary based on the tissue and type of assay; thus it is imperative to follow the fixation protocol that best suits your experiment (see Gage [1] for review).

Fixed tissue that has been properly perfused, fixed, and postfixed can be immersed in a cryoprotective solution and stored frozen until it is time to section with a cryostat at a desired thickness. Sectioned sample slices are then placed on slides for desired staining (e.g. fluoro-jade B, silver stain) or immunostaining with specific antibodies. Immunostaining for IHC can be performed using several different techniques, such as the direct method or indirect method. The direct method is performed with a labeled primary antibody that immediately reacts with the antigen of interest, but this tends to be less selective and sensitive when compared with the more popular indirect method. The indirect method with streptavidin-biotin will be used in this section as an IHC example with colorimetric staining using a chromogen (compound that produces a color product after chemical reaction). First, using the frozen sections, incubate slides in hydrogen peroxide (concentration and duration depends upon the frozen tissue type) to inactivate endogenous peroxidase activity. After washing slides with buffer, then block nonspecific binding with secondary antibody-based serum; it should be noted that if nonspecific binding is not a concern, this step can be omitted. After blocking (if blocking is used), blot excess serum solution from the slide and incubate with the primary antibody, which is an antibody that is selective for the specific antigen of interest (e.g. epidermal growth factor receptor [EGFR], extracellular signal-regulated kinase [ERK], protein kinase B [Akt/PKB], glial fibrillary acidic protein [GFAP]). It is important to note that the concentration and duration of antibody incubation is antibody, tissue, and technique (IHC, Western blot, ELISA, etc.) specific and appropriate antibody titration is necessary prior to the full IHC staining protocol. The slide can also be incubated with Triton X-100 to improve antibody penetration. After primary antibody incubation, wash the slide with buffer and incubate with the biotinylated secondary antibody (or HRP-tagged secondary antibody, which can then proceed to substrate addition). After incubation, rinse with buffer, incubate with conjugated streptavidin-horseradish peroxidase (HRP), and rinse again. Finally, incubate the slide with substrate chromogen, such as 3,3'-diaminobenzidine (DAB), which converts to a brown precipitate in the presence of HRP and hydrogen peroxide that can be visualized with a microscope. If an increase in signal output is needed, a fluorescent label instead of a chromogen can be used. In this method, the secondary antibody is biotinylated, and then a streptavidin–fluorophore complex is added to bind to the biotinylated secondary antibody and produces a fluorescent signal, detected using a fluorescence microscope (Figure 6.2).

IHC is implemented in the diagnosis of certain conditions, but it is especially useful in classifying cancer types. Oftentimes, IHC is confused with a similar technique, immunocytochemistry (ICC). ICC examines antigens of interest in individual cells from culture or smears after the surrounding matrix has been removed. The primary antibody binds to the antigen of interest within the cell, and the secondary antibody is conjugated with a fluorophore that enables its

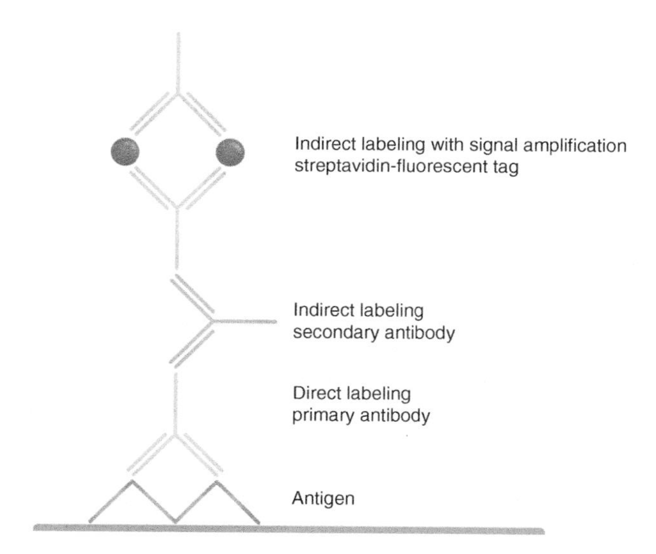

Figure 6.2 Illustration of indirect immunohistochemistry methods using streptavidin-biotin.

visualization using a fluorescent microscope. The subcellular location of the antigen of interest can be determined using this technique.

6.3.1.4 FRET

Epifluorescence microscopy can also be used to identify biomolecular interactions and measure the distance between interactions of interest (e.g. protein–protein interactions), determine the location of biomolecules of interest (e.g. membrane proteins), and measure the distances between different domains on a single protein. This information can be determined over time to elucidate valuable signal transduction information in both time and space. The most common method used to determine this information is FRET. FRET is a photophysical phenomenon that involves the non-radiative energy transfer between two fluorophores. The excited donor fluorophore transfers this non-radiative energy to another acceptor molecule. The acceptor molecule does not need to be fluorescent. In order for FRET to occur, the donor and acceptor molecules need to be physically close to each other, a distance that has been shown to be less than 10 nm (Figure 6.3).

There are many different types of fluorescent labels available for FRET-based assays. The most well-known classes are genetically encoded labels such as green fluorescent protein (GFP), cyan fluorescent protein (CFP),

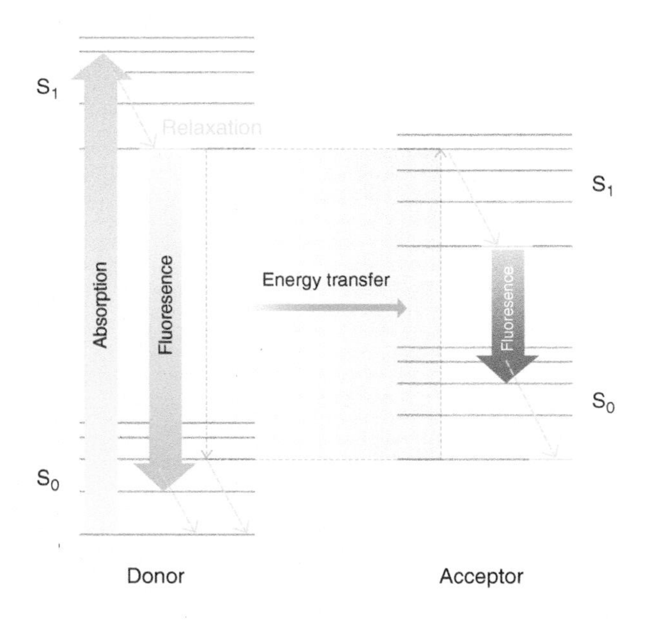

Figure 6.3 A resonance energy transfer Jablonski diagram.

yellow fluorescent protein (YFP), mCherry, and monomeric variant of EosFP (mEOS). These fluorescent proteins are advantageous for FRET assays due to their ability to measure dynamic processes both in space and over time in living cells via microscopy as well as microtiter plate fluorescent assays for *in vitro* studies. The donor and acceptor molecules need to be tagged to the targets of interest. For fluorescent protein tags, the target protein and fluorescent reporter protein can be fused together through recombinant cloning at the N- or C-terminus. Some disadvantages of fluorescent protein tags are pH sensitivity, limited labeling sites, and the potential for oligomerization, which can be overcome by using appropriate controls or by using mutant variants of fluorescent proteins that have been developed specifically to reduce this effect. Another disadvantage is that fluorescent proteins are rather large tags. Their size may interfere with a target protein's endogenous activity, such as conformational changes, and also inhibit the target protein's ability to interact with its corresponding substrate protein.

6.3.2 Enzyme-Linked Immunosorbent Assay (ELISA)

As mentioned in Chapter 5, ELISA is a common method used to measure the concentration of the protein of interest in matrices such as serum, blood, and even feces. The ability of ELISAs to quantitatively measure analytes of various classes, from cytokines to kinases, has made them an invaluable tool for signal transduction research. There are several types of standard ELISAs that will be discussed in this section – competitive, sandwich, and direct cellular – as well as the multiplexable bead-based suspension array assay, which is a derivative of the traditional sandwich ELISA.

6.3.2.1 Competitive ELISA

Competitive ELISAs are used to quantify antigens in a given sample medium. To perform a competitive ELISA, the wells of a microtiter plate are pre-coated with purified antigen for the desired target of the assay. To the competitive assay control wells, an enzyme–antibody conjugate, where the antibody is specific for the antigen of interest and the enzyme is used to measure the well response, is added to provide the control response if no antigen is present in a standard/sample well. To the remaining wells, standards or samples (containing unknown amounts of antigen) are loaded into the antigen pre-coated wells along with the enzyme–antibody conjugate. All wells are incubated for a set amount of time and washed to remove any unbound enzyme–antibody conjugate. The substrate corresponding to the enzyme attached to the antibody is added to initiate a detectable reaction, whether that be using absorbance or fluorescence detection, and measured with a microplate reader. The decrease in signal indicates the presence of antigen in a standard or sample-containing well. In order to quantify the amount of antigen present in sample wells, a standard curve must be used. The competitive ELISA is suitable for studies where a few milligrams or more of purified protein/antigen are readily available and cost effective. However, if that is not possible for the antigen of interest, a sandwich ELISA would be a more advantageous immunosorbent assay. It is also important to note that "enzyme–antibody" conjugate can also be substituted for a biotinylated antibody and the "substrate" would be streptavidin–fluorophore conjugate for fluorescence detection.

6.3.2.2 Sandwich ELISA

Sandwich ELISAs are one of the more common and advantageous types of immunosorbent assays for antigen detection due to their increased sensitivity over assays where the antigen is directly bound to the microtiter plate, such as competitive ELISA. They are widely available in kit form from many manufacturers, and each kit comes with specific protocols and most reagents necessary to complete the assay, making these assays very simple and reproducible. To detect the antigen of interest, the potential antigen-containing sample is added to respective microtiter wells coated with a capture antibody selective for the desired antigen and incubated for a set amount of time determined by the kit manufacturer, or if the assay was developed in a laboratory, for the optimized amount of time. After incubation, any unbound antigen is washed away and a secondary/detection antibody is added, which is an enzyme–antibody conjugate where the antibody is specific for the antigen of interest and the enzyme is part of the detection. The sandwich (capture antibody, antigen, secondary/detection antibody) is incubated, the wells are again washed to remove any unbound secondary antibody, and the substrate for the enzyme–substrate detection reaction is added, and the solution color change (absorbance) or fluorescence can be measured using a plate reader. Just as with a competitive ELISA, sandwich ELISA may also use biotinylated antibodies instead of enzyme–antibody conjugates.

6.3.2.3 Direct Cellular ELISA

Direct cellular immunosorbent assays are useful for investigating antigens on a cellular surface, such as receptor proteins. Unlike traditional competitive or sandwich ELISAs that typically use cell lysates or tissue homogenates, direct cellular ELISA allows for the analysis of intact cell samples. In direct cellular ELISA, cell suspensions are incubated with enzyme–antibody conjugates, where the antibody is specific for cell surface molecules (e.g. receptors). After incubation, the suspension is centrifuged and the supernatant is discarded, leaving the pellet of cells with bound antibody conjugate attached. The pellet is resuspended in wash buffer, mixed, and centrifuged again. After centrifugation, the supernatant is again discarded. The pellet is then resuspended in substrate-containing buffer, and the response is measured via absorbance or fluorescence. The direct cellular ELISA can also be amended to an indirect cellular ELISA, where a biotinylated antibody is used instead of the enzyme–antibody conjugate and a streptavidin–alkaline phosphatase/HRP conjugate as the substrate for the detectable reaction.

Selecting the appropriate enzyme–substrate reaction for the ELISAs discussed in this section (competitive, sandwich, cellular) can be challenging. The selection depends on cost of reagents, desired time to prepare the reagents (some require the preparation of several buffers and others arrive ready to use), the instrumentation available in the laboratory space used to carry out the assay, and desired detection limit of the assay, whereby some enzyme–substrate reactions are more sensitive than others.

6.3.2.4 Multiplex Suspension Array Assays

The dynamic and coordinated intracellular network response to external stimuli involves many different signaling molecules at any given time. The potentially large number of key protein targets responding to a given insult poses a significant experimental challenge if only one protein is measured at a time (single target/singleplex). Advances in numerous bioanalytical methods have allowed for multiple proteins to be measured per sample analyzed, such as multiplexable ELISAs. Traditionally, Western blot, which will be discussed later on in this chapter, and immunoprecipitation have been the assays of choice for phosphoprotein analyses. While Western blotting has been the gold standard for phosphoprotein determinations, it is considered a low-throughput method. To overcome the limitations of single-target experiments, a multiplexed and high-throughput assay for phosphoprotein analysis has been developed by Luminex Corp., commonly referred to as suspension bead-based array or bead-based (microsphere) sandwich ELISA. This assay is a modification of a traditional ELISA, whereby the capture antibody (Ab) is immobilized to magnetic (or nonmagnetic) polystyrene microspheres. The assay workflow proceeds as follows: (i) capture Ab beads are added to the assay plate wells, (ii) add the analyte(s)-containing sample and incubate, (iii) add biotinylated detection Ab and incubate, (iv) add fluorescent reporter (e.g. streptavidin–phycoerythrin complex) and incubate, and (v) analyze samples via suspension array.

Since this is a multiplexable method, 100+ different Ab-bead combinations can be assayed per well. The beads are coded with two dyes that can be excited by the same wavelength but emit different wavelengths. Depending on the ratio of the two encapsulated dyes, the bead region (code) can be measured by the instrument. In order to differentiate between bead dyes and fluorescent reporter dye, the suspension array platform utilizes a dual laser system: a red solid-state laser (635 nm) to excite dyes encapsulated within the microspheres for bead-type identification and an Nd:YAG laser (532 nm) to excite the phycoerythrin dye (fluorescent reporter). The fluorescent reporter is bound to the analyte–bead complex for quantitation of the captured analyte. All emission intensities are detected with a photomultiplier tube. This method is high throughput and requires low sample volume. Currently, Luminex-manufactured magnetic microspheres (sold under the name MagPlex) have 500 unique bead regions, allowing for 500 different antibody assays to be performed per well of a 96-well plate.

For a smaller, simpler, and more affordable version of the suspension array system, Luminex introduced the MAGPIX, which utilizes magnets and sophisticated CCD camera technology instead of flow cytometry. MAGPIX uses the same magnetic microspheres as discussed previously (antibody-coated magnetic spheres containing identification dyes). Once all relevant reagents and incubations have been carried out (similar to the flow cytometry suspension array), the contents of a sandwich ELISA bead-containing well are removed by an instrument probe and driven with fluidics to a chamber, and a magnet attracts the beads from the chamber. The beads then assemble in a monolayer fashion and the bead-encapsulated dye is excited by a red light-emitting diode (LED) light, and the CCD camera takes an image to determine the code/identity of the beads, which corresponds to the immobilized antibody on the bead. The fluorescent reporter (for the detection antibody) is excited by a green LED light, and the CCD camera takes an image to determine the presence/absence of an antigen.

Both the MAGPIX and bead-based suspension array systems can measure not only phosphoproteins but also cytokines, chemokines, and kinases (e.g. total ERK protein in addition to phosphorylated ERK). These biomolecules can be quantified using either of these systems by including standard curves in your analyses. While the ability to multiplex is extremely beneficial for signal transduction research, this also presents a "multiplexable problem" for quantitation, where the linear range of the standard curve for each analyte/antigen to be measured in a given assay run may not cover the range of responses in your samples. Therefore, the sample well protein concentration may be suitable for "most" of the analytes measured (possibly 16/20), but there will be some that do not fit into their corresponding standard curves. To overcome these issues, it may be wise to then analyze certain analytes as a singleplex assay after interpolating the results from a large multiplex assay.

6.3.2.5 Electrochemiluminescence (ECL) Array

A recently developed antibody-based protein and phosphoprotein method is the electrochemiluminescence (ECL) array. The ECL array combines ECL and sandwich ELISA in a multispot array to make this method multiplexable (4–10 analytes/well). For this assay, the capture antibody is immobilized to the plate surface, the analyte is bound, and the detection antibody completes the "sandwich." The detection antibody is tagged with ruthenium(II) trisbipyridine-(4-methyl sulfone) (Ru(bpy)). After required incubation steps to form the sandwich complex, tripropylamine (TPA) is added within five minutes of reading on the specialized plate reader. Once in the plate reader, a potential is applied to the electrode/well, and the $Ru(bpy)_3^{2+}$–TPA reaction is initiated.

The simplified chemical reaction sequence below follows the assay from potential applied to measurable orange emission of 610 nm [2]:

(1) $Ru(bpy)_3{}^{2+} \rightarrow Ru(bpy)_3{}^{3+} + e^-$
(2a) $Ru(bpy)_3{}^{3+} + TPA \rightarrow Ru(bpy)_3{}^{2+} + TPA^+$
(2b) $TPA \rightarrow TPA^+ + e^-$
(3) $TPA^+ \rightarrow TPA + H^+$
(4) $Ru(bpy)_3{}^{3+} + TPA \rightarrow [Ru(bpy)_3{}^{2+}]^*$
(5) $[Ru(bpy)_3{}^{2+}]^* \rightarrow Ru(bpy)_3{}^{2+}$ + light (610 nm)

6.3.3 Gel Electrophoresis

Gel electrophoresis is a multifunctional technique useful for separating DNA, RNA, and proteins based on their size and charge. It can be used as a standalone method or as a precursor for electroblotting or mass spectrometry (MS). As an electrophoretic technique, an electric potential is applied to the gel where one end is negatively charged (cathode) and the other is positive (anode). Since DNA and RNA are negatively charged, they will migrate toward the positively charged end of the gel. Conversely, proteins are positively charged and will migrate toward the negatively charged end of the gel. The composition of the gel affects the separation of the sample mixture. Agarose gel is one type of porous medium used for gel electrophoresis that is particularly well suited for larger biomolecules, such as DNA and large proteins. The other popular gel used for this technique is acrylamide. The acrylamide is used at various concentrations to form a cross-linked polyacrylamide polymer, and the desired concentration depends on the size of biomolecules to be separated for a given protocol. The agarose or acrylamide gel acts as a size filter, allowing smaller molecules to traverse the gel further than larger molecules. This results in longer DNA strands or larger proteins to be found closest to the top of the gel and the shorter DNA strands or small proteins to travel the furthest distance. The separation of proteins by size is beneficial but may still result in many proteins appearing in one band, providing poor separation. To improve separation, 2D gel electrophoresis can be used, where a mixture of proteins is separated by isoelectric point in the first dimension. Then, in the second dimension (perpendicular to the first separation), molecules are separated on their size. By using 2D gel electrophoresis, unknown, complex proteins can be identified in a sample. Additionally, 2D gels can be used with an additional analytical method, where protein bands of interest can be excised, digested, and analyzed using electrospray ionization (ESI) MS.

6.3.4 Western Blot

In Chapter 5, we discussed the utility and methodology of Western blot analyses for cellular signaling experiments. While Western blots have been the workhorse for specific protein or phosphoprotein analyses, they suffer from being relatively low throughput, requiring long incubation times; are labor intensive; and are unable to simultaneously measure many analytes. Traditional Westerns can only measure one analyte of interest at a time and are limited to the number of lanes available on a gel after ladder and standard lanes have been accounted for, which makes it a relatively low-throughput technique. To overcome some of these limitations, a relatively new technology has been developed – the simple Western. The simple Western is an automated, gel-free, and transfer blot-free method to determine relative quantitation of a protein of interest in a sample. With the simple Western, you can get information from your sample using a size-based or charge-based platform. In this chapter, we will discuss the size-based immunoassay format, which is the most similar to a traditional Western blot in terms of information received; however, it is important to know that there are other types of information that can be utilized from this instrument.

The simple Western immunoassay can be succinctly conceptualized as capillary electrophoresis coupled with an ELISA. The starting material required for analysis is total protein (derived from whole tissue homogenate or cell culture lysates, among other mediums). All of the reagents required to run the assay come in a convenient kit with easy-to-follow instructions to streamline benchtop sample preparation into about an hour. After the plate is prepared per manufacturer's instruction, the microtiter plate is loaded into the simple Western instrument. The capillary collects a preset amount of stacking and separation matrix buffers and then loads the prepared sample. The proteins in the sample are then separated based on size (i.e. high and low molecular weight proteins) within the capillary. The separated proteins are immobilized in the capillary and the pseudo-ELISA begins. The desired proteins are identified with a primary mono- or polyclonal antibody, such as anti-ERK (total) or phosphoprotein targets, such as anti-phospho-ERK (threonine 202/tyrosine 204), and detected with a secondary antibody, such as a HRP-labeled secondary antibody, followed by chemiluminescent detection. The signal is then detected and quantitated (Figure 6.4). As long as the appropriate controls are loaded in the same plate, you can calculate relative protein or relative phosphoprotein response of the analyte of interest. Depending on the instrument used, 11–96 samples can be analyzed at one time. Additionally, the method and all incubation steps are automated within the instrument, saving time and preventing potential laboratory errors. Since there is no actual blotting step using the simple Western assay, any issues that are typically caused by poor protein transfer are eliminated. While this method is not highly multiplexable (i.e. no more than five analytes can be measured from one sample well as long as the molecular weight bands do not overlap and there is little nonspecific binding by the various antibodies), there is the option to run different antibodies on the sample plate (e.g. four sample wells

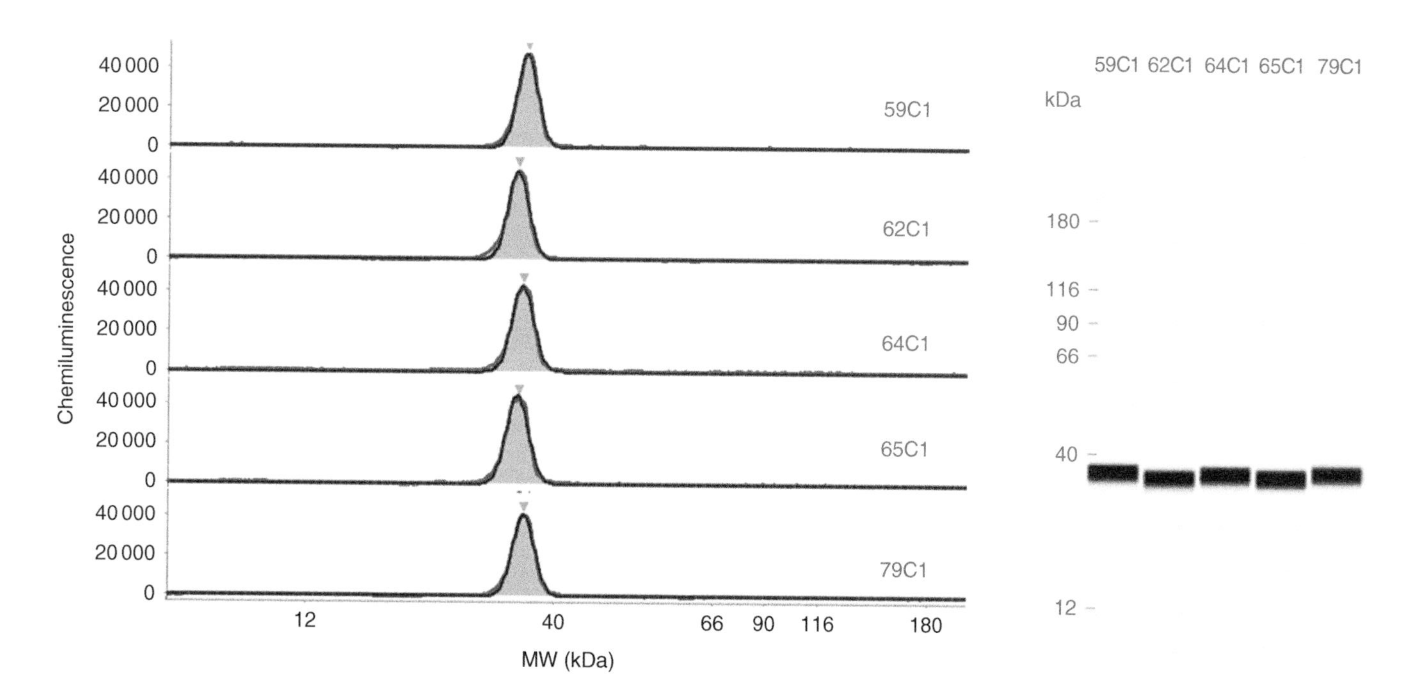

Figure 6.4 Example image of a simple Western and electropherogram that would result from using an antibody for blotting.

measuring anti-ERK, four sample wells measuring anti-phospho-ERK, four sample wells measuring anti-MEK1, etc.).

6.3.5 Protein Nuclear Magnetic Resonance (NMR)

Commonly thought of as a technique for organic chemists and small molecule analyses, nuclear magnetic resonance (NMR) is a valuable method for determining PTMs and their amino acid target sites (see Theillet [3] for review). There exist many atomic nuclei that act as though they are spinning around an imaginary rod/axis. Due to the nuclei's positive charge, the spinning nuclei act like small magnets that can interact with external magnetic force. In NMR, when these nuclei are placed between the poles of a strong magnet, their magnetic fields align with (parallel) or against (antiparallel) the external field. Once in a parallel or antiparallel orientation, the nuclei are irradiated with electromagnetic radiation of the necessary frequency, and the lower energy nuclear state spin-flips to the higher energy state. The action of the spin-flip means that the nuclei are in resonance with the applied electromagnetic radiation. The variation in absorption bands of NMR spectra is due to each nucleus' electromagnetic environment, where nuclei can be "shielded" from the applied field by surrounding electrons. Structural information can be determined from these shielding effects, or chemical shift, for small molecules and even proteins. High-resolution NMR has the capability of detecting a wide array of PTMs, such as phosphorylation, acylation, alkylation, and glycosylation, in solution. The development of 2D heteronuclear correlation NMR methods using ^{1}H, ^{13}C, ^{15}N, and ^{31}P nuclei, such as ^{1}H–^{15}N correlation for phosphorylated serines and threonines, has dramatically impacted the ability to identify structural information in these complex samples. Protein kinases can have multiple endogenously phosphorylated sites, which increases the complexity of the NMR spectrum. The tumor suppressor protein p53, which is important for DNA repair activation and apoptosis, is one such protein with multiple phosphorylation sites where ^{1}H–^{15}N correlation is advantageous. In Figure 6.5a, the phosphate addition to serine and threonine residues is shown, and Figure 6.5b shows the superimposed NMR spectra of treated and untreated N-terminal disordered transactivation domain (N-term TAD) of human p53 when combined with (green) or without (black) HeLa cell extracts and the spectral shifts of multiple phosphorylation sites [3]. NMR has also been used to determine the multiple phosphorylation sites of Tau, which has been aggressively investigated for its role in neurodegenerative diseases, such as Alzheimer's disease (AD). Elucidating the critical signaling events and proteins responsible (e.g. ERK) for potential deleterious phosphorylation of specific Tau residues is important for understanding the underlying mechanisms and conditions leading to AD [4].

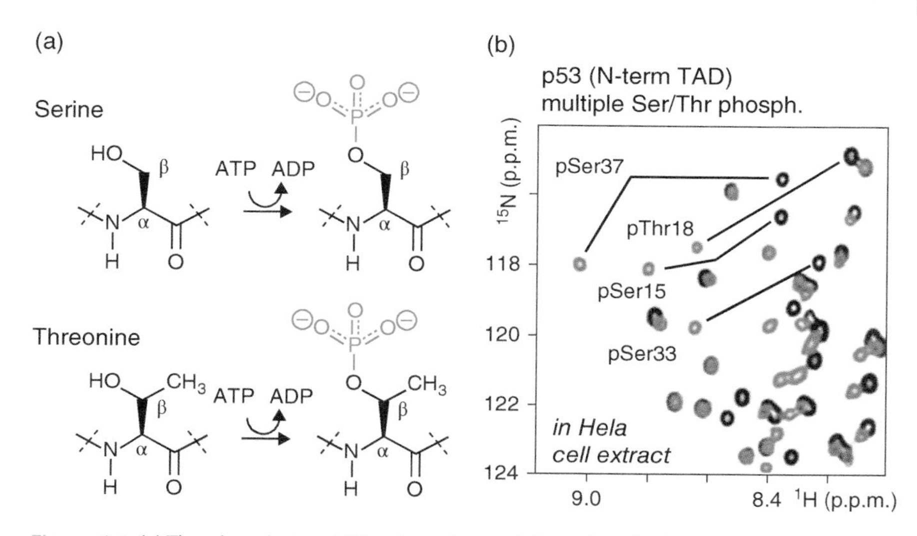

Figure 6.5 (a) The phosphate addition to serine and threonine. (b) Superimposed 2D NMR spectra of treated N-term TAD of p53 when combined with HeLa cell extracts (green) and without (black). *Source:* Reprinted with permission from *Springer Nature Journal of Biomolecular NMR*, Theillet et al. [3].

6.4 Techniques to Generate Large Datasets for Signal Transduction Network Analysis

The highly dynamic and interconnected nature of signaling networks has made it increasingly difficult to elucidate and predict network responses to endogenous or xenobiotic stress. However, with advances in computational and network biology, high-throughput experimental techniques such as -omics investigations using MS, multiplex bead-based ELISA suspension array systems, microchip arrays, next-generation sequencing, and public databases have greatly improved signaling research. With these technological advances, many scientific fields can benefit such as tracing cancer pathobiology and developing new treatments as well as developing toxicological risk assessment approaches for determining potential toxic outcome of a xenobiotic or mixture using mechanistic information can be attempted.

6.4.1 '-omics Using Mass Spectrometry

Proteomic technologies enable the simultaneous analysis/detection of a large number of proteins and can be used to characterize different pathophysiological conditions *in vitro* and *in vivo*. A major advantage of proteomics over antibody-based techniques is that antibodies for specific analytes are unnecessary; therefore identification of proteins that were not known to be significant to the experimental conditions before analysis is possible. Traditional proteomic approaches have focused on profiling strategies, including 2D gel electrophoresis combined with MS, liquid chromatography separation combined with mass spectrometry (LC-MS), desorption electrospray ionization (DESI), and matrix-assisted laser desorption ionization (MALDI) MS.

There are two main approaches to proteomics: "top down" and "bottom up." In top-down proteomics, intact proteins are injected into the mass spectrometer, where specific proteins can be selected in trapping-type instruments, such as Fourier transform ion cyclotron resonance (FT-ICR), linear ion quadrupole trap, or Orbitrap. By injecting the intact protein into the mass spectrometer, both the intact and fragment ions can be measured. Fragmentation for top-down proteomics can be carried out using electron-capture dissociation (ECD) or electron-transfer dissociation (ETD). Top-down proteomics allows for complete protein sequence coverage, characterization of different proteoforms, and molecular changes via genetic variation or addition of small molecules from PTMs (for a review, see Catherman [5]). In bottom-up proteomics, proteins in complex samples, such as tissue or cell culture lysates, are chemically or enzymatically digested into peptide fragments before further analysis using MS or LC-MS. Top-down proteomics is best suited for specific questions about a protein or PTM on a protein, whereas bottom-up proteomics is best suited for large complex samples and potential proteins of interest are unknown prior to analysis. Bottom-up suffers from several limitations regarding peptide sequence coverage, peptide fragments appearing in several different protein sequences, and low-abundance protein information may be lost in the "noise" of other high-abundance protein's peptides in the sample matrix. Top-down approaches for PTM identification suffer from the need for high-resolution and mass-accuracy mass spectrometers, which many laboratories do not have available. High-resolution and mass-accuracy spectrometers are required for intact protein PTM experiments because they have the ability to distinguish small changes on the protein, such as acetylation (mass change of ~42 Da) or phosphorylation (mass change of ~80 Da). For further reading and in-depth discussions about proteomics principles and techniques, see Twyman [6].

6.4.1.1 Separation Techniques

Classic proteomic approaches for bottom-up analyses include protein separation by sodium dodecyl sulfate-polyacrylamide gel electrophoresis (SDS-PAGE) or 2D gel electrophoresis and protein identification by MS. The major advantage of 2D gel is to discover the unrecognized proteins of interest without interference of a large number/the entire complex sample of proteins. When using 2D gel electrophoresis, desired protein bands can be excised from the gel, digested, and analyzed using traditional ESI MS.

For a more high-throughput method for separating proteins in complex sample matrix, LC is often utilized. The protein-containing sample can be separated based on an assortment of any physical properties, such as hydrophobicity (reversed-phase liquid chromatography [RPLC]), hydrophilicity (hydrophobic interaction liquid chromatography [HILIC]), mass/size (size exclusion), or charge (ion-exchange chromatography). The stationary phase selected is dependent on experiment needs. After protein-containing samples are loaded onto the LC (e.g. from tissue homogenates or cell lysates), fractions are collected so that many vials of separated proteins are saved for further preparation before MS. Once LC fractions are collected, the protein-containing fractions are chemically and/or enzymatically digested into smaller peptide fragments prior to MS (or additional LC-MS) [7]. There are many different types of sample preparations, but the most common is as follows: (i) neutralize pH of the fractions to remove acetonitrile and any trifluoroacetic acid, (ii) reduce disulfide bonds with dithiothreitol (DTT), (iii) alkylate any free cysteine residues with iodoacetamide (this prevents protein aggregation and precipitation during sample preparation), (iv) digest the unfolded protein sample chemically (e.g. cyanogen bromide) or enzymatically (e.g. trypsin, chymotrypsin, Lys-C protease, Asp-N protease) to obtain peptide fragments cleaved before/after at certain amino acid residues, and (v) desalt samples, if necessary, with C-18 solid-phase extraction cartridges or reverse osmosis membrane filtration [7].

6.4.1.2 Phosphoprotein Enrichment for Phosphoproteomics: IMAC, MOAC, and SMOAC

Identification of protein phosphorylation sites has been exceedingly difficult via MS. To improve phosphoprotein identification, enrichment protocols have been employed for phosphoproteomics. Immobilized metal ion affinity chromatography (IMAC) is one such enrichment method. In IMAC, the negatively charged phosphate groups of phosphoproteins/peptides interact with positively charged metal ions, such as Al^{3+}, Fe^{3+}, or Ga^{3+}. The bound phosphopeptides are released from the column stationary phase with ethylenediaminetetraaceticacid (EDTA), alkaline buffers (pH 10–11), or phosphate-containing buffers. IMAC enrichment is both sensitive and selective for peptide mixtures that have been purified or significantly fractionated prior to enrichment. This method is limited when it comes to working with complex samples, such as whole-cell extracts, but can be overcome with significant prefrationation. Additionally, lowering the pH of the mobile phase/sample is suggested for IMAC because it can reduce the nonspecific binding to IMAC resin and improve phosphopeptide selectivity.

Another phosphoprotein enrichment method is with metal oxide affinity chromatography (MOAC), specifically with $Al(OH)_3$ or TiO_2 resin. For example, the TiO_2 resin attracts the oxygen of phosphate groups with a high degree of specificity when in the presence of 2,5-dihydroxybenzoic acid (DHB) in the mobile phase/loading buffer. This method has been applied to a more user-friendly, simple, and rapid protocol by using TiO_2 beads, of which there are many commercially available kits. A recently developed method for enhanced phosphoprotein enrichment is serial metal oxide affinity chromatography (SMOAC), where washes from TiO_2 enrichment are put through IMAC enrichment, further increasing the number of phosphopeptides available for MS analysis.

6.4.1.3 Quantitation with Chemical Tags: iTRAQ and TMT

Qualitative information, such as identification of proteins or PTM of proteins, in a complex sample matrix is a major advantage of using MS for proteomics and phosphoproteomics. However, quantitative information about a sample is desperately needed to compare treated against control samples at the protein level. To solve this issue, chemical tags added to peptide fragments, such as isobaric labeling (iTRAQ) or tandem mass tagging (TMT), is used for MS quantitation (for step-by-step protocol, see Zhang and Elias [8]). After peptide preparation (i.e. digested peptide fragments), samples are labeled with iTRAQ or TMT reagents, so that each sample (control, treatment 1, treatment 2, etc.) is labeled with a different isotope or mass reporter tag. For iTRAQ, kits are available for 4–8-plex experiments, and TMT kits are available in 2–11-plex. With differentially labeled sample peptide fragments, treatment groups can be analyzed at one time and compared to obtain relative protein abundance information.

6.4.2 RNA Sequencing (RNA-Seq)

RNA sequencing (RNA-Seq) is a "big" dataset generating technique commonly used in a wide variety of biological studies, such as identification of molecular drivers of a given disease or exposure condition. While this technique offers a large and rich dataset, it is still quite expensive. This method requires good-quality isolated RNA for cDNA library preparation. After obtaining isolated RNA, the type of RNA that is desired can be selected: whole RNA or mRNA (filtered for keeping 3′ polyadenylated tail discarding ribosomal RNA). Desired RNA is then reverse transcribed to cDNA [9]. The cDNA library is fragmented and size selected and then loaded into a sequencer (e.g. Illumina platforms) to generate 100s of millions of paired reads. The type of analysis used for RNA-Seq data is based on the use of a reference genome. If your sample of interest is generated in a species with an incomplete or unknown genome, *de novo* analysis is used to interrogate the transcriptome. The most commonly used transcriptome assembly algorithms use de Bruijn graphs such as the programs Velvet, Trinity, and Bridger [10, 11]. If your sample has a known genome, the software analysis platforms most commonly used are TopHat and Cufflinks [12].

From these large datasets, differential expression analysis, false discovery rate, and p-values for comparison of differentially expressed genes can be used to identify new genes, pathways, or cellular processes of interest for a given disease or exposure condition, which would otherwise remain unknown without this high-throughput shotgun/whole transcriptome technique.

6.5 Bioenergetics

Kinase signaling is an energy-demanding process, and its reliance on phosphorylation results in the consumption of substantial amounts of available ATP. ATP production governs ATP-consuming processes, such as signal transduction in mammalian cells, and this production is primarily driven by oxidative phosphorylation within mitochondria. Mitochondria are the energy production hub of a cell via production of ATP via phosphorylation and, of equal importance, key mediators in kinase signal transduction, regulating cell survival, proliferation, differentiation, and death. Due to their critical role in many cellular processes, mitochondria are also susceptible to exposure effects. Additionally, irreversible processes leading to cell death primarily rely on two mitochondria-related phenomena: (i) the inability to reverse mitochondrial dysfunction, resulting in ATP depletion, and (ii) the disturbance of membrane function (both mitochondrial membrane and plasma membrane). Therefore, maintenance of mitochondrial bioenergetics and integrity is crucial to cellular fate.

6.5.1 Oxygen Consumption

Intracellular molecular oxygen is a key component of cellular homeostasis and mitochondrial oxidative phosphorylation via the electron transport chain (ETC). After xenobiotic exposure, early cellular changes associated with mitochondrial bioenergetics and cellular respiration can be monitored using real-time oxygen consumption assays. Determining key time points related to perturbations in cellular respiration can further the mechanistic understanding of xenobiotic exposure and eventual cell survival/death. High-throughput real-time assays for oxygen consumption that do not perturb endogenous intracellular activity are limited. The most popular method for cellular oxygen consumption is the Clark electrode; however, it is a very low-throughput method (one sample at a time). The development of oxygen-sensitive extracellular probes that can reproducibly measure discrete changes in oxygen consumption over time (with comparable sensitivity to the Clark electrode) has significantly advanced mitochondrial and toxicological research. One such probe was developed by Luxcel Corp. (Cork, Ireland), manufactured as MitoXpress. The MitoXpress probe is an extracellular phosphorescent platinum-coproporphyrin dye with a long emission time and stable phosphorescent signal that can be used for 24-hour kinetic measurements in a 96-well plate format, making it high throughput. For this assay, the MitoXpress probe is quenched by molecular oxygen (O_2). A decrease in extracellular O_2 concentration (increase in cellular oxygen consumption) is measured as an increase in signal (less probe is quenched by O_2), whereas an increase in extracellular O_2 concentration is measured as a decrease in signal (more O_2 to quench the probe; less oxygen is consumed by the cell).

Another relevant and useful tool to measure oxygen consumption related to bioenergetics is the Seahorse bioanalyzer technology. The Seahorse bioanalyzer measures oxygen consumption rate in a microtiter plate by exploiting the ETC biochemistry. For *in vitro* oxygen consumption rate measurement, a prepared cell culture plate with the optimized amount of cells/well and desired exposure conditions is used with the patented sensor cartridge. The sensor cartridge has four compartments: (i) open for toxicant injection if desired or exposures can be done by the experimenter prior to the assay, (ii) oligomycin, (iii) carbonyl cyanide 4-trifluoromethoxyphenylhydrazone (FCCP), and (iv) rotenone/antimycin A. After following the manufacturer-provided instructions for reagent and cartridge preparation, the cell-containing plate with the cartridge is loaded, and the basal respiration is measured for 20 minutes. After 20 minutes, oligomycin is injected where it inhibits complex V of the ETC and the decrease in oxygen consumption rate is correlated to cellular ATP production via mitochondrial respiration. Next, FCCP is added to act as a proton uncoupler and disrupt the mitochondrial membrane potential, which allows uninhibited electron flow into the mitochondria and measurement of maximum respiration via complex IV. Finally, the last injection is a solution of rotenone (complex I inhibitor) and antimycin A (complex III inhibitor), which essentially inhibits all mitochondrial respiration, and therefore oxygen consumption due to non-mitochondrial respiration can be determined.

6.5.2 Reactive Oxygen Species (ROS) Fluorescent Probes

Fluorescent probes are a fast and sensitive means to detect cellular reactive oxygen species (ROS) generation and oxidative activity *in vitro*, but all suffer from a selectivity point of view. Among these, 2′,7′-dichlorofluorescin diacetate (DCFDA) has been commonly used as a probe for oxidative stress, especially for the intracellular production of H_2O_2, which has been suggested to be a suitable marker of the overall oxidative stress of the cell [13]. Fluorescent probes specifically targeting ROS produced transiently, such as hypochlorite acid (HOCl), have been developed and applied to real-time high-throughput assays [14–16]. HOCl may cause tissue damage in the host tissue and play an important role in modulating the signaling pathway of ischemia/reperfusion injury, inflammation, and atherosclerosis [14]. Myeloperoxidase (MPO) catalyzes the production of HOCl from hydrogen peroxide (H_2O_2) and chloride anion (Cl−) during the respiratory burst of the neutrophil, which is a transient reaction and can be missed if the assays are not performed at the critical time. In a study from Liu et al. [15], quantum dots (QDs), a novel and sensitive type of probe for HOCl based on a fluorescence quenching mechanism, were used in a rapid assay for the screening of HOCl scavengers and MPO inhibitors. The time course curve of QD fluorescence quenching by neutrophil-like cells after phorbol 12-myristate 13-acetate (PMA) stimulation or H_2O_2 indicates the different mechanisms of HOCl generation by H_2O_2 or PMA treatments. The addition of H_2O_2 quenched the QD fluorescence much faster than the fluorescence quenching caused by PMA stimulation because H_2O_2 was directly used as a substrate by intracellular MPO to generate HOCl, whereas PMA treatment triggers a "respiratory burst," leading to an extracellular release of MPO, which is a gradual process that needs activation of nicotinamide adenine dinucleotide phosphate (NADPH) oxidase and superoxide dismutase (SOD). Therefore, the QD assay with H_2O_2 added to cells is suitable for the evaluation of intracellular MPO inhibitors and HOCl scavengers, while the QD assay with PMA stimulation of cells is more suitable for screening of extracellular MPO inhibitors and HOCl scavengers.

6.5.3 ATP Assays

As described previously, ATP is a critical biomolecule for cell survival. While real-time assays for ATP generation are limited, endpoint assays for ATP are widely used and extremely sensitive. Additionally, ADP/ATP ratios have been used to measure cell viability, apoptosis, and necrosis. The predominant endpoint ATP assay is the D-luciferin-luciferase assay, which requires cell lysis for *in vitro* applications. The luciferase assay is a bioluminescent assay where the substrate D-luciferin is converted into oxyluciferin, by the luciferase enzyme in the presence of ATP. The conversion from luciferin to oxyluciferin emits a stable, relatively long-lived light emission (~30 minutes). This assay is very sensitive, ranging from 10^{-13} mol to greater than 10^{-6} mol ATP, and can be used for a wide variety of sample matrices, such as soil, milk, plasma, and cell culture.

6.5.4 Nicotinamide Adenine Dinucleotide (NADH) Assay

Nicotinamide adenine dinucleotide (NADH) is a vital component of mitochondrial function, energy metabolism, and oxidative stress. As a critical component of the ETC and bioenergetics, early cellular effects and perturbations can be monitored via changes in cellular NADH. One of the most useful features of NADH is that it strongly absorbs at 340 nm; therefore, real-time kinetic measurements of cellular NADH can be obtained without the assistance of any fluorescent tag or probe. An example of an NADH assay will be further discussed in Chapter 8.

6.5.5 Mitochondrial Membrane Potential

The irreversible processes leading to cell death primarily rely on two mitochondria-related phenomena, one of which is the disturbance of mitochondrial membrane function. Therefore, maintenance of mitochondrial bioenergetics and integrity is crucial to cellular fate. The loss of mitochondrial membrane integrity, measured as the loss of mitochondrial membrane potential, is a trademark of apoptosis and is also implicated in necrosis. One popular and versatile assay for measuring mitochondrial membrane potential is the JC-1 assay. This assay uses the cationic cyanine dye JC-1 (5,5′,6,6′-tetrachloro-1,1′,3,3′-tetraethylbenzimidazolylcarbocyanine iodide), where mitochondria of healthy cells are stained red. The intact mitochondrial outer membrane potential, bearing a negative charge, allows the lipophilic dye, which has a delocalized positive charge, to pass through to the mitochondrial matrix and accumulate. Once the dye reaches a critical concentration within the mitochondrial matrix, the dye accumulates as J-aggregates and emits a red fluorescence upon excitation. In the presence of lower mitochondrial potentials, the JC-1 dye cannot properly accumulate in the mitochondrial matrix to form J-aggregates and instead emits the green fluorescence of its monomeric form. The JC-1 dye can be used in flow cytometry, epifluorescence microscopy, and fluorescence microplate readers.

6.6 Relating Signaling to Cellular Outcome Using Relevant Assays

Perturbations in cellular signaling cascades can be measured by the techniques discussed briefly in this chapter. However, relating these perturbations to overall cellular outcome – apoptosis, necrosis, or proliferation – in a high-throughput manner is also critical to unraveling cellular response to external stimuli. There are many commercially available assays for determining cellular outcome, and each of them has their advantages and limitations. Many of these techniques were already discussed in Chapter 5, such as trypan blue, erythrosin, crystal violet, and neutral red. Following the previous section on bioenergetics, in this chapter, we will briefly touch on metabolic viability assays via 3-(4,5-dimethylthiazol-2-yl)-2,5-diphenyltetrazolium bromide (MTT)-type, alamar blue, and lactate dehydrogenase (LDH) assays, as well as cell death via plasma membrane permeability.

6.6.1 MTT/MTS/WST Assays

Metabolic viability assays include the colorimetric MTT, 3-(4,5-dimethylthiazol-2-yl)-5-(3-carboxymethoxyphenyl)-2-(4-sulfophenyl)-2H-tetrazolium

(MTS), 2,3-bis-(2-methoxy-4-nitro-5-sulphophenyl)-2H-tetrazolium-5-carboxanilide (XTT), and water-soluble tetrazolium salts (WST). While these assays are used to determine viability, most are included as an orthogonal measurement of cytotoxicity or for initial screening of appropriate dosing ranges. For a more accurate measure of cell death, plasma membrane degradation is appropriate, such as the ethidium homodimer-1 (EthD-1) dye-based assay. However, due to its low cost, robustness, and ease of use, MTT is commonly included in studies to estimate viability. Briefly, the MTT reagent 3-[4,5-dimethylthiazol-2-yl]-2,5-diphenyltetrazolium bromide is a yellow water-soluble salt that is converted into insoluble formazan crystals in the presence of succinate dehydrogenase within the mitochondria of metabolically active cells via cleavage of the tetrazolium ring. The insoluble purple formazan crystals can be solubilized by dimethyl sulfoxide (DMSO) or detergent for quantification via absorbance using 96-well plate spectrophotometry. Similar results can be obtained using the MTS assay, but the solubilizing step is removed. MTS coupled with phenazine methosulfate (PMS) can produce soluble formazan that can be read at wavelengths ranging from 490 to 500 nm. Since the MTS and PMS are not removed, there is a greater probability that the background could impact the quality of the results. Unlike MTT and MTS, WST (both WST-1 and WST-8) is reduced outside of the cell, so it works well with treatments that affect mitochondrial activity where MTT use is not advised. Similar to MTS, WST does not have a solubilization step, and it is not as toxic to the cells as MTT. The plate can be read at an absorbance of 450 nm [17]. A major setback for MTT assay is that some compounds can influence the mitochondrial enzymes that are responsible for converting the MTT dye to formazan, but not ultimately kill the cells. It is important that researchers appropriately qualify their analyses and know the difference between viability (that may alter metabolic regulation) and cytotoxicity (cell death).

6.6.2 LDH Assay

Cytotoxicity can be determined by measuring the leakage of LDH, a cytosolic protein found in almost all cell types, into the extracellular space. LDH is an enzyme responsible catalyzing the reaction of lactate into pyruvic acid (and pyruvic acid back to lactate) with concomitant conversion of NAD^+ and NADH. The LDH cytotoxicity assay is commonly used for *in vitro* research and available as a commercial assay from many manufacturers. The commercially available assays exploit the concomitant NADH generation from the LDH-catalyzed reaction of lactate to pyruvate and uses a secondary reaction where a substrate (e.g. tetrazolium salts) requires NADH for conversion to an optically measurable product (formazan dyes). For example, the commercially available LDH assay uses 3-(4-iodophenyl)-2-(4-nitrophenyl)-5-phenyltetrazolium chloride (INT) that is reduced to a red formazan dye in the presence of NADH.

6.6.3 Resazurin Assay (Alamar Blue)

The resazurin reduction assay has been implemented in bacterial and yeast experiments for over 50 years, and within the last 20 years, it has been accepted as an appropriate cell viability test in mammalian cell lines. The assay works so that the resazurin or alamar blue dye is reduced to a pink fluorescent dye (resorufin) in the culture medium, which is stimulated by cellular activity. This reductive process is believed to be a result of the use of oxygen during metabolism in the living cells. Alamar blue is prepared in a 10% solution with NaCl/propidium iodide (PI) and is added to each well of a 96-well plate. A laser scanning confocal microscope can be used with an excitation of 568 nm and an emission of 585 nm. Alamar blue dye is believed to provide little toxicity to the cells in culture, making it a useful assay to measure cell viability [18].

6.6.4 Cell Death: Plasma Membrane Degradation Assay

The "point of no return" for which a cell decides to die has yet to be determined [19]. However, certain cellular features implicate a cell has died; one such feature is the loss of plasma membrane integrity. A simple, reproducible, and high-throughput assay for plasma membrane degradation is the EthD-1 assay, commonly referred to as a "dead assay." The EthD-1 dye is impermeable for cells with an intact plasma membrane. However, when the plasma membrane is permeabilized, the EthD-1 can penetrate the cell and intercalate with DNA nucleic acids. When EthD-1 is bound to strands/segments of DNA, the dye emits a strong red fluorescence, whereas in the absence of available DNA segments, the dye has a very-low-intensity fluorescent signal. This dye is useful not only for *in vitro* plate reader-based assays but also for fluorescence microscopy.

6.7 Summary

This chapter focuses on techniques that are frequently relied on for cellular signaling network investigations in molecular biology, pathology, toxicology, pharmacology, and oncology laboratories. These techniques range from high-content analyses of distinct intracellular targets to high-throughput assays capable of measuring hundreds to thousands of proteins in a given sample, as well as small biomolecules (e.g. ATP or O_2), mRNA, and proteins. While this chapter covers a wide range of techniques relevant to cellular signal transduction research, it is important to note that new techniques are constantly emerging to deliver quantitative and qualitative information for more targets, faster data collection, and less sample and monetary cost. Additionally, sophisticated data analysis and statistical platforms are every bit as critical to signal transduction analyses as the measurement techniques themselves.

References

1 Gage, G.J., Kipke, D.R., and Shain, W. (2012). Whole animal perfusion fixation for rodents. *J. Vis. Exp.* 65: e3564.

2 Sun, C., Lu, W., Gao, Y., and Li, J. (2009). Electrochemiluminescence from $Ru(bpy)_3^{2+}$ immobilized in poly(3,4-ethylenedioxythiophene)/ poly(styrenesulfonate)–poly(vinyl alcohol) composite films. *Anal. Chim. Acta* 632 (2): 163–167.

3 Theillet, F.X., Smet-Nocca, C., Liokatis, S. et al. (2012). Cell signaling, post-translational protein modifications and NMR spectroscopy. *J. Biomol. NMR* 54 (3): 217–236.

4 Qi, L., Ke, L., Liu, X. et al. (2016). Subcutaneous administration of liraglutide ameliorates learning and memory impairment by modulating tau hyperphosphorylation via the glycogen synthase kinase-3β pathway in an amyloid β protein induced alzheimer disease mouse model. *Eur. J. Pharmacol.* 783: 23–32.

5 Catherman, A.D., Skinner, O.S., and Kelleher, N.L. (2014). Top down proteomics: facts and perspectives. *Biochem. Biophys. Res. Commun.* 445 (4): 683–693.

6 Twyman, R.M. (2014). *Principles of Proteomics*, 2e. New York: Garland Science, Taylor & Francis Group, LLC.

7 Gundry, R.L., White, M.L., Murray, C.I. et al. (2010). Preparation of proteins and peptides for mass spectrometry analysis in a bottom-up proteomics workflow. *Curr. Protoc. Mol. Biol.* 10 (25): 1–29.

8 Zhang, L. and Elias, J.E. (2017). Relative protein quantification using tandem mass tag mass spectrometry. *Methods Mol. Biol.* 1550: 185–198.

9 Griffith, M., Walker, J.R., Spies, N.C. et al. (2015). Informatics for RNA sequencing: a web resource for analysis on the cloud. *PLoS Comput. Biol.* 11 (8): e1004393.

10 Ye, Y. and Tang, H. (2016). Utilizing de Bruijn graph of metagenome assembly for metatranscriptome analysis. *Bioinformatics* 32 (7): 1001–1008.

11 Chang, Z., Li, G., Liu, J. et al. (2015). Bridger: a new framework for *de novo* transcriptome assembly using RNA-seq data. *Genome Biol.* 16 (1): 30.

12 Trapnell, C., Roberts, A., Goff, L. et al. (2012). Differential gene and transcript expression analysis of RNA-seq experiments with TopHat and Cufflinks. *Nat. Protoc.* 7 (3): 562–578.

13 Wang, H. and Joseph, J.A. (1999). Quantifying cellular oxidative stress by dichlorofluorescein assay using microplate reader. *Free Radic. Biol. Med.* 27 (5–6): 612–616.

14 Yan, Y., Wang, S., Liu, Z. et al. (2010). CdSe-ZnS quantum dots for selective and sensitive detection and quantification of hypochlorite. *Anal. Chem.* 82 (23): 9775–9781.

15 Liu, Z., Yan, Y., Wang, S. et al. (2013). Assaying myeloperoxidase inhibitors and hypochlorous acid scavengers in HL60 cell line using quantum dots. *Am. J. Biomed. Sci.* 5 (2): 140–153.
16 Marsche, G., Semlitsch, M., Hammer, A. et al. (2007). Hypochlorite-modified albumin colocalizes with RAGE in the artery wall and promotes MCP-1 expression via the RAGE-Erk1/2 MAP-kinase pathway. *FASEB J.* 21 (4): 1145–1152.
17 Tominaga, H., Ishiyama, M., Ohseto, F. et al. (1999). A water-soluble tetrazolium salt useful for colorimetric cell viability assay. *Anal. Commun.* 36 (2): 47–50.
18 O'Brien, J., Wilson, I., Orton, T., and Pognan, F. (2000). Investigation of the Alamar Blue (resazurin) fluorescent dye for the assessment of mammalian cell cytotoxicity. *Eur. J. Biochem.* 267 (17): 5421–5426.
19 Kroemer, G., El-Deiry, W.S., Golstein, P. et al. (2005). Classification of cell death: recommendations of the Nomenclatures Committee on Cell Death. *Cell Death Differ.* 12 (Suppl. 2): 1463–1467.

Suggested Reading

Buchser, W., Collins, M., Garyantes, T. et al. (2012). Assay development guidelines for image-based high content screening, high content analysis and high content imaging. In: *Assay Guidance Manual [Internet]* (ed. G.S. Sittampalam, N.P. Coussens, H. Nelson, et al.). Bethesda: Eli Lilly & Company and the National Center for Advancing Translational Sciences.

Engvall, E. and Perlmann, P. (1971). Enzyme-linked immunosorbent assay (ELISA) quantitative assay of immunoglobulin G. *Immunochemistry* 8 (9): 871–874.

Galluzzi, L., Aaronson, S.A., Abrams, J. et al. (2009). Guidelines for the use and interpretation of assays for monitoring cell death in higher eukaryotes. *Cell Death Differ.* 16 (8): 1093–1107.

Garcia-Ojalvo, J. and Martinez Arias, A. (2012). Towards a statistical mechanics of cell fate decisions. *Curr. Opin. Genet. Dev.* 22 (6): 619–626.

Giuliano, K.A., DeBiasio, R.L., Dunlay, R.T. et al. (1997). High content screening: a new approach to easing key bottlenecks in the drug discovery process. *J. Biomol. Screen.* 2 (4): 249–259.

Ishikawa-Ankerhold, H.C., Ankerhold, R., and Drummen, G.P. (2012). Advanced fluorescence microscopy techniques- FRAP, FLIP, FLAP, FRET and FLIM. *Molecules* 17 (4): 4047–4132.

Immunohistochemistry (IHC) vs. Immunocytochemistry (ICC) (2018). Thermo Fisher Scientific. https://www.thermofisher.com/us/en/home/life-science/protein-biology/protein-biology-learning-center/protein-biology-resource-library/pierce-protein-methods/immunohistochemistry-immunocytochemistry.html (accessed 5 February 2018).

Lichtman, J.W. and Conchello, J.A. (2005). Fluorescence microscopy. *Nat. Methods* 2 (12): 910–919.

Mahmood, T. and Yang, P.-C. (2012). Western Blot: technique, theory, and trouble shooting. *N. Am. J. Med. Sci.* 4 (9): 429–434.

Olympus Microscopy Resource center (2018). Olympus Corporation. http://www.olympusmicro.com/index.html (accessed 5 February 2018).

Sorkin, A. and Von Zastrow, M. (2002). Signal transduction and endocytosis: close encounters of many kinds. *Nat. Rev. Mol. Cell Biol.* 3 (8): 600–614.

Taylor, D.L. (2007). Past, present, and future of high content screening and the field of cellomics. *Methods Mol. Biol.* 356: 3–18.

Tsuchiya, M., Wong, S.T., Yeo, Z.X. et al. (2007). Gene expression waves: Cell cycle independent collective dynamics in cultured cells. *FEBS J.* 274 (11): 2878–2886.

Walker, J.M. (2005). *The Proteomics Protocols Handbook.* Totowa, NJ: Humana Press Inc.

7

Computational Methods for Signal Transduction

A Network Approach

Giovanni Scardoni[1], Gabriele Tosadori[1], John Morris[2], Sakshi Pratap[3], Carlo Laudanna[4], and Alice Han[5]

[1] *Center for Biomedical Computing, University of Verona, Verona, Italy*
[2] *Resource for Biocomputing, Visualization and Informatics, University of California, San Francisco, CA, USA*
[3] *Birla Institute of Technology & Science, Goa, India*
[4] *Department of Pathology and Diagnostics, University of Verona, Verona, Italy*
[5] *Chem Bio & Exposure Sci Team, Pacific Northwest National Laboratory, Richland, WA, USA*

7.1 Introduction

Systems biology is a field that aims to investigate the properties of complex biological responses by utilizing models to describe molecular interactions and adaptations [1]. For instance, molecular information (e.g. changes in gene or protein expression, metabolite formation) can be compiled to form pathways and networks that capture the biological response. Signal transduction networks describe how information is processed and disseminated among various biological molecules, principally proteins, but can also involve enzymes and metabolites. Cell signaling largely relies on protein–protein interactions (PPIs) and signaling cascades to regulate essential cellular activities. As the name suggests, PPIs encompass all interactions that take place between specific proteins, and elucidating the network of protein communication is a current endeavor of immense interest and importance. Recent efforts to map the human interactome (PPIs that occur in a cell) have mapped around 14 000 binary interactions [1], which is only a small fraction of the estimated 650 000 interactions thought to exist [2]. On a smaller scale, pathways (e.g. MAPK pathway) can be used to organize groups of biomolecules that are associated by a specific function, also known as functional modules [3].

Once pathways and networks are formed, different computational approaches can be employed to extract information to illustrate a global depiction of cellular signaling and function. The size and complexity of a

Cellular Signal Transduction in Toxicology and Pharmacology: Data Collection, Analysis, and Interpretation, First Edition. Edited by Jonathan W. Boyd and Richard R. Neubig.

network absolutely influence the overall interpretation of the biological response; investigating both small (i.e. few nodes or proteins) and large (i.e. many nodes or proteins) networks can offer alternative perspectives of specific cellular function. For instance, small networks portray a very focused set of interactions that can be modeled as dynamic systems that generate emergent properties [1]. Emergent properties are prompted when the interaction of individual components leads to new functions, such as feedback loops [4]. By evaluating these properties, it is possible to identify how networks change, or are perturbed, under various experimental conditions (e.g. protein inhibition). These dynamic analyses are only feasible for simple pathways composed of well-defined inputs and outputs with a discrete number of molecules. Alternatively, large networks are intricate sets of interconnected molecular events that require algorithms and/or computational power to reduce their complexity and obtain meaningful data for specific pharmacological or toxicological endpoints.

Computational techniques are essential for proper analysis of both small and large networks because they facilitate the management of complex temporal dynamics along with the substantial amount of information contained within a dataset. Such computational techniques are implemented through algorithms, software, and databases. Several platforms, dedicated software, libraries, and resources are freely available to draw appropriate representations and investigate both biological and mathematical properties of the network [5–7]. For instance, databases offer high-throughput quantitative datasets that are shared by research groups, including whole interactomes for different organisms, or those associated with diseases [8]. Software and algorithms, on the other hand, can be specifically designed to perform many different operations to uncover properties and characteristics of the modeled processes.

One algorithmic approach is to use network analysis, which can be used to explore very complex datasets. An advantage of network analysis over other computational techniques is its ability to quickly add new or updated information. However, once a network is created, it simply represents the *known* interactions between the actors; further biological information (e.g. quantitative data, dynamics, or molecular modifications) is not necessarily integrated within that network. Hence, the design of the network requires consideration of several key aspects, such as cellular processes of interest, the biological context, and the known experimental data. It is of critical importance to recognize that the field of network analysis is composed of mathematical and biological aspects that possess their own methodologies and difficulties. This consideration is of utmost importance to the field, as the networks are not only used to prove a hypothesis or an experimental result, but they can also drive new experimental insights. In summary, there are many advantages and disadvantages to approaching complex signal transduction datasets with network analysis. The specifics of proper use of these computational techniques will be discussed individually throughout this chapter.

7.2 Network Construction

7.2.1 Introduction to Network Construction

The first step of network analysis is to construct an appropriate graphical network representing the biological process of interest. A typical network consists of nodes (any element within the network, e.g. proteins) and edges (relationships or interaction between nodes) (Figure 7.1). The construction of a network can be performed by using different approaches that depend on the available datasets, the cellular process under investigation, the starting hypothesis, and the expected outputs and results. The dataset is usually composed of a group of proteins, genes, or posttranslational modifications involved in an integrated biological function. Many databases are available and are used to infer different kinds of information about biological interactions and processes. In the field of signaling networks, it is possible to query PPI databases and pathway-based databases that provide information about single molecules and whole biological processes; some examples are STRING, BioGRID, IntAct, MIPS, PINA, MINT, DIP, Pathway Commons, PathCards, Reactome, CST Pathway, GeneMANIA, WikiPathways, PSICQUIC, and PID. Furthermore, there are a number of specific databases like Gene Ontology, KEGG, RCSB PDB, and HPRD that describe roles, tasks, and structure of the actors involved in interaction networks; these can be used in a blended approach for enhanced modeling efforts. In summation, when constructing a network, many aspects must be considered, starting from all the available information.

7.2.2 Network Construction from a Probe

A standard approach of network construction is to begin with a set of proteins or genes (called probes). Starting from a probe typically requires some preliminary experimental data that drive the network construction, which can yield unexpected interactions between unforeseen actors and even unpredictable results. Probe-based constructions allow researchers to investigate specific

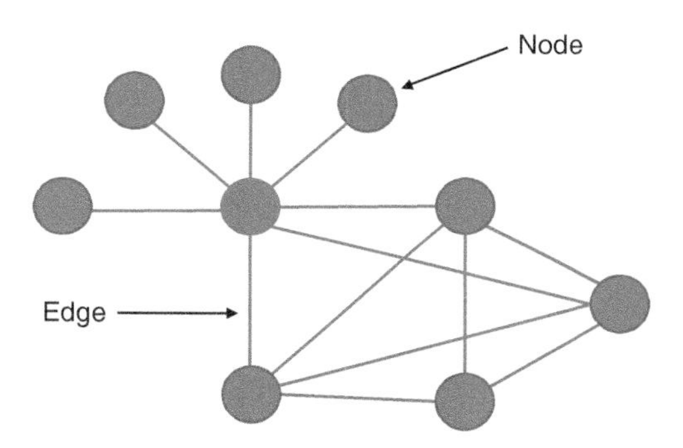

Figure 7.1 A network of nodes (any element in the network) and edges (connections between nodes). *Source:* Reproduced with permission of American Chemical Society.

experimental results in a general context that may not always be reproducible in a wet lab, thus improving the predictive ability while reducing time and costs. In this sense, it is possible to design a complex network that comprises hundreds of biological actors. It is, however, very important to note that when building a network of interactions, the probe values (e.g. level of activation or expression) should be derived from the same experiment. If this is not possible, it is important to exercise caution that nodes/edges are both reproducible and representative of the sample population. These can be collected by the investigator or assembled from literature (ensuring consistency with data collection between samples). Additionally, there are practical considerations that must be exercised when assembling a network from probes derived from literature databases: if the primary probe filter that is required refers to the names of the actors that are investigated, the standardization of the names can become a challenge because various databases use different nomenclature that may prevent the researcher from finding corresponding information about a specific probe (e.g. molecule); synonyms can also present a problem since a molecule could have a standard name and several other synonyms, which can make the query phase more difficult than anticipated. However, once the naming is organized, the database will return all available information about the probe interactions that are related to the molecules of interest. There are several kinds of probe interactions, such as physical interactions, co-expression interactions, interactions that are inferred from the literature by using text-mining algorithms, or experimentally derived interactions. This wealth of information can be further processed by using filters offered only by databases, which select interactions that are considered pertinent by user-defined parameters (e.g. disease specific, animal species). Some argue that experimentally derived interactions are considered more reliable than an interaction obtained by a text-mining algorithm, but such an assumption is not always correct, as the source of the information and experimental design must be taken into account. If the objective is to assemble as many interactions as possible, then it is favorable to include *all* data sources. Nonetheless, careful selection must be performed to incorporate only actual and reliable interactions, thus obtaining a high level of precision by discarding non-validated interactions. Finally, it is important to note that each interaction can be directed (e.g. the direction of the signal transduction matters) or undirected (Figure 7.2), and weighted or unweighted (Figure 7.3), depending on the available data; such information could be useful in depicting a specific process into a highly detailed model.

7.2.3 Mapping Methodology

Another approach to network construction that does not utilize databases, but instead requires an existing network (e.g. interactome), is described as a mapping methodology. To construct a network by using mapping methodology,

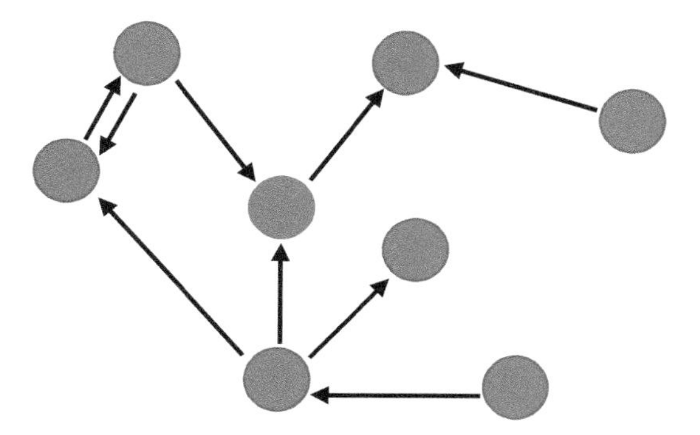

Figure 7.2 A directed network. *Source:* Reproduced with permission of American Chemical Society.

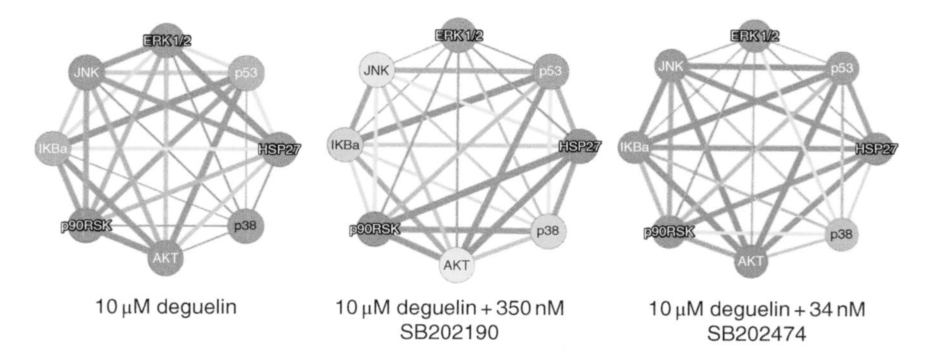

Figure 7.3 Network graphs constructed from protein responses following various chemical exposures. Edge color and thickness are scaled to represent relationships between proteins. Red thick lines correspond to short distances and green thin lines correspond to long distances. Node colors are additionally scaled to represent the weight for each node. Nodes with high values are shown in red and nodes with low values are shown in green. *Source:* Reprinted (adapted) with permission from Currie et al. [9]. Copyright (2014) American Chemical Society.

several computational techniques have been developed: extracting the *neighbors* of the probe nodes or extracting the *shortest paths* connecting them. First, the *neighbors* method starts by identifying the subnetwork that contains the target nodes of the probe (Figures 7.4–7.7). An ideal network will explore new nodes as well as connections among the original nodes associated with the probe. For instance, "first neighbors," or nodes that share a direct interaction, can be identified. A drawback of using the first neighbors method is that the resulting network can omit information about other potential, more distant interactors. This method, however, has the advantage of being quite small and manageable. The dimension of the subnetwork depends on the number of nodes that constitute the probe, so the higher the number of nodes, the bigger the resulting

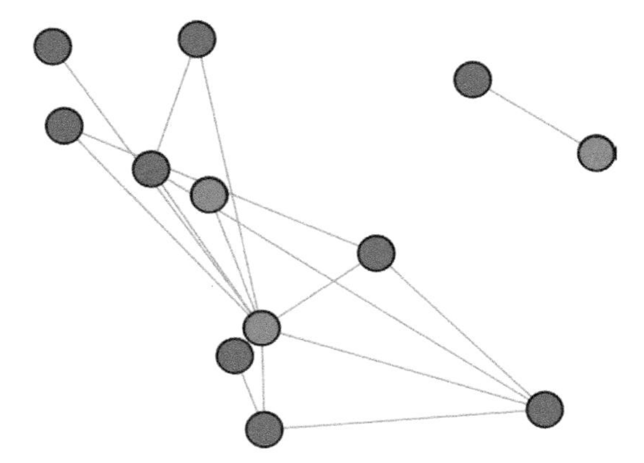

Figure 7.4 The nodes belonging to the probe are colored in red. The three red nodes were mapped into a bigger network, and the network that considered only the first neighbors was extracted. The results show 2 connected components: one is composed of 10 nodes and the other one of only 2 nodes. Analyzing the two-node network is useless, even if it contains a node belonging to the probe, while the other one, which contains two red nodes, can be investigated.

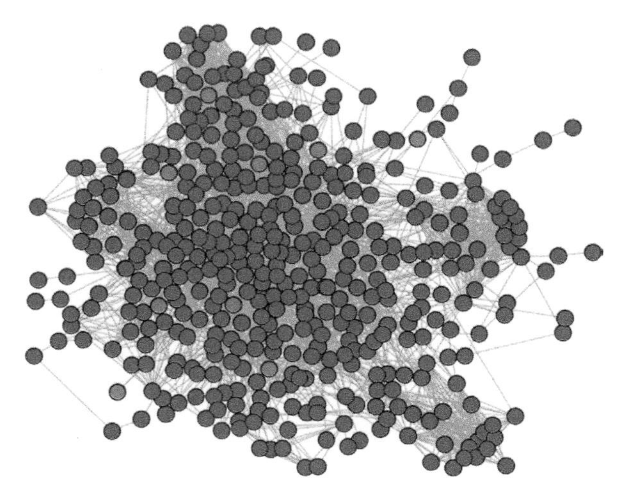

Figure 7.5 The nodes belonging to the probe (in this case, selected by picking some random nodes) are colored in red.

network. This analysis is also beneficial if the network is a single connected component, meaning all the nodes that form the network are able to communicate with all the other nodes (as it is impossible to investigate the mathematical properties of a node that is disconnected and does not have neighbors). The *shortest path* method, on the other hand, means that once the probe is mapped, the network

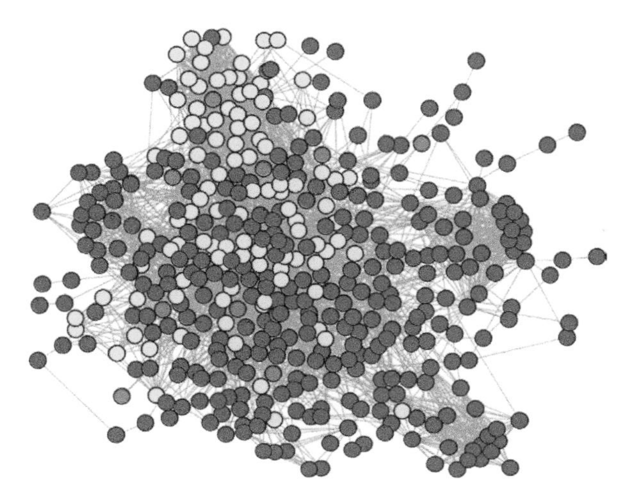

Figure 7.6 The first neighbors of the probe are colored in yellow.

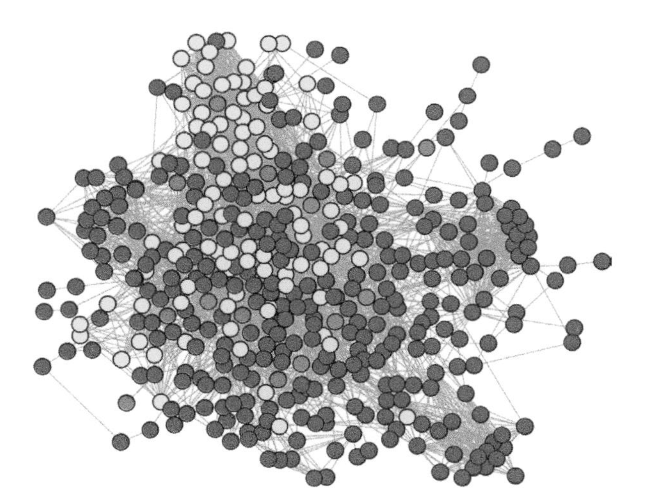

Figure 7.7 The figure shows how it is possible to split a network starting from the selected probe (red in Figure 7.5). When selecting the first neighbors, the network will be composed of more nodes like in Figure 7.6, in which the yellow nodes are the first neighbors of the nodes in the probe. In this figure, nodes that are found to be part of some of the shortest paths that connect the nodes associated with red probes are shown in green. All the colored nodes are found to be part of a short path. Yellow nodes are also first neighbors, while green nodes are not first neighbors.

is used to determine the shortest paths that connect the nodes to the probe by means of all the vertices that constitute the network. The paths that connect the nodes may involve a different number of interactors, compared with the first neighbors methods. Usually this technique leads to bigger sets of nodes, which requires higher computational power for retrieval of the paths of interest.

7.2.4 Small Networks

Small networks are useful for describing functional modules or basic systems that carry out a specific task. There are many molecular pathways known to be involved in specific cellular functions, and such processes can be represented as networks. The number of nodes involved in a pathway is usually very controlled, thus allowing the simulation of a specific cellular behavior. Furthermore, the study of emergent properties could be of interest to understand the dynamics of such systems and to investigate how different conditions impact the connectivity and communication within a network. Another useful method of analysis that can be applied to small networks is to compare the structure of different networks. When obtaining a network from probes, one possible scenario is to obtain several networks related to the same process. For example, you can compare networks of responses categorized into diseased and healthy subjects/samples. Such networks are usually different in structure, protein involvement and activation, or expression level. It would be advantageous to compare these networks to distinguish between healthy and sick patients or between patients with different subtypes of the same pathology. This type of approach, however, cannot be applied to large networks due to the complexity of computation that can require large amounts of time and processing power to be executed.

7.2.5 Large Networks

A large network represents the interaction between different functional modules that communicate by the means of other nodes potentially involved in different biological processes. The interactome, which represents all the interactions that take place in a cell or disease network, identifies all the actors that are known to be part of the pathological process. These big networks require different approaches of analysis, as the high number of nodes that are investigated, as well as the biological complexity, makes these networks difficult to interpret. Therefore, it is important to reduce the complexity of such systems as much as possible in order to extract useful information, and different mathematical tools are available to tackle such challenges. Biological processes that are modeled as graphs inherit all the properties that come from such a configuration, making it possible to investigate functional modules, recurrent patterns, or the role of a node within the network through evaluation of graphical parameters. In other words, different properties (global and local) of the graph can be investigated. These parameters will be introduced and further discussed in the next sections.

7.3 Facing the Network Analysis

A few concepts regarding the benefits, outcomes, and constructions of networks have been introduced in the previous sections, but a deep understanding of their roles and potential biological meaning is necessary to interpretation of the network, and will be discussed in the remainder of this chapter. Biological networks are best evaluated and described by parameters known as node centralities, network centralities, and clusters. Node centralities identify the most important nodes in the network, network centralities provide a global description of the network, and clusters identify particularly dense regions of nodes that are considered important for network functionality due to their structure (e.g. high numbers of edges). One of the most common and well-known properties of biological networks is called scale-freeness [10]; a network is scale-free if a relative change of one node results in a proportional relative change in a second node. This property is associated with an interesting behavior of biological networks, called preferential attachment, which states that nodes with a higher number of neighbors or edges (degree) have a higher probability to attract new interactions, compared with nodes with a lower degree [11]. To better understand the meaning of the most common network parameters, it is useful to think about biological networks as road maps where the towns are the proteins and the roads connecting the towns are the interactions (also known as edges) between the proteins. Consider the map of Spain in Figure 7.8. It is challenging to identify the most important nodes on the road map. For example, Madrid (in the red circle) is an important node, as it has a lot of neighbors. It is also central in the sense that it is not far from most of the towns on the map. We can conclude that Madrid is an important (central) node in the network. On the other hand, there are other towns that are important, even if they do not have the geographically privileged position of Madrid. For instance, Valencia (blue circle) is not in the center of the map, and many towns are closer to Madrid than to Valencia. Valencia is, however, still an important connecting node between the northeast (between France and the south of Spain). Blocking the traffic in Valencia will result in blocking most of the traffic from the north to the south and vice versa. So, Valencia is an important connecting node, even if it is not in the "center" of the map. Therefore, we can argue that Madrid corresponds to a high degree and a high closeness node, while Valencia corresponds to a node with high betweenness and stress value. These parameters, and many others, are discussed in detail throughout the following sections. Each one reflects a particular topological property that can be interpreted as a particular biological function in a protein network, depending on the context and aim of the analysis.

Figure 7.8 Road maps as networks. This map has been produced using Google maps.

7.3.1 Centralities Definition and Description

Centralities are indices used to describe the role of a node in a network. Some centralities are global parameters describing the whole network. Others are local parameters describing the importance of single nodes.

Preliminary Definitions [12]

- Let $G = (N,E)$ for a directed (if edges have direction) or an undirected (if edges have no direction) graph, with $n = |N|$ vertices.
- $\deg(v)$ indicates the degree of the vertex.
- $\text{dist}(v,\omega)$ is the shortest path between v and ω.
- σ_{st} is the number of shortest paths between s and t.
- $\sigma_{st}(v)$ is the number of shortest paths between s and t passing through the vertex v.

Important Concepts to Keep in Mind

- Vertex = nodes; edges = arches.
- The distance between two nodes, $\text{dist}(v,\omega)$, is the shortest path between the two nodes.
- All calculated scores are computed, giving higher values a positive meaning, where positive refers to node proximity. Thus, independent of the calculated node centrality, higher scores indicate proximity, and lower scores indicate remoteness of a given node v from the other nodes in the graph.

7.3.2 Global Parameters

7.3.2.1 Diameter (ΔG)

The diameter is the maximal distance (using shortest paths between nodes) among all the distances calculated between each pair of vertices in a graph. It indicates how far the two most distant nodes are within the network. It can describe graph compactness, regarding the overall proximity between nodes. A high graph diameter indicates that the two nodes determining that diameter are very distant, implying little graph compactness. However, it is possible that two nodes are very distant, thus giving a high graph diameter, but several other nodes are not. Therefore, a graph could have high diameter and still be rather compact or have very compact regions. Thus, a high graph diameter can be misleading in terms of evaluation of graph compactness. In contrast, a low graph diameter is much more informative and reliable. A low diameter may indicate that all the nodes are in close proximity and the graph is compact. In quantitative terms, high and low are better defined when compared to the total number of nodes in the graph. A low diameter of a very big graph (with hundreds of nodes) is much more meaningful in terms of compactness than a low diameter of a small graph (with few nodes) (Figure 7.9).

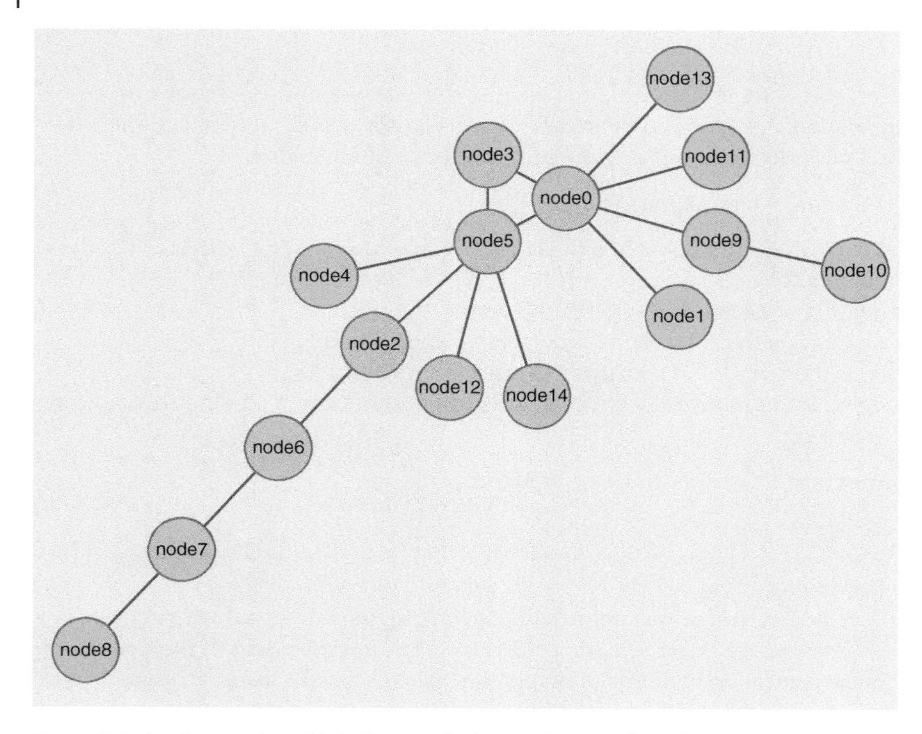

Figure 7.9 A network where high diameter is due to a low number of nodes.

The diameter of a biological network (e.g. protein signaling network) can be interpreted as the overall easiness of the proteins to communicate and/or influence their reciprocal function. It could also be a sign of functional convergence. A large protein network with low diameter may suggest that the proteins within the network had a functional coevolution. The diameter should be carefully weighted if the graph is not fully connected (that is, there are isolated nodes).

7.3.2.2 Average Distance

$$A_v D_G = \frac{\sum_{i,j \in N} \text{dist}(i,j)}{n(n-1)}$$

where n is the number of nodes in G.

The average distance is the average shortest path of a graph, and it is calculated by taking the sum of all shortest paths between pairs of vertices and dividing by the total number of nodes. Being an average, it can be somewhat more informative than the diameter, and it can also be considered a general

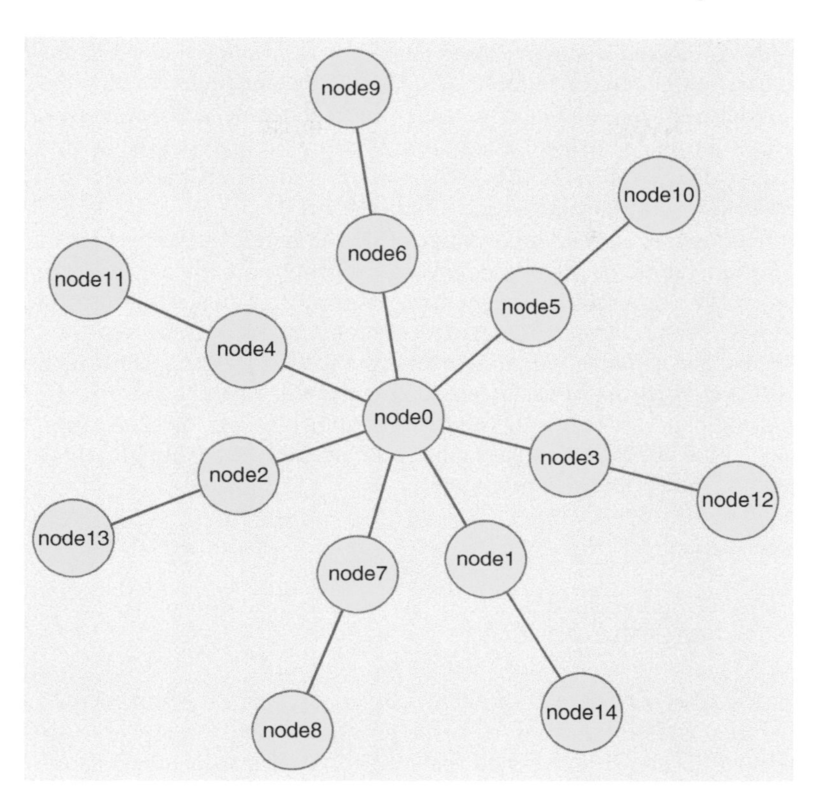

Figure 7.10 A network with low diameter and average distance. The network is "compact."

indicator of network navigability. A high average distance indicates that the nodes are distant (disperse), implying little graph compactness. In contrast, a low average distance indicates that all the nodes are in proximity and the graph is compact (Figure 7.10). Like the diameter, high and low values are better defined when compared to the total number of nodes in the graph.

The biological meaning of the average distance is similar to the diameter. A large protein network with low average distance may suggest that the proteins within the network have a tendency to generate functional complexes and/or modules (although additional centrality indices should be also calculated to support that indication).

7.3.3 Local Parameters

7.3.3.1 Degree

The degree is the simplest topological index, corresponding to the number of nodes adjacent to a given node v, where "adjacent" means directly connected.

Those directly connected nodes are also called first neighbors of v. Thus, the degree also corresponds to the number of adjacent incident edges. In directed networks, we define in-degree, when the edges target the node, and out-degree, when the edges target its adjacent neighbors. Nodes with high degree (highly connected) are called hubs and hold together several nodes with lower degrees. Natural networks usually contain a high number of hubs.

The degree allows an immediate evaluation of the regulatory relevance of the node. For instance, in signaling networks, proteins with a very high degree interact with several other signaling proteins, thus suggesting central regulatory roles. Groups of very-high-degree proteins are likely to be represent regulatory hubs. For instance, signaling proteins encoded by oncogenes, such as HRAS, SRC, or TP53, are hubs. Depending on the nature of the protein, the degree could indicate a central role in amplification (kinases), diversification and turnover (small GTPases), signaling module assembly (docking proteins), gene expression (transcription factors), etc.

7.3.3.2 Eccentricity

$$C_{\text{ecc}}(v) = \frac{1}{\max\{\text{dist}(v,\omega) : \omega \in N\}}$$

The eccentricity of a node v is calculated by computing the shortest path between a node v and all other nodes in the graph; then the longest of those shortest paths is chosen. Once this path is identified, its reciprocal is calculated. By doing that, an eccentricity with higher value assumes a positive meaning in terms of node proximity. Indeed, if the eccentricity of node v is high, this means that all other nodes are in proximity. In contrast, if the eccentricity is low, this means that there is at least one node (and all its neighbors) that is far from node v. Of course, this does not exclude that several other nodes might be much closer. Thus, eccentricity is a more meaningful parameter if its value is high. Notably, high and low values are more significant when compared to the average eccentricity of the graph. Average eccentricity is calculated by averaging the eccentricity values of all nodes in the graph.

The eccentricity of a node in a biological network, for instance, a protein signaling network, can be interpreted as the ease of a protein to be functionally reached by all other proteins in the network. Thus, a protein with high eccentricity, compared with the average eccentricity of the network, will be more easily influenced by the activity of other proteins (i.e. the protein is subject to a more stringent or complex regulation) or, conversely, could easily influence several other proteins. In contrast, a low eccentricity, compared with the average eccentricity of the network, could indicate a marginal functional role (although this should also be evaluated with other parameters and contextualized to the network annotations).

7.3.3.3 Closeness

$$C_{\text{clo}}(\upsilon) = \frac{1}{\sum_{\omega \in N} \text{dist}(\upsilon, \omega)}$$

The closeness of a node υ is calculated by computing the shortest path between the node and all other nodes in the graph, summing these shortest paths, and then calculating the reciprocal of the sum. Typically, high and low values are more meaningful when compared to the average closeness of the graph – calculated by averaging the closeness values of all nodes in the graph. Notably, high values of closeness should indicate that all other nodes are in proximity to node υ. In contrast, low values of closeness should indicate that all other nodes are distant from node υ. However, a high closeness value can be determined by the presence of few nodes very close to node υ, with others much more distant, or by the fact that all nodes are generally very close to υ. Likewise, a low closeness value can be determined by the presence of few nodes very distant from node υ, with other much closer, or by the fact that all nodes are generally distant from v. Thus, the closeness value should be considered as an average tendency to node proximity or isolation. Closeness should be always compared to the eccentricity. A node with high eccentricity and high closeness is very likely to be central in the graph (Figure 7.11).

The closeness of a node in a biological network, for instance, a protein signaling network, can be interpreted not only as the probability of a protein to be functionally relevant to several other proteins but also with the possibility to be irrelevant for few other proteins. Thus, a protein with high closeness, compared with the average closeness of the network, will be easily central to the regulation of some proteins with the possibility that other proteins will not be influenced by its activity. It should be noted that exploring proteins with low closeness (compared with the average closeness of the network) may appear less relevant for a specific network, but these could still hold immense value as intersecting boundaries with other networks. Accordingly, a signaling network with a very high average closeness is more likely organizing functional units or modules, whereas a signaling network with very low average closeness will behave more likely as an open cluster of proteins connecting different regulatory modules.

7.3.3.4 Radiality

$$C_{\text{rad}(\upsilon)} = \frac{\sum_{\omega \in N} \left(\Delta G + 1 - \text{dist}(\upsilon, \omega) \right)}{n-1}$$

The radiality of a node υ is calculated by computing the shortest path between the node υ and all other nodes in the graph. The value of each path is then

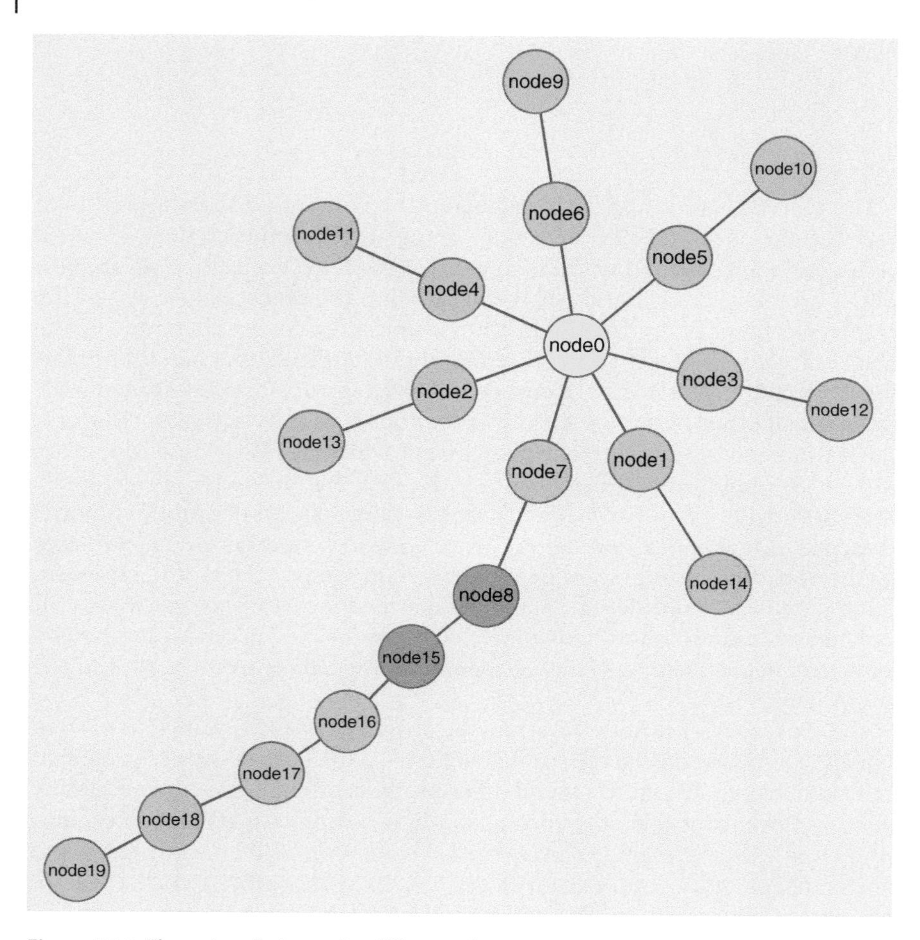

Figure 7.11 The network shows the difference between eccentricity and closeness. The values of eccentricity are node 0 = 0.14, node 8 = 0.2, and node 15 = 0.2. The closeness values are node 0 = 0.021, node 8 = 0.017, and node 15 = 0.014. In this case node 0 is closer than node 8 and node 15 to the most of nodes in the graph. The eccentricity value of node 0 is smaller than the value of node 8 and node 15, but this is due to the small number of nodes. If these nodes represent proteins, this probably means that node 0 is fundamental for most of reaction in the network and that node 8 and node 15 are important only for reactions between a few proteins.

subtracted by the value of the diameter +1 ($G+1$), and the resulting values are summed. Finally, the obtained value is divided by the number of nodes ($n-1$). As the diameter is the maximal possible distance between nodes, subtracting the shortest paths between the node v and its neighbors from the diameter will give high values if the paths are short and low values if the paths are long.

Overall, if the radiality is high, this means that the node is generally closer to the other nodes, with respect to the diameter. Conversely, if the radiality is low, the node is peripheral. Also, high and low values are more meaningful when compared to the average radiality of the graph calculated by averaging the radiality values of all nodes in the graph. Similar to closeness, the radiality value should be considered as an average tendency toward node proximity or isolation, instead of an assessment of an individual node. The radiality should be compared to the closeness and to the eccentricity: a node with high eccentricity, high closeness, and high radiality is a consistent indication of a high central position in the graph.

The radiality of a node in a biological network, for instance, a protein signaling network, can be interpreted as the probability of a protein to be functionally relevant for several other proteins, but with the possibility to be irrelevant for few other proteins. Thus, a protein with high radiality, compared with the average radiality of the network, is central to the regulation of some proteins, while other proteins may not be influenced by its activity. It may also be of interest to analyze proteins with low radiality, compared with the average radiality of the network, as these proteins, although less relevant for that specific network, are possibly behaving as intersecting boundaries with other networks. Accordingly, a signaling network with a very high average radiality is more likely organizing functional units or modules, whereas a signaling network with very low average radiality will behave more likely as an open cluster of proteins connecting different regulatory modules. All these interpretations should be accompanied with the evaluation of eccentricity and closeness.

7.3.3.5 Centroid Value

$$C_{\text{cen}}(\upsilon) = \min\{f(\upsilon,\omega) : \omega \in N\{\upsilon\}\}$$

where $f(\upsilon, \omega) = \gamma_\upsilon(\omega) - \gamma_\omega(\upsilon)$ and $\gamma_\upsilon(\omega)$ is the number of vertices closer to node υ than to ω. The centroid value is one of the most complex node centralities. It is computed by focusing on pairs of nodes (υ,ω) and systematically counting the nodes that are closer (in terms of shortest path) to υ or to ω (Figure 7.12). The centroid value provides a centrality index that is always weighted with the values of all other nodes in the graph. Indeed, the node with the highest centroid value is also the node with the highest number of neighbors, not only first, compared with all other nodes. In other terms, a node υ with the highest centroid value is the node with the highest number of neighbors separated by the shortest path to υ. The centroid value suggests that a specific node has a central position within a graph region characterized by a high density of interacting nodes. Again, high and low values are more meaningful when compared

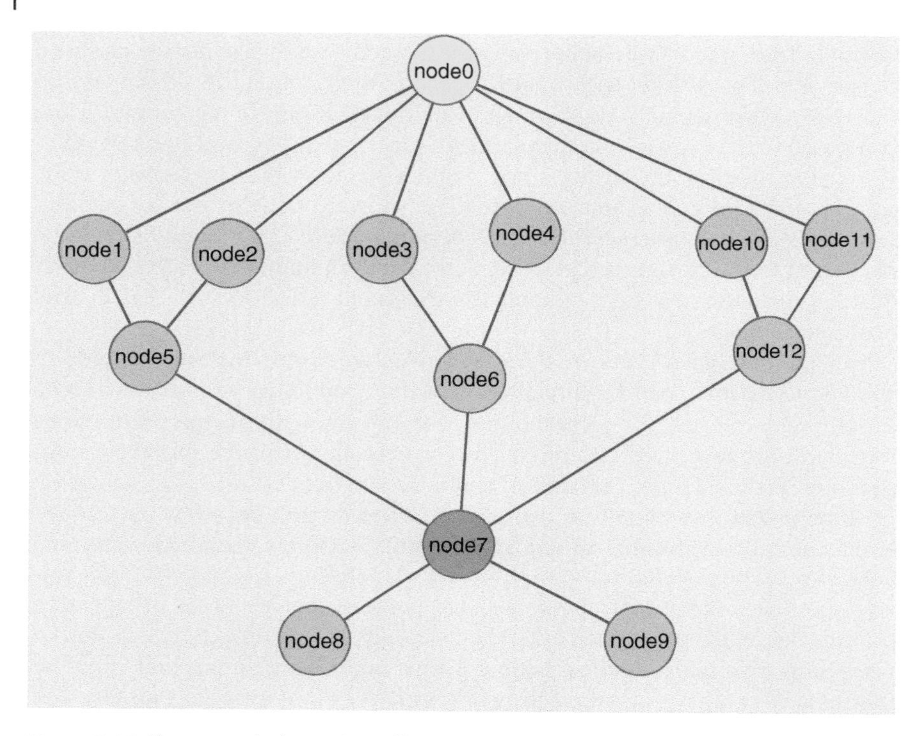

Figure 7.12 The network shows the difference between centroid and closeness. Here, node 0 has the highest centroid value (centroid = 1, closeness = 0.04), and node 7 has highest closeness value (centroid = −1, closeness = 0.05).

to the average centrality value of the graph G calculated by averaging the centrality values of all nodes in the graph.

The centroid value of a node in a biological network, for instance, a protein signaling network, can be interpreted as the probability of a protein to be functionally capable of organizing discrete protein clusters or modules. Thus, a protein with high centroid value, compared with the average centroid value of the network, may be possibly involved in coordinating the activity of other highly connected proteins, altogether devoted to the regulation of a specific cell activity (for instance, cell adhesion, gene expression, proliferation, etc.). Accordingly, a signaling network with a very high average centroid value is more likely organizing functional units or modules, whereas a signaling network with very low average centroid value will behave more likely as an open cluster of proteins connecting different regulatory modules. It can be useful to compare the centroid value to algorithms detecting dense regions in a graph, indicating protein clusters, such as MCODE [13].

7.3.3.6 Stress

$$C_{\text{str}}(\upsilon) = \sum_{s \neq \upsilon \in N} \sum_{t \neq \upsilon \in N} \sigma_{st}(\upsilon)$$

Stress is calculated by measuring the number of shortest paths passing through a node. To calculate the stress of a node υ, all shortest paths in a graph are calculated, and then the number of shortest paths passing through υ is counted. A stressed node is a node traversed by a high number of shortest paths. Notably, and importantly, a high stress value does not automatically imply that the node υ is critical to maintain the connection between nodes whose paths are passing through it. Indeed, it is possible that two nodes are connected by means of other shortest paths without passing through the node υ. High and low values are more meaningful when compared to the average stress value of the graph.

The stress of a node in a biological network, for instance, a protein signaling network, can indicate the relevance of a protein as functionally capable of holding together communicating nodes. The higher the stress value, the higher the relevance of the protein in connecting regulatory molecules. Due to the nature of this centrality, it is possible that the stress simply indicates a molecule heavily involved in cellular processes but not relevant to maintain the communication between other proteins.

7.3.3.7 S.-P. Betweenness

$$C_{\text{spb}}(\upsilon) = \sum_{s \neq \upsilon \in N} \sum_{t \neq \upsilon \in N} \delta_{st}(\upsilon)$$

where

$$\delta_{st}(\upsilon) = \frac{\sigma_{st}(\upsilon)}{\sigma_{st}}$$

The S.-P. betweenness is a node centrality index. It is similar to stress, but this parameter provides a more elaborate and informative centrality index. The betweenness of a node n is calculated by considering couples of nodes ($\upsilon 1$; $\upsilon 2$) and counting the number of shortest paths linking $\upsilon 1$ and $\upsilon 2$ passing through a node n. Then, the value is related to the total number of shortest paths linking $\upsilon 1$ and $\upsilon 2$. Thus, a node can be traversed by only one path linking $\upsilon 1$ and $\upsilon 2$, but if this path is the only one connecting $\upsilon 1$ and $\upsilon 2$, the node n will score a higher betweenness value. Thus, a high S.-P. betweenness score means that the node, for certain paths, is crucial to maintaining node connections. To know the number of paths for which the node is critical, it is also necessary to calculate the stress. Thus, stress and S.-P. betweenness can be used to gain complementary information (Figure 7.13). Further information could be gained by

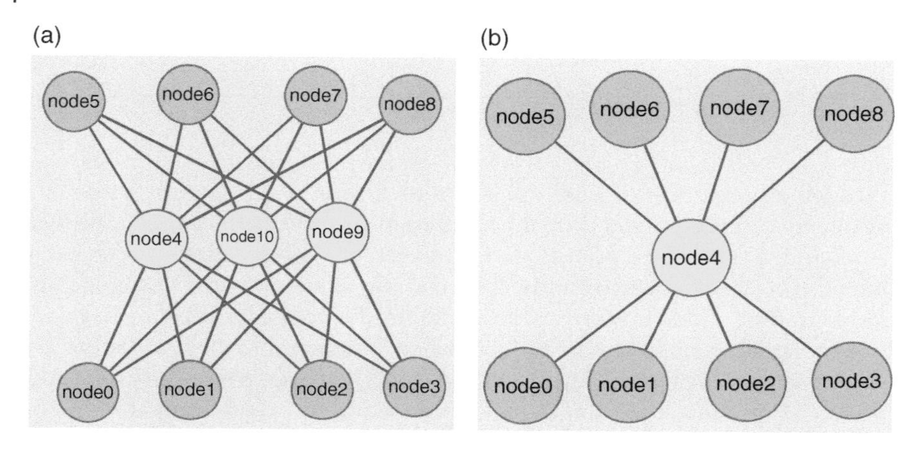

Figure 7.13 Betweenness vs. stress. In figure (a), node 4, node 10, and node 9 present high values of stress (=56) and the same value of betweenness (=18.67). In figure (b), node 4 presents the same value of stress observed in figure (a) but higher value of betweenness (=56). This is because the number of shortest paths passing through node 4 is the same in the two networks. However, in the second network, node 4 is the only node connecting the two parts of the network. In this sense, betweenness is more precise than stress in describing how the node is fundamental to the network. If we remove node 4 in figure (a), the connection between the nodes in the network does not change significantly. If we remove node 4 from figure (b), then the network is completely disconnected.

referencing the S.-P. betweenness to node couples, thus quantifying the importance of a node for two connected nodes. Again, high and low values are typically more meaningful when compared to the average S.-P. betweenness value of the graph G, calculated by averaging the S.-P. betweenness values of all nodes in the graph.

The S.-P. betweenness of a node in a biological network, for instance, a protein signaling network, can indicate the relevance of a protein as functionally capable of holding together communicating proteins. The higher the value, the higher the relevance of the protein as organizing regulatory molecule. The S.-P. betweenness of a protein effectively indicates the capability of a protein to bring in communication between distant proteins. In signaling modules, proteins with high S.-P. betweenness are likely crucial to maintain functionally and coherence of signaling mechanisms.

7.3.3.8 Eigenvector

The eigenvector centrality assigns relative scores to all nodes in the network based on the concept that connections to high-scoring nodes contribute more to the score of the node in question than equal connections to low-scoring nodes. The definition is a recursive definition: a high eigenvector value means that a node has several neighbors with high eigenvector value, which can be simplified to the degree of connectedness of a node within a module. So, the eigenvector centrality can be viewed as a sort of weighted degree, where not only the number of the neighbors is important but also the score (the eigenvector itself) of the neighbors.

Similar to the degree, eigenvector allows an immediate evaluation of the regulatory relevance of the node. A protein with a very high eigenvector is a protein interacting with several important proteins, thus suggesting a central regulatory role for a module. A protein with low eigenvector can be considered a peripheral protein, interacting with few noncentral proteins.

7.3.3.9 Bridging Centrality

The bridging centrality of a node is the product of the bridging coefficient (BC) and the betweenness centrality (Btw), shown by the formula BC*Btw. The BC of a node is high if the node has neighbors with high degree. A node has high bridging centrality if it is a connecting node (due to the Btw component) of high-degree nodes (due to BC component). This can suggest that the node is a connecting node between clusters or between dense regions of the network.

The Btw component of the bridging centrality suggests that in a biological network such as a PPI network, a protein with high bridging centrality value is functionally capable of holding together communicating proteins. The BC component may also suggest that such communicating proteins are regulatory proteins, since they interact with many other proteins. Thus, a protein with high bridging centrality is a protein possibly bringing in communication sets of regulatory proteins. In the context of directed networks, a high value of bridging centrality may indicate that a protein is regulated by many proteins at once or is regulating the overall architecture of many clusters.

7.3.3.10 Edge Betweenness

The edge betweenness centrality is an edge centrality index. It is the edge version of the node betweenness centrality. Given an edge E, edge betweenness is calculated considering couples of nodes ($v1$, $v2$) and counting the number of shortest paths linking $v1$ and $v2$ that pass through E. Then, the value is related to the total number of shortest paths linking $v1$ and $v2$. Thus, an edge can be traversed by only one path linking $v1$ and $v2$, but if this path is the only path connecting $v1$ and $v2$, the edge will score a higher betweenness value. Thus, a high edge betweenness score means that the edge, for certain paths, is crucial to maintaining node connections.

Since the edge betweenness is an edge score of centrality, with the edge indicating the functional relationship between two nodes (for instance, a kinase-triggered phosphorylation), computation of this centrality moves the focus from node relevance to a biochemical process relevance, thus indicating that a specific biochemical reaction (or any other type of functional relationship) has a central role in the network functional organization. The higher the edge betweenness value, the higher the relevance of the interaction as organizing regulatory process. In signaling modules, interactions with high edge betweenness are likely prevalent to maintain function and coherence of signaling mechanisms. In a metabolic network, a high edge betweenness value is related to the relevance of a reaction in energy and mass processing. Overall, edge betweenness quantifies the capability of a certain reaction to be an information (signaling) or mass (metabolic) transit reaction.

7.3.3.11 Normalization and Relative Centralities

Once centralities have been computed, the question that arises is: what does it mean to have a centrality of, for example, 0.4 for a node? This clearly depends on different parameters, such as the number of nodes in the network, the maximum value of the centrality, or the topological structure of the network. In order to compare centrality scores between the elements of a graph or between the elements of different graphs, a normalization of centrality values is needed. A common normalization method applicable to most centralities is to divide each value by the maximum centrality value or by the sum of all values.

7.3.4 Clusters

Another interesting characteristic of networks is clustering: a cluster is a group of nodes that are highly connected. To identify nodes with high connectivity, the clustering coefficient parameters can be used. The clustering coefficient of a node i is the fraction of the number of triangles connected to i over the number of possible triangles connected to i (Figure 7.14). Even if the clustering coefficient is a mathematical property that can be easily calculated, the notion of cluster remains quite indefinite. Given a graph, there can be several different solutions to find clusters, since, by definition, a "high density region" and a "group of highly connected nodes" are notions that are not rigorous by themselves. So, depending on the chosen algorithm and on the parameters set for

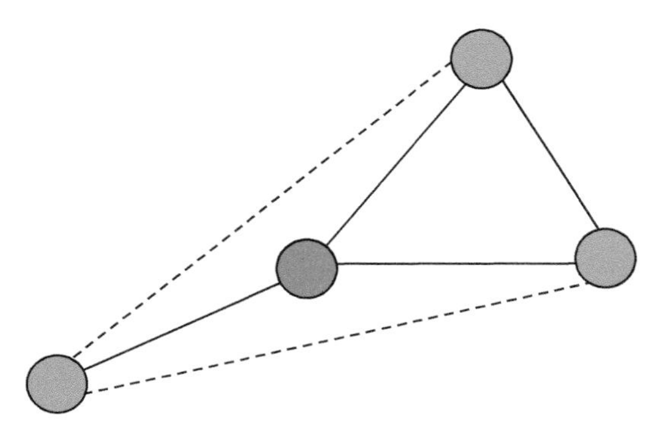

Figure 7.14 The clustering coefficient of the red node is 1/3. It has degree equals to 3 so there are three possible triangles that can be formed. Only the one with the solid line is realized.

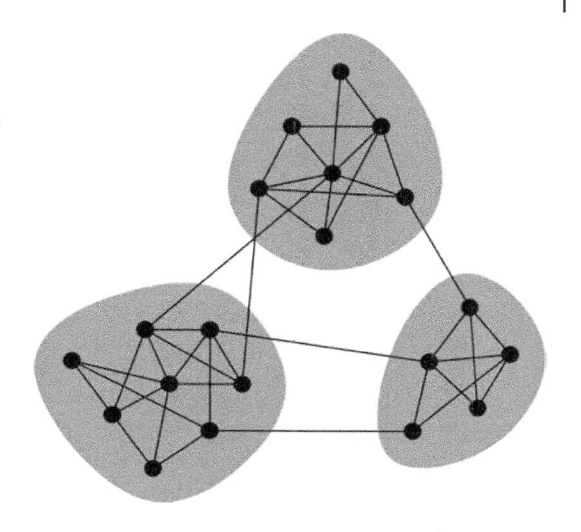

Figure 7.15 A possible clustering solution for a graph with three clusters. Even if it appears like a good solution, one or more nodes could be moved from one cluster to another without affecting the "quality" of the solution, producing an alternative option as valid as the previous one.

such algorithm (if any), we can find one large cluster or more than one cluster. Sometimes, there are nodes that are a part of two different clusters or on the border between two clusters; this phenomenon can be caused by the data or the clustering algorithm. This scenario is similar to the identification of a mostly sunny or a partly cloudy day. If there are no clouds, we all agree that it is a sunny day; if the sky is completely covered by the clouds, everybody agrees that it is a cloudy day. However, most days have some clouds and some sun, so how do we distinguish the states? It depends on the observational data (how many clouds, and in what spatial orientation), but it also depends on the discretion and knowledge of the observer. Figure 7.15 shows a possible clustering solution for a graph. But if we move one node from one cluster to a near one, we obtain another good solution. Depending on our knowledge of the biological process described by the network, we can decide to move one node to one cluster to another or to consider them as connecting nodes between different clusters. Biological networks are usually structured in one or more clusters, which are formed of many proteins that are similar in structure and interactions. Often in signal transduction, this manifests itself as a multi-functional protein, which means that there exists a group of proteins that have more than one interacting partner and are capable of carrying out similar functions. If one of these protein types is missing or damaged for some reason, its function can be easily replaced by a similar one. This is an evident evolutionary advantage of the clustering choice of the protein network evolution, and this partially demonstrates why clustering regions correspond to crucial functions of certain biological process. Such crucial functions require the robustness that is guaranteed by the clustering structure.

7.4 Employing Centrality Analysis to Evaluate Stressed Biological Systems

This chapter introduced a number of network parameters and offers a means to interpret networked responses. It is important to highlight that the utilization of these centrality parameters is rising in current research aimed at investigating complex biological systems under stressful conditions. For instance, perturbations of biological pathways or networks are well known to occur as result of toxic chemical exposures. As such, toxicity testing is shifting from common endpoint investigations to pathway-based approaches, where the degree of perturbation is monitored. For instance, Currie et al. induced mitochondrial stress in human hepatocytes (HepG2) through exposure to deguelin, a complex I inhibitor. Relative phosphorylation responses of various proteins were measured and formed into a network, and parameters describing the network perturbations revealed the dynamic relationship between key proteins – p38, c-Jun N-terminal kinase (JNK), and extracellular signal-related kinase 1/2 (ERK1/2) – under conditions of mitochondrial stress. The network revealed shifts in the relationships between these MAPK pathways at different doses of deguelin (Figure 7.16) [9]. Network centrality analysis has also been used to investigate cellular networks involved in injury and repair. Intra- and extracellular regulators of healing mechanisms, such as cytokines, signaling proteins, and growth factors, have been described to possess significant roles in facilitating optimal recovery. Han et al. explored a collection of 30 responses composed of cytokines, proteins, phosphorylated proteins, and caspase-3, measured in skeletal muscle tissue following a traumatic injury. A specific centrality parameter, radiality, identified responses that were substantially amplified and significant to the integrated biological tissue response. A spatial and temporal understanding of the injury response was acquired by tracking changes in radiality associated with key actors (Figures 7.17 and 7.18). Notably,

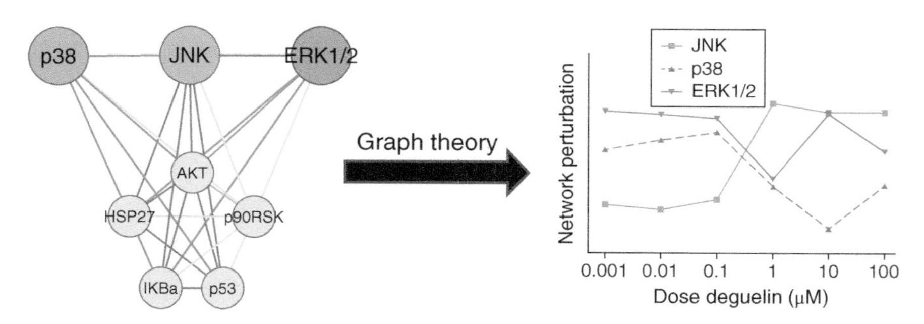

Figure 7.16 Construction of a protein signal transduction network resulting from a chemical exposure to deguelin (complex I inhibitor). Derived centrality parameters suggest signaling perturbations, revealing shifts within the dynamic molecular relationship. *Source:* Reprinted (adapted) with permission from Currie et al. [9]. Copyright (2014) American Chemical Society.

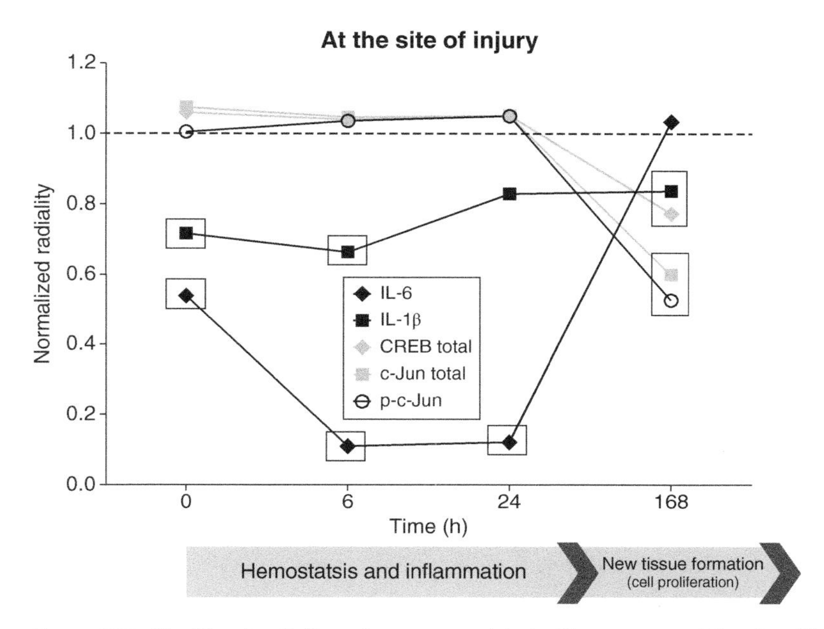

Figure 7.17 Significant radiality outcomes associated with responses at the site of fracture. Nodes that possessed significant radiality outcomes (significance threshold: average radiality ± standard deviation) were plotted over time. Five nodes (IL-6, IL-1β, CREB, c-Jun, p-c-Jun) from networks describe responses at the site of fracture that were significant (in boxes below graph) at varying times.

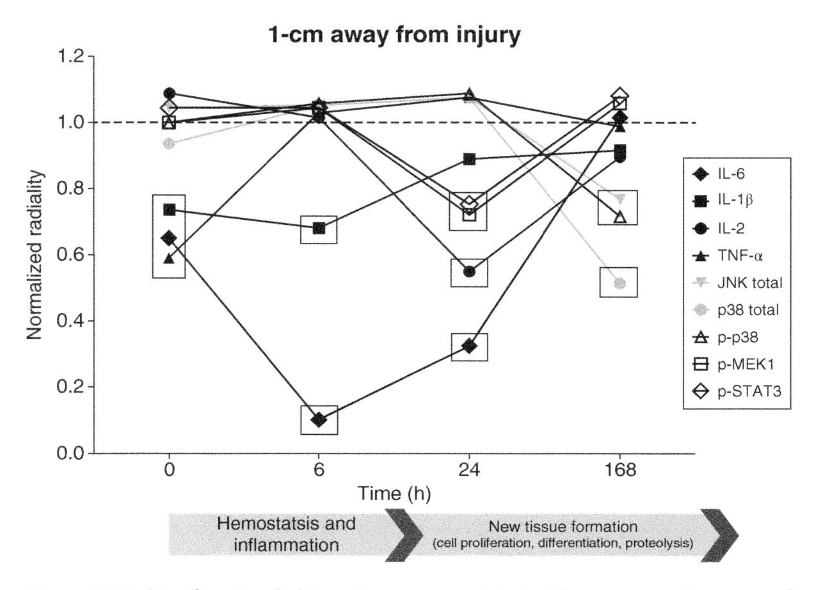

Figure 7.18 Significant radiality outcomes associated with responses 1 cm away from the site of fracture. Nodes that possessed significant radiality outcomes (significance threshold: average radiality ± standard deviation) were plotted over time. Nine nodes (IL-6, IL-1β, IL-2, TNF- α, JNK, p38, p-p38, p-MEK1, p-STAT3) from networks associated with injury 1 cm away from the fracture were significant (in boxes below graph) at varying times.

distinct radiality profiles were observed between tissue samples subjected to varying extents of injury, demonstrating the advantages of using a network analysis approach to extract significant information from a complex dataset [14].

7.5 Interference Notion: How to Perform Virtual Knockout Experiments on Biological Networks

In complex networks, single nodes are not isolated entities but rather are functionally related to other nodes. A node can affect the function of a group of nodes, but, at the same time, it may have no influence on other nodes in the network. Conversely, a node can be affected by some nodes more than others. This reciprocal influence between nodes depends on their function as well as on the network's structure. A node can directly affect the function of its neighbors and indirectly affect the function of distant nodes in an undirected way, based on cascades of interactions that depend on the topology of the network. The centrality-based notions of interference and robustness allow evaluation of the influence between nodes or group of nodes, reciprocally affecting their own functionality. Conceptually, they are based on a sort of virtual knockout experiment on the network: a node is removed from the network, and the effects of such removal on the network structure are analyzed (Figure 7.19).

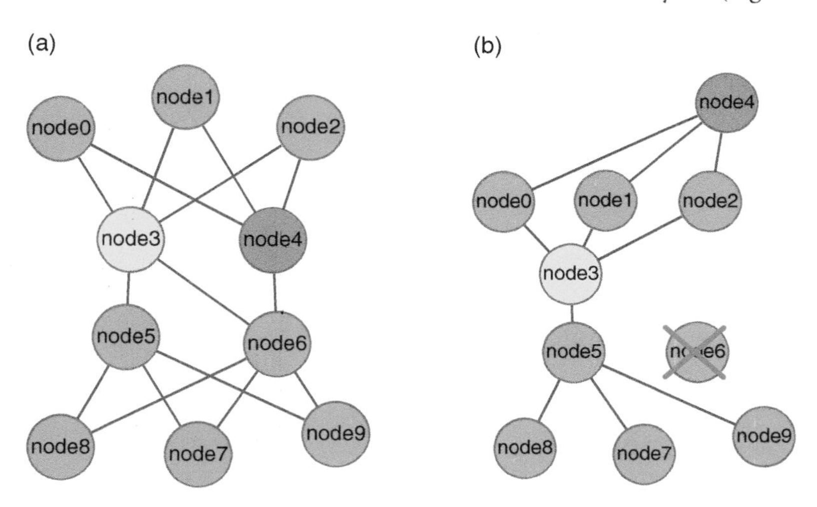

Figure 7.19 Positive and negative interference. In figure (a), node 3 and node 4 are the nodes connecting the top of the network with the bottom. In figure (b), node 6 has been removed: node 4 becomes a peripheral node, and its betweenness decreases. The presence of node 6 is important for node 4 to play a central role (positive interference). At the same time, node 3 and node 5 become fundamental connections between the top and the bottom. Their betweenness values increases. The presence of node 6 in the network on the left damages the "central role" of node 3 and node 5 (negative interference).

From a node-centered perspective, centralities are advantageous parameters to evaluate in order to detect the effects of a single node alteration. As the centrality value of a node is strictly dependent on the network structure and on the properties of other nodes in the network, the consequences of a node deletion are well captured by the variation on the centrality values of all the other nodes. Notably, this kind of approach can model common situations in which nodes are removed or added from/to a physical network. In a biological network, one or more nodes (genes, proteins, metabolites) are possibly removed from the network due to gene deletion, pharmacological treatment, or protein degradation. For instance, in the case of a pharmacological treatment, it is possible to infer side effects of a drug by looking at the topological properties of nodes in a drug-treated network, meaning a network in which a drug-targeted node (protein) was removed. Similarly, we can simulate the consequences of gene deletions, which implies loss of coding genetic material and corresponding encoded proteins, thus resulting in the removal of one or more nodes from the network. If a node (A), upon removal from the network of a specific node (B) or of a group of nodes, *decreases* its value for a certain centrality index, its interference value is positive. This means that this node (A), topologically speaking, takes advantage (is positively influenced by) of the presence in the network of the node (B) or of the respective group of nodes. Thus, removal of node (B) or of that group of nodes from the network negatively affects the topological role of the node (A). If a node (A), upon removal from the network of a specific node (B) or of a group of nodes, *increases* its value for a certain centrality index, its interference value is negative. This means that this node (A), topologically speaking, is disadvantaged (is negatively influenced) by the presence in the network of the node (B) or of that group of nodes. Thus, removal of node (B) or of that group of nodes from the network positively affects the topological role of node (A). This is called negative interference.

7.5.1 Integrating Experimental Dataset into a Topological Analysis

As described in the previous sections, when constructing a network from an experimental probe, we can utilize expression or activation levels of each protein. These data are important and should be considered in a bioinformatics approach involving both lab and *in silico* experiments. The first method is to plot topological analysis and lab experiment analysis in a graph as seen in Figure 7.20. This can be a good first step, but it is not enough to capture all the information included in combined virtual and real protein networks. The next two sections of this chapter will explain how experimental data can be included in the computation of network centralities in order to perform a more powerful and informative analysis. All the features of this section are provided by the CentiScaPe app for Cytoscape.

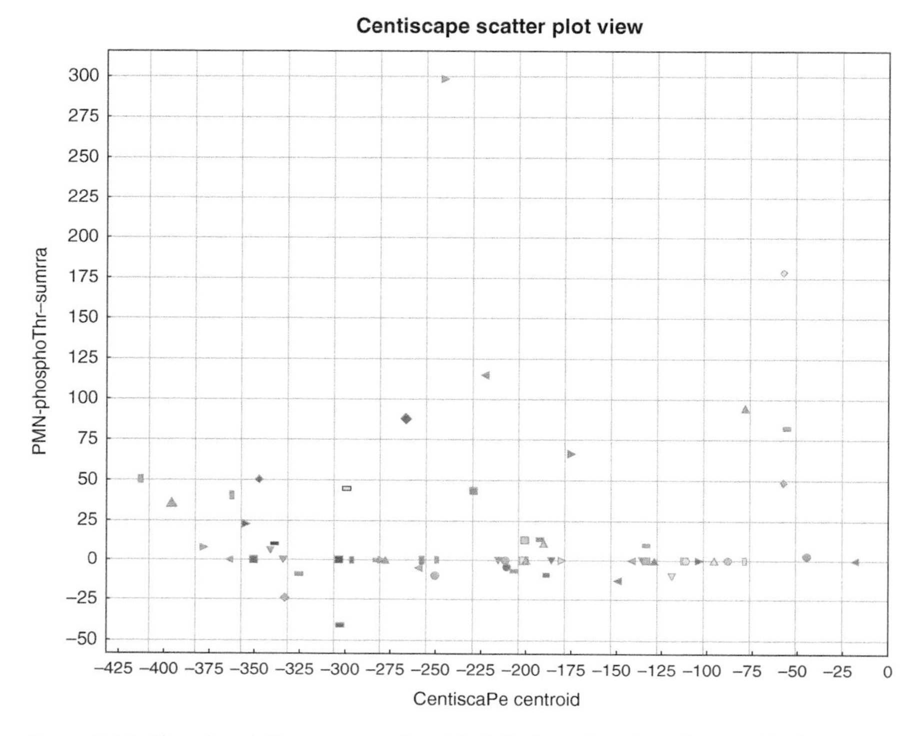

Figure 7.20 Phosphorylation response (y-axis) plotted as a function of centroid value (x-axis). In the top right of the plot, proteins with high topological and experimental levels are observed.

7.5.2 Integrating Expression or Activation Levels as Nodes Attributes

The expression or activation levels are simply numbers describing the importance of a protein in the context it was extracted. The higher the activation or expression level obtained experimentally, the higher is its importance in that particular context. Consider the network in Figure 7.21 and the respective activation level. The network is modified according to such values, and then the computation of the centralities (or any other parameter) is performed with the new network.

7.5.3 Edge Attributes as Distance in a Computation

In the classical definition of centralities, the shortest paths between the nodes are computed as binary constants: 1 for the presence and 0 for the absence of an edge connecting two nodes. However, providing a numerical value to each edge between nodes (called a weight), which describe different features (depending on the context), can be valuable. For example, a weight between nodes u and v

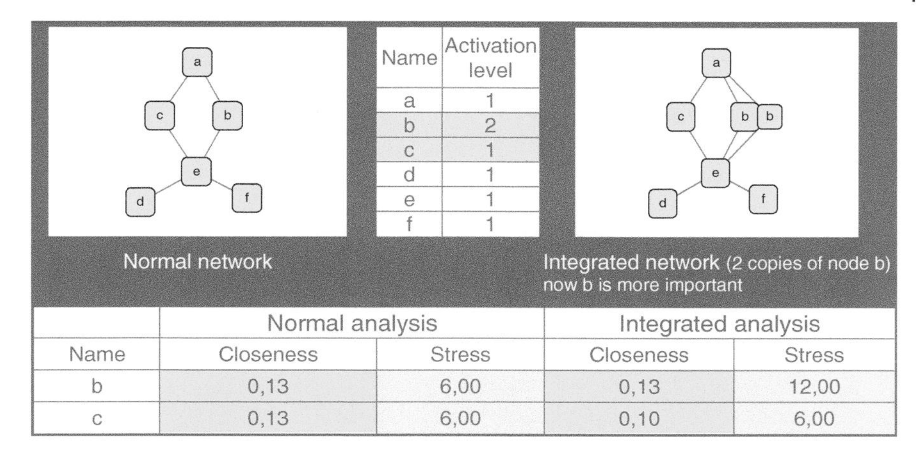

Name	Activation level
a	1
b	2
c	1
d	1
e	1
f	1

	Normal analysis		Integrated analysis	
Name	Closeness	Stress	Closeness	Stress
b	0,13	6,00	0,13	12,00
c	0,13	6,00	0,10	6,00

Figure 7.21 Integration of experimental dataset in the network structure. In this example, node b has an activation level of 2, allowing us to create two copies of node b (as seen in the network on the right). Computation takes into account that node b is more relevant, and the centrality results reflect the new structure.

can be viewed as a cost that is necessary to travel from u to v, as a distance, or as the strength of the interactions between the two vertices. Therefore, considering weights within the shortest path computations often leads to different results from the computation without the weights. Consider the network (A → B → C) in Figure 7.22 where dist(A,B) = 2, dist(B,C) = 3, and dist(A,C) = 7. The shortest path from A to C is not the direct one (=7), but rather the one passing through B (2 + 3 = 5). Since edge weights modify the result of shortest path-based centrality computations, it is critical that they have been properly assigned to the edges in accordance with the biological process being studied.

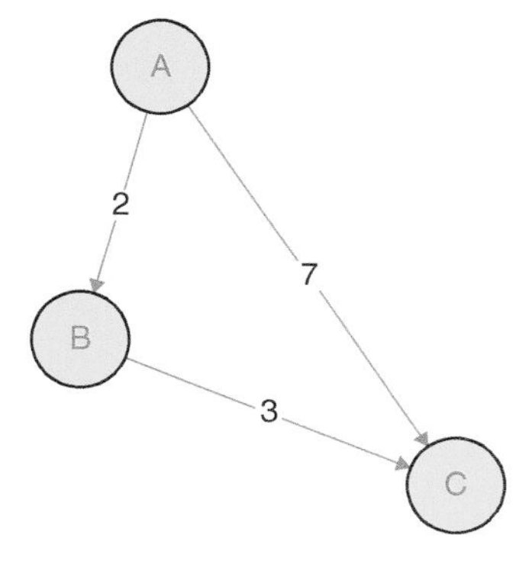

Figure 7.22 The shortest path computation can change when weighted distances are used. The shortest path from A to C is A → C if the weights are not considered. If the weights of the edges are considered as a distance, the path (A → B → C) (=5) is shorter than the simple path A → C (=7).

7.6 Network Analysis Software

7.6.1 Cytoscape and Its Apps

Cytoscape (http://www.cytoscape.org) is an open-source package for analysis, visualization, and integration of network-oriented data (Figure 7.23). It is most commonly used for biological networks, but it has been used in a variety of other contexts as well. Cytoscape is written in Java (currently requires Java 8) and runs on all of the major platforms

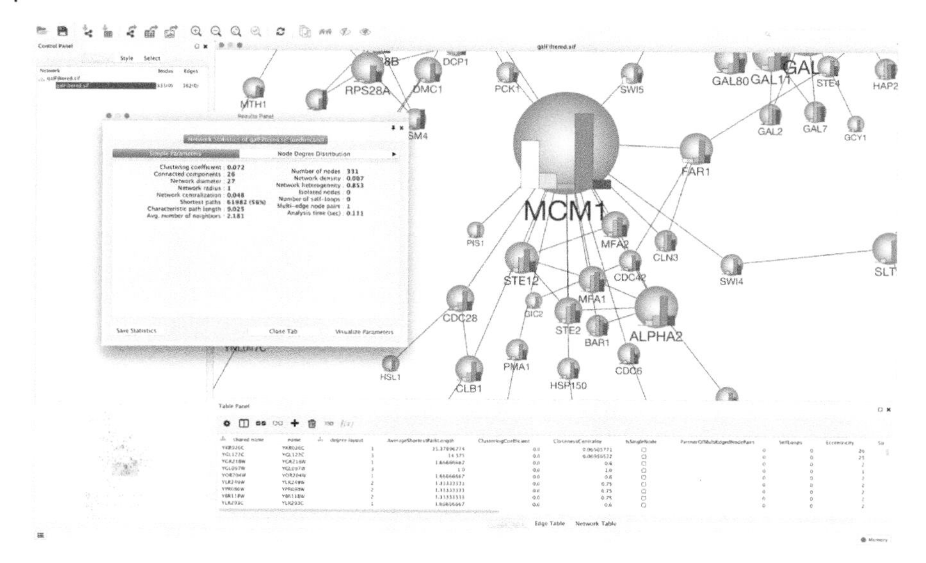

Figure 7.23 Screenshot of Cytoscape 3.3 showing part of a yeast interaction network. The bar chart shows several network parameters (betweenness centrality, closeness centrality, clustering coefficient, radiality, and topological coefficient). The statistics for the entire network are shown in the Network Analyzer results panel.

(Windows, Mac OS, and Linux). Cytoscape's strengths are its ability to integrate data from a variety of sources and map that data into the network; a wide variety of visualization options, including graphs, charts, and custom graphics; and its app ecosystem, which includes a public repository that allows contributions from many partners (http://apps.cytoscape.org), and a variety of apps (currently 256 apps) are available for a variety of network visualization, integration, and analysis tasks. Cytoscape has a limited set of network analysis capabilities, relying mostly on apps to provide sophisticated analysis functionality. Capabilities delivered with Cytoscape (version 3.3) include: importing network data and tabular data from public repositories, network formats, spreadsheets, or delimited text; mapping tabular data to a variety of network visual attributes, including node and edge colors, sizes (thickness), node charts and graphs; having tools: (i) to merge networks or tables together; (ii) to calculate simple network properties including node degree, various centralities, scale-free properties, radius, diameter, connected components, clustering coefficient, etc.; (iii) to position nodes and edges using diverse layout algorithms; (iv) to map visual network attributes from tabular data; (v) to enable the user to save and restore sessions and move sessions between computers (even if they have different operating systems); and finally (vi) to load and manage apps.

7.6.1.1 structureViz/RINalyzer

structureViz and RINalyzer are twos Cytoscape apps that work together to provide tools to link the fields of network biology and structural biology. structureViz provides a link between Cytoscape and UCSF Chimera, a package for molecular visualization and analysis [15]. structureViz provides the ability to link nodes in a network to the corresponding molecular structure and to create residue interaction networks from protein structures where nodes represent amino acid residues and edges represent bonds and contacts. RINalyzer provides tools to analyze residue interaction networks.

7.6.1.2 CentiScaPe

CentiScaPe is a Java app for Cytoscape that computes specific centrality parameters describing the network topology. It is a project of the Center for Biomedical Computing (CBMC), University of Verona, Italy. It computes average distance and diameter as global parameters. Computed node centralities are degree, stress, betweenness, radiality, closeness, centroid value, eccentricity, eigenvector, and bridging centrality. Edge betweenness (for edges) is also computed. Centralities are computed for directed, undirected, and weighted networks. Support for multiple network analysis and the integration of experimental dataset in the network structure are provided.

7.6.1.3 PesCa

The PesCa [16] app for the Cytoscape network analysis environment is specifically designed to help researchers infer and manipulate networks based on the shortest path principle. PesCa offers different algorithms, allowing network reconstruction and analysis starting from a list of genes, proteins, and, in general, a set of interconnected nodes. The app is useful in the early stages of analysis (i.e. to create networks or generate clusters based on shortest path computation), but it can also help further investigate relationships. It is suitable for situations requiring connection of a set of nodes that do not share links, such as isolated nodes in subnetworks. Overall, the plug-in enhances the ability of discovering interesting, and not very obvious, relations between high dimensional sets of interacting objects.

7.6.1.4 Interference

Interference is a Cytoscape app that evaluates the topological effects of single- or multiple-node removal from a network. Interference allows virtual node knockout experiments, meaning it is possible to remove one or more nodes from a network and analyze the consequences on network structure via variations of the node centrality values. As the centrality value of a node is strictly dependent on the network structure and on the properties of other nodes in the network, the consequences of a node deletion are well captured by the variation on the centrality values of all the other nodes. The Interference approach can model common situations in which real nodes are removed or added from/to a network. In biological networks, oftentimes one or more nodes (genes, proteins, metabolites) are removed from the network (or limited in their activity) because of gene deletion, pharmacological treatment, or protein degradation. For example, Interference can be used to simulate pharmacological treatment, since one can potentially predict side effects of the drug by looking at the topological properties of nodes in a drug-treated network (where a drug-targeted node was removed from the network). Specifically, Interference computes the centrality values of the network, removes the node(s) of interest, and computes the centralities in the new network (the one in which the node(s) has been removed). The results are simplified to an evaluation of the differences between the centralities in the two networks: one with the node(s) of interest still present and one in which the node(s) has been removed. Some nodes will increase, whereas others will reduce, their individual centrality values. This may provide insight on the functional, regulatory, or relevance of specific node(s). Interference plug-in computes the interference values for the following centralities: stress, betweenness, radiality, closeness, centroid value, and eccentricity.

7.6.1.5 clusterMaker2

One of the most common analyses for networks and network-related attributes is to classify nodes based on their topological distance from other nodes or based on the similarity of node attributes (Figure 7.24). clusterMaker2

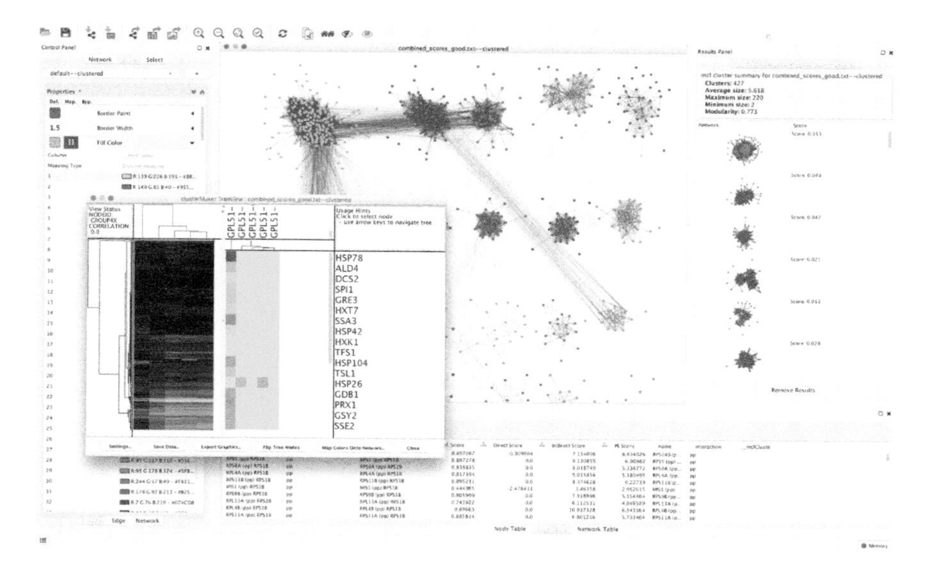

Figure 7.24 Screenshot of clusterMaker2 showing the results of performing an MCL cluster to partition a yeast interaction network as well as a hierarchical cluster of the expression fold change resulting from a heat shock experiment.

(http://apps.cytoscape.org/apps/clustermaker2) is a Cytoscape app that provides a wide variety of clustering algorithms (unsupervised classifiers) and visualizations for Cytoscape. It currently supports three kinds of clustering algorithms: node attribute algorithms, network partition algorithms, and post-clustering filters (applied after network partitioning).

In addition, clusterMaker2 provides different ways to visualize the data: Create New Network from Attribute and Create New Network from clusters for network partition algorithms and JTree TreeView, JTree KnnView, and JTree HeatMap view for node attribute data. Network partition algorithms include both discrete algorithms (e.g. Affinity Propagation, Connected Components, Community Clustering [GLay], MCODE, MCL, Spectral Clustering of Protein Sequences [SCPS], and Transitivity Clustering) and fuzzy algorithms (Fuzzy C-Means Clustering and a Cluster Fuzzifier) that utilize a discrete algorithm as a starting point to select cluster centroids. Post-cluster filters include Best Neighbor, Cutting Edge, Density, and Haircut filters. Node attribute cluster algorithms include HOPACH, PAM, Hierarchical, k-Means, k-medoid, AutoSOME, and the ability to create a correlation network by taking the feature vector distance between nodes as edge weights. Future versions will add support for PCA. Many of the implementations are parallelized and will take advantage of multicore machines.

7.6.1.6 chemViz

chemViz is another Cytoscape app to extend Cytoscape's analytic capabilities. It adds the ability to show the 2D structures of small molecules on nodes, calculate and display a wide variety of chemical descriptors in a table, display chemical properties and links to common chemical databases in the results panel, and calculate and display chemical similarity networks.

7.6.2 Other Tools

7.6.2.1 Gephi

Gephi is an open-source platform that allows you to visualize and explore different kinds of networks and graphs. It can be used for networks up to 50 000 nodes and 1 000 000 edges. The tool is fast, easy to learn, and comprehensive, providing almost every feature needed to study the different attributes of a network. Graphs can be easily manipulated, allowing you to filter the edges or nodes and change their size, label, or color. It also provides various algorithms to change the network layout and style. It even has the provision to cluster your network into different subnetworks based on certain characteristics. Gephi has built-in capabilities to calculate many of the essential network node and edge characteristics as well as an active community developing plug-ins to assist in performing many specialized tasks. It is a general tool that can be used for many kinds of network visualization, ranging from twitter feeds to geographical connectivity. An additional feature is that you can export any visualization at any time as a .png or .svg to refer to a particular data state later.

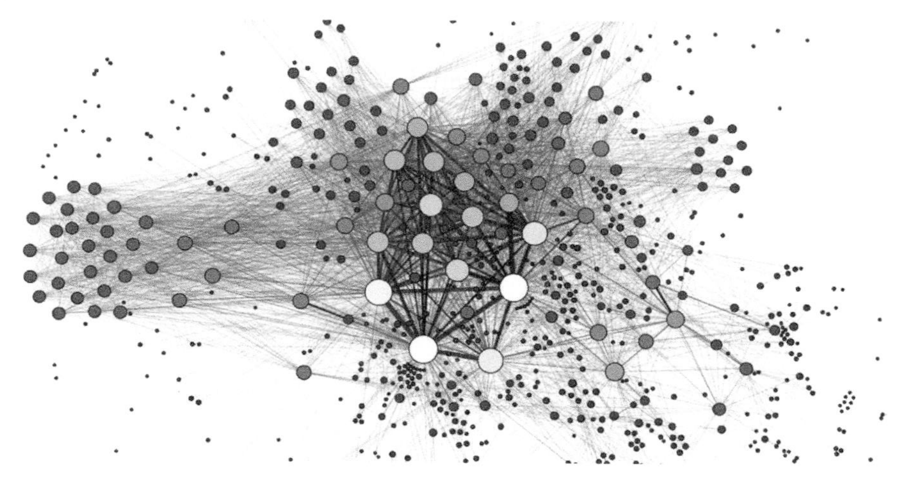

Figure 7.25 Screenshot of a network visualized with D3.

7.6.2.2 D3.js

D3.js is one of the most popular JavaScripts used for visualizing data on the Web browser. It is an extremely powerful tool that uses HTML, CSS, and SVG to render the graphs and tables. D3 enables you to create several different kinds of visualization (Figure 7.25). It has rich features and support for creating interactive graphs as well as other types of charts (bar/line, etc.) that can be used to visualize and explore the different aspects of the data. Despite the power of D3, there is a trade-off: the whole system must be built from the ground up. D3 does not have a predefined set of graphs or views, but it has features that allow manipulation of data, creation of styles, transitions, and interactions to create these views. While there are a number of tutorials and example for D3.js, the additional steps required for ground-up manipulation of visuals might be a daunting task for those more comfortable with plug-and-play graphics. However, D3.js is Web based, and sharing the visualization is relatively easy, especially in cases where a particular network is being studied by multiple users. Applications like Cytoscape and Gephi have provisions that let you export the graph visualization data in the json format directly compatible with certain views of D3, thus allowing you to use the power of both.

7.6.2.3 VisANT

VisANT runs as a Java applet on the browser and is designed to specifically study biological interactions and relations (Figure 7.26). The tool is designed to be used by scientists and researchers with at least some prior knowledge of biological networks, possible interactions, and access to different available datasets with their accompanying characteristics. The system is integrated

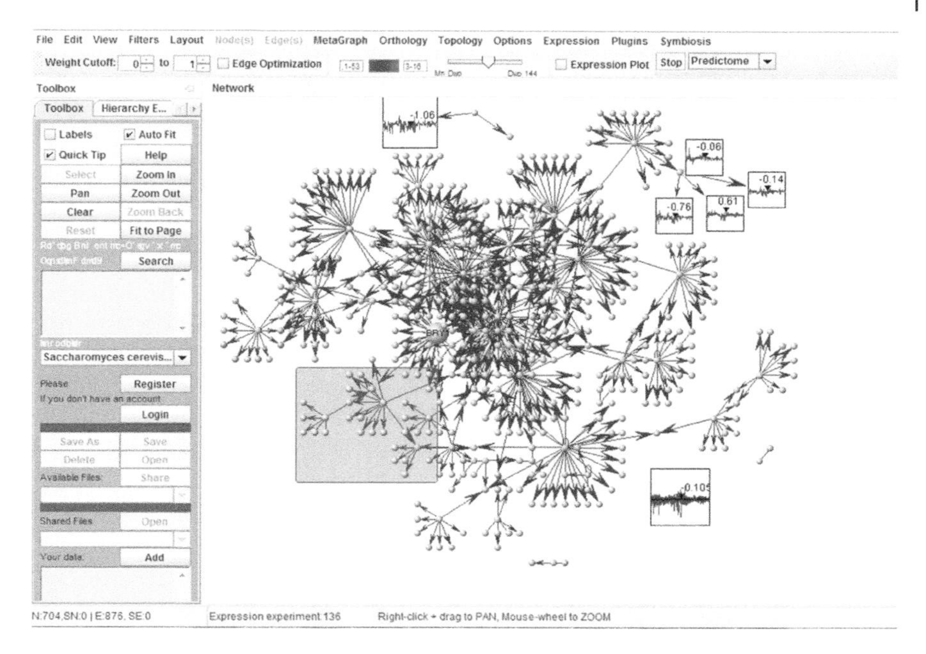

Figure 7.26 Screenshot of VisANT.

with many standard databases like GenBank and KEGG. VisANT uses the Predictome database that contains both putative and experimental interactome data to provide services like name resolution and interaction predictions, which eases integration of data from different sources. VisANT uses its server-side database to supplement user's tasks. For example, if a user-defined dataset contains interactions between genes or proteins, the user can test how that dataset relates to available similar datasets in the database. The user can also query network data for functional information.

VisANT is a tool for integrative research and visualization of data in the context of sequence, pathway, structure, and associated annotations. It has the provisions to support large networks, and according to the developers, it has been tested to work well with over 226k edges and nodes. Pathway data are provided by the KEGG database. KEGG pathway, which is currently available for metabolic pathways in the system, can be easily queried in VisANT using its KEGG pathway ID or chemical compounds. The COG database is used to provide homology information for relationships between species. Annotation information is drawn from KEGG and the Gene Ontology, and cross-referencing of genes and proteins to GenBank and Swiss-Prot is provided. If interactions or associations are discovered by multiple methods, the corresponding edges will be segmented with different colors corresponding to the methods. VisANT also allows analysis of user-defined networks that can be imported in the node/edge lists, BioPAX, PSI-MI, or the VisANT network specification format, visML. It supports weighted, directed, and undirected, as well as combination of these networks. For any network in the workspace, VisANT provides a means to visually manipulate and filter the data and also change its visual properties, like the color, width, and shape of the nodes and edges. It can be used to analyze a network to study complex features as well as characteristics of networks like shortest paths, node distributions, network motif identification such as directed loops or feedforward loops, and clustering coefficients. Changing the layout and alignment of the network can help to interpret key features at different levels. Users can customize, modify, save, and share network views. VisANT has plug-ins to supplement complex analysis and specialized functions not available by default. Additionally, anyone can develop the Java-based plug-ins that can be used for manipulating the network data model, server-side database, or the user interface.

7.7 Conclusions

The primary goal in using network analytics to interpret signal transduction data is to reduce complexity in order to obtain a global understanding of the processes involved in the cellular network. In this sense, networks can be used to infer statistically relevant results, reveal common characteristics or conserved patterns among various conditions, and even model diseases on a molecular signaling level. All of the available tools mentioned in this chapter such as databases, network construction methods, and network analytics allow researchers to uncover specific functional modules that can be potentially involved in biological mechanisms or new interesting targets to investigate. While a network approach is only one manner of data analysis, there are definite advantages in treating these intricate molecular responses as a group or subgroups of interconnected components in the overall biological response. Finally, like any other analytical or statistical method, great consideration must be paid to all aspects that may impact the overall outcome. For example, how the network is constructed, the number and types of elements in network, and types of relationships among the nodes are only a few factors that may significantly impact apparent node centralities.

References

1 Rolland, T., Tasan, M., Charloteaux, B. et al. (2014). A proteome-scale map of the human interactome network. *Cell* 159 (5): 1212–1226.
2 Stumpf, M.P., Thorne, T., de Silva, E. et al. (2008). Estimating the size of the human interactome. *Proc. Natl. Acad. Sci. U. S. A.* 105 (19): 6959–6964.

3 Hartwell, L.H., Hopfield, J.J., Leibler, S., and Murray, A.W. (1999). From molecular to modular cell biology. *Nature* 402 (6761 Suppl): C47–C52.
4 Bhalla, U.S. and Iyengar, R. (1999). Emergent properties of networks of biological signaling pathways. *Science* 283 (5400): 381–387.
5 Yu, D., Kim, M., Xiao, G., and Hwang, T.H. (2013). Review of biological network data and its applications. *Genomics Inform.* 11 (4): 200–210.
6 Summer, G., Kelder, T., Radonjic, M. et al. (2016). The Network Library: a framework to rapidly integrate network biology resources. *Bioinformatics* 32 (17): i473–i478.
7 Ji, Z., Yan, K., Li, W. et al. (2017). Mathematical and computational modeling in complex biological systems. *Biomed. Res. Int.* 2017: 5958321.
8 Goh, K.I., Cusick, M.E., Valle, D. et al. (2007). The human disease network. *Proc. Natl. Acad. Sci. U.S.A.* 104 (21): 8685–8690.
9 Currie, H.N., Vrana, J.A., Han, A.A. et al. (2014). An approach to investigate intracellular protein network responses. *Chem. Res. Toxicol.* 27 (1): 17–26.
10 Barab'si, A.L. (2009). Scale-free networks: a decade and beyond. *Science* 325 (5939): 412–413.
11 Barabasi, A.L. and Oltvai, Z.N. (2004). Network biology: understanding the cell's functional organization. *Nat. Rev. Genet.* 5 (2): 101–113.
12 Scardoni, G., Petterlini, M., and Laudanna, C. (2009). Analyzing biological network parameters with CentiScaPe. *Bioinformatics* 25 (21): 2857–2859.
13 Bader, G.D. and Hogue, C.W. (2003). An automated method for finding molecular complexes in large protein interaction networks. *BMC Bioinformatics* 4 (2): https://doi.org/10.1186/1471-2105-4-2.
14 Han, A.A., Currie, H.N., Loos, M.S. et al. (2017). The impact of cytokine responses in the intra- and extracellular signaling network of a traumatic injury. *Cytokine* 106: 136–147.
15 Morris, J.H., Huang, C.C., Babbitt, P.C., and Ferrin, T.E. (2007). structureViz: linking Cytoscape and UCSF Chimera. *Bioinformatics* 23 (17): 2345–2347.
16 Scardoni, G., Tosadori, G., Pratap, S. et al. (2015). Finding the shortest path with PesCa: a tool for network reconstruction. *F1000Res* 4.

8

A Toxicological Application of Signal Transduction

Early Cellular Changes Can Be Indicative of Toxicity

Julie Vrana Miller[1], *Nicole Prince*[2], *Julia A. Mouch*[2], *and Jonathan W. Boyd*[3]

[1]*Cardno ChemRisk, Pittsburgh, PA, USA*
[2]*Department of Orthopaedics, West Virginia University School of Medicine, Morgantown, WV, USA*
[3]*Department of Orthopaedics and Department of Physiology and Pharmacology, West Virginia University School of Medicine, Morgantown, WV, USA*

8.1 Introduction

The early cellular changes initiated by external stimuli (whether that be chemical, physical, or biological agents) can offer a host of information about the adaptive response and adverse effects related to individual chemical and mixture exposures. Notable examples of early cellular changes are perturbations of dynamic intracellular signaling networks and alterations in cellular bioenergetics, such as increased/decreased oxygen consumption or electron transport chain (ETC) uncoupling. Due to the interconnectivity of various effector proteins and biomolecules, activity at a distinct intracellular location can have consequences at distal locations. Furthermore, a true understanding of biological response to chemical exposures necessitates a better understanding of cellular changes in response to a range of concentrations (especially low-dose exposures) and is essential to toxicity testing and chemical risk assessment. Finally, the ability to measure the rapid and dynamic cellular responses to exposure(s) is critical for an enhanced understanding of toxicity. The experimental conditions and assays capable of capturing relevant mechanistic information for toxicity testing will be discussed in this chapter. A better understanding of the mechanistic components related to chemical response has numerous implications across many fields, such as risk assessment for toxicology and target molecule/pathway analysis for drug development and pharmacology. With mechanistic data collected from various doses, intracellular proteins and biomolecules, and cellular endpoints,

Cellular Signal Transduction in Toxicology and Pharmacology: Data Collection, Analysis, and Interpretation, First Edition. Edited by Jonathan W. Boyd and Richard R. Neubig.

better toxicity prediction models can be developed for individual chemical and mixture risk assessment.

8.2 Classification of Toxic Agent and Exposure Effects: A Toxicological Perspective

Humans are continuously exposed to a plethora of chemicals. These chemicals can be traced to a variety of sources, such as environmental, pharmaceutical, or industrial. With the large scope of individual chemicals humans can be exposed to on a daily basis, this raises a question: what classifies a chemical or agent as "toxic?" A toxic agent, whether it be chemical, biological, or physical, can be classified by their use (e.g. pharmaceutical, pesticide, additive, etc.), source (e.g. man-made, plant or animal toxin, etc.), target organ(s) (e.g. brain, liver, heart, etc.), and effects (e.g. cardiotoxicity, carcinogenicity, immunogenicity, etc.) [1]. A primary tenet of toxicology states that all chemical agents are toxic, but it is the dose that determines if an agent is toxic or safe [2]. While the classification of what constitutes an agent as toxic, and the threshold thereof, can be ambiguous, the experimental characterization of an agent that incorporates chemical properties and biological exposure effects can be very useful for toxicological risk assessment.

Appropriately characterizing biological effects of a single agent exposure can be a daunting task. Toxicological risk assessment quantitatively determines the possible effects of a xenobiotic on human health [1]. Xenobiotic exposures can provoke adverse, deleterious, or dangerous effects on an organism [3]. Conversely, some effects can be beneficial, such as pharmaceutical side effects of antihistamines (drowsiness) or oral contraceptives (decrease acne severity). To properly address the adverse or beneficial nature of a chemical exposure, the dose and time course of toxicity/adaptation should be properly elucidated.

8.2.1 Dose–Response for Chemical Exposure Toxicity Testing and Risk Assessment

A central concept for toxicity studies and risk assessment is the dose–response relationship [4]. Dose–response relationships are described by dose–response curves, in which the response variable can be any desired effect to be measured, such as cell death, survival, or cellular oxygen consumption; this response is measured across a range of doses to determine the doses required to elicit a response. There are many types of dose–response models used in toxicity testing, but the most dominant is the threshold model [5]. The threshold model has been used in many scientific disciplines, such as biology, pharmacology, and toxicology, and has been the primary model for regulatory agencies, such as the US Federal Drug Administration (FDA) and Environmental Protection Agency (EPA) [5]. In the threshold model, depending on the effect measured and assay sensitivity, there exists a dose below which the probability of a measured response for a sample or individual compared to control is zero [6]. This threshold dose is also referred to as the no-observed-adverse-effect level (NOAEL) [7]. An alternative model, the hormesis model, has also proven useful for low-dose (below the NOAEL) risk assessment [8] and has seen a recent surge in interest due to advances in molecular toxicology testing [4]. The hormesis model is a biphasic dose–response relationship and can be succinctly described as low-dose activation followed by high-dose inhibition, which can appear as a U-shaped or J-shaped dose–response curve [9].

The renewed interest in hormesis as a valid model for toxicity testing has opened the door for low-dose biochemical and molecular toxicology research [9]. It is important to note that the hormetic dose–response relationships should also consider time in toxicity testing. This is due to the fact that hormesis responses may be a compensatory response that follows the initial disruption in homeostasis, resulting in the characteristic low-dose stimulatory response [10]. Biological systems are highly coordinated and dynamic [11]; the exclusion of temporal response in risk assessment modeling would ignore the ability of an organism to adapt and respond to a low-dose exposure. Therefore, toxicological risk assessment needs to be inclusive of the spatiotemporal aspect of biological response post-exposure as well as a wide range of doses, including low doses.

8.2.2 Chemical Mixtures

Current chemical exposure risk assessment is primarily carried out for single xenobiotics [12]. However, in reality, humans are continuously exposed to a vast number of components, whether they be chemical, physical, or biological agents, at various doses, and through a variety of exposure routes on a daily basis [13]. Understanding and ultimately predicting the possible combined effects of a given mixture exposure is necessary for risk assessment toxicology. There are two types of mixtures: simple and complex. Simple mixtures contain a small number of different chemicals, and the composition is known. Complex mixtures contain hundreds to thousands of individual chemicals, of which the composition (dose or constituents) is not known [14]. Experimentally determining all possible mixture combinations, or even binary mixtures, for a range of doses at different time points is not physically or financially possible [12]. Therefore, adequate models capable of predicting mixture responses are necessary for various sectors, such as pharmacological adverse interaction risk assessments and environmental exposure risk assessments. The two most commonly used and accepted mixture prediction models are Loewe additivity (dose addition) and Bliss independence (response addition) [15].

Dose addition is typically used when two or more chemical agents have a similar mechanism of action [16]. Dose addition is based on the theory that two chemicals in a mixture act as a dilution of each other [17, 18]. Traditional dose additivity can be described as shown in Eq. (8.1):

$$\Sigma \frac{[\alpha']_i}{E[\alpha]_i} = 1, \tag{8.1}$$

where α' is the dose of the chemical i when administered as a mixture producing response E and α is the concentration of the chemical agent, i, required to produce the response effect E when administered alone (single exposure). Equation (8.1) can be used for n number of agents as a mixture. If the overall expression is <1, the mixture is considered synergistic. Conversely, if the overall expression is >1, the mixture is considered antagonistic [19].

Response addition (Bliss independence) is typically used for two (or more) chemicals that do not have the same mechanism of action, that is, the organism will respond to each chemical agent independently, as though the other agent(s) are not present [15, 20]. The prediction of mixture effects using response addition is equivalent to the conditional sum of independent chemical effect probabilities [21]. Response addition can be described by expression (Eq. 8.2)

$$f(\alpha_{\text{mix}}) = 1 - \Pi\left[1 - F(\alpha_i)\right], \tag{8.2}$$

where $F(\alpha_i)$ is the response effect produced by chemical i at dose α. Since this is a probabilistic model, $F(\alpha_i)$ cannot be greater than 1. Dose addition and response addition models are generalizations of nonlinear regression models, such as a Gompertz growth curve or Hill plot, and can be easily compared with observed mixture dose–response curves [19, 21].

8.2.3 Mode of Action Versus Mechanism of Action

Regulatory guidelines for mixture risk assessments rely heavily on the similarity (or dissimilarity) of two chemical components' mechanistic information for model selection of a given mixture [22–27]. In this discussion of prediction models for mixture toxicity, the term "mechanism of action" is used loosely. The term "mechanism" of action is often used interchangeably with "mode" of action [16]. However, these terms have specific definitions but are often defined differently depending upon the literature article cited [28, 29]. Traditionally, "mechanism of action" refers to the series of molecular events from the absorption of an effective dose of an agent to the eventual biological response [30]. To fully describe the mechanism of action for a chemical, each component outlined in Figure 8.1 would need to be determined experimentally.

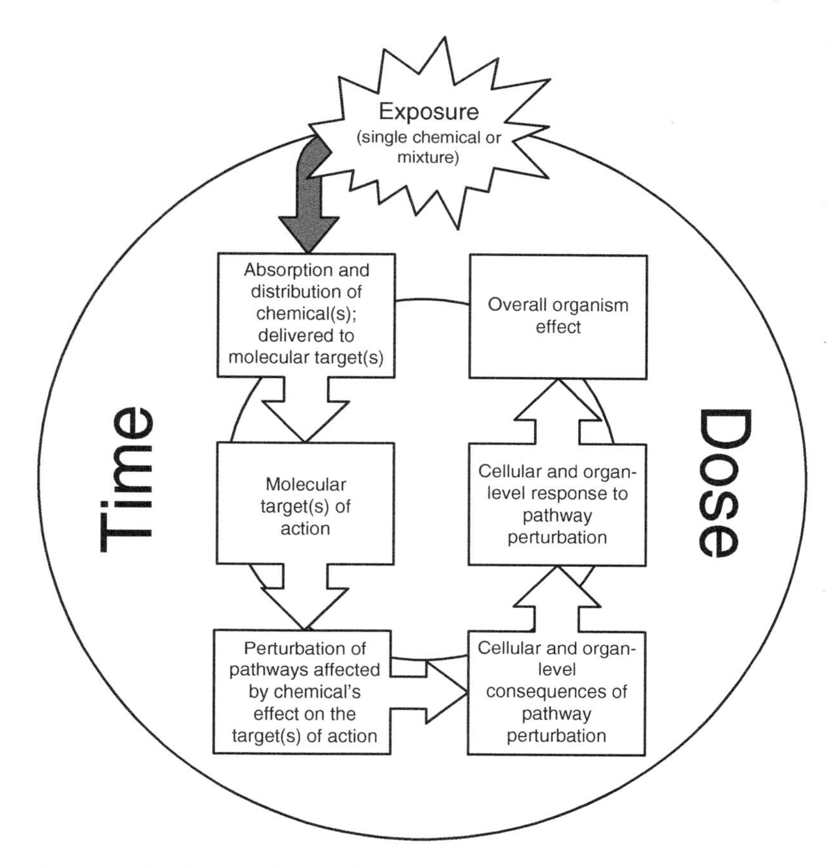

Figure 8.1 Mechanism of action. Schematic outlining the components necessary to elucidate the mechanism of action for an exposure. All steps must include the temporal and dose components to fully understand mechanism of action for a given chemical or mixture exposure.

The term "mode of action" is a more generalized means to describe a chemical's action on a given organism [31]. The term "mode of action" describes the type of observed response of an organism exposed to a given chemical or mixture and may only refer to the significant events or components of the mechanism necessary for producing a particular biological response or toxic effect [32]. Regardless of the terminology, the suitability of either model for a given experiment or effect prediction remains debatable when thorough mechanisms or modes of action are unknown, especially for low-dose studies [29, 33]. Often, the models fail to predict mixtures toxicity when an observed mixture acts synergistically [34, 35] and antagonistically [36, 37] or with dose-dependent subtle interactions (e.g. synergy for low-dose mixtures and antagonist for high-dose mixtures) [38, 39]. To address these challenges, future model development should incorporate low-dose (i.e. below the NOAEL) mixtures mechanistic effects on the appropriate spatiotemporal scale.

8.3 Early Cellular Changes Post-exposure

Understanding and ultimately predicting potential biological effects and health outcomes from single chemical or mixture exposure remains an arduous task for risk assessment. The predominant approach for toxicity testing relies heavily on whole animal studies evaluating observable apical responses, such as clinical effects or pathologic changes from high-dose exposures [29]. While many of these studies have been thorough, the amount of time and resources required to carry out this low-throughput methodology has left this approach unable to meet the demands of current toxicology needs [40]. The enormous backlog of chemicals waiting to be evaluated for toxic outcome has inspired a paradigm shift in toxicity testing, proposed by the US National Research Council (NRC) report, *Toxicity Testing in the 21st Century: A Vision and a Strategy* [40, 41]. The NRC proposed a transition from *in vivo* low-throughput animal toxicity testing to an *in vitro* high-throughput approach utilizing well-designed mechanistic information-based assays. This *in vitro* approach would take advantage of early cellular perturbations post-exposure associated with toxicity endpoints in human cell lines and tissues to elucidate mechanistically relevant information regarding the mode(s) of action for a potential xenobiotic or mixture [42]. Additionally, the high-throughput nature of this approach would allow for testing on a wide range of doses, especially low doses, which is not currently possible with traditional whole animal studies [41]. New risk assessment approaches would include a suite of assays in order to cast a wide net on early cellular changes, such as changes in cellular bioenergetics, and various pathway perturbations, such as alterations in posttranslational modifications (PTMs) post-exposure, to fully understand the cellular response, whether that be adaptation after exposure to a new homeostatic state or cell death [43].

Recent advances in systems toxicology have opened the door for the collection and analysis of large amounts of mechanistic data across a wide dosing range and time scale. Incorporating toxicodynamic factors for risk assessment can aid in our understanding of xenobiotic toxicity. The term toxicodynamics can be succinctly described as what the toxicant does to the body (as opposed to toxicokinetics, which describe what happens to the toxicant once it is in the body). Toxicodynamic analyses interrogate the spatiotemporal interaction of a xenobiotic with biological targets, the corresponding disruption of molecular pathways and bioenergetics, and downstream effects after exposure [44].

Marrying toxicokinetic and toxicodynamic approaches for risk assessment can offer a wealth of knowledge regarding xenobiotic exposure and the biological perturbations associated with exposure and subsequent effects (Figure 8.2). Perturbations of biological processes by xenobiotic exposure can elicit early cellular changes, leading to an adaptive stress response for continued survival or adverse response leading to toxicity [45]. Early cellular changes corresponding to toxicodynamics will be highlighted in this chapter, specifically intracellular signaling perturbations (e.g. PTMs after chemical exposure) and alterations in cellular bioenergetics.

8.3.1 Intracellular Signaling Perturbations Associated with Exposure

The toxicodynamic components of early cellular changes, biological perturbations, and their associated responses yield valuable mechanistic information for toxicity risk assessment. Early and late effects associated with chemical exposure are mediated by plasma membrane receptor proteins acting as sensors for their downstream signaling pathways [41]. Signaling pathways are not

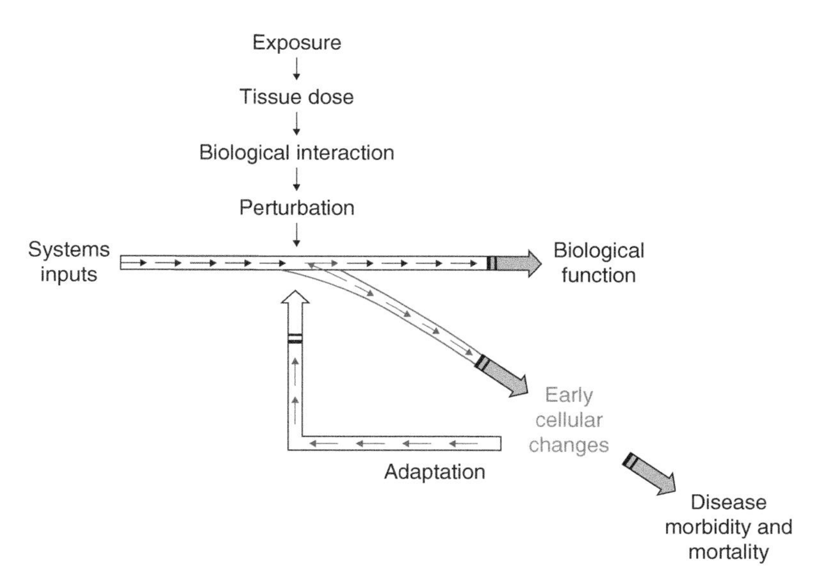

Figure 8.2 Early cellular changes related to biological outcome. Toxicokinetic components of exposure involve the absorption of the chemical exposure, biotransformation, and distribution, resulting in the tissue dose. The chemical then interacts with molecular targets, perturbing endogenous pathways and processes, resulting in early cellular changes, which contributes to the toxicodynamic response post-exposure. Early cellular changes can lead to adverse effects (toxicity) or adaptive response and survival. *Source:* Adapted from Andersen and Krewski [43]. Reproduced with permission of Oxford University Press.

static, linear pathways that simply transmit signals, but rather are responsible for encoding and integrating both internal and external cues [46]. Exposure to chemical insults can perturb the dynamic and highly coordinated signaling pathways responsible for normal biological function and maintenance [47]. Additionally, chemical insult at one molecular target and pathway can propagate throughout the signaling network due to pathway crosstalk and the interconnectivity of various signaling cascades [11]. As such, cellular spatiotemporal signaling dynamics are responsible for integrating and interpreting intra- and extracellular cues to make cellular fate decisions, such as proliferation, differentiation, or programmed cell death (apoptosis) [11, 48, 49].

Signaling networks are primarily regulated by PTMs. As discussed in Chapter 2, PTMs are vital to signaling coordination and diversification of proteins for various functions [50, 51]. PTMs can modify proteins after translation via complex molecules (glycosylation and isoprenylation), peptides or proteins (ubiquitylation and SUMOylation), chemical groups (acetylation, methylation, and phosphorylation), and cleavage (proteolysis) [52].

Phosphorylation is a PTM that can provide key insights into the early and rapid cellular changes that result from xenobiotic exposures. The rapid toxicodynamic response of cells to xenobiotics is primarily coordinated by these signal transduction networks, which follow a simple framework: the phosphorylation/dephosphorylation cycle mediated by kinases and phosphatases [53, 54]. Kinases are intimately involved in the regulation of signaling events relevant to cellular death and survival processes [55]. Xenobiotics can interfere with kinase signaling via activation (e.g. overstimulation) or inhibition [44]. Kinases can be activated/deactivated via the addition or removal of a phosphate group at serine, threonine, and tyrosine residues (—OH containing amino acids) [50], shown in Figure 8.3. The phosphorylation/dephosphorylation of a protein acts as a molecular switch to activate or deactivate a protein. In the catalytic domain of protein kinases, the —OH group on serine, threonine, and tyrosine residues acts as a nucleophile, attacking the γ-phosphate on adenosine triphosphate (ATP), resulting in the transfer of a phosphoryl group to the protein [56]. In this scheme, magnesium is a critical component for protein phosphorylation, acting as a supportive chelator. The covalent bonding between phosphate groups and protein kinases typically induces a conformational change, aiding or preventing protein–protein (enzyme–substrate) interactions [57]. It is this protein–protein/enzyme–substrate interaction where critical cellular information can be transmitted throughout the signaling network.

Under xenobiotic stress conditions, many pathways can be perturbed, including survival and death pathways. The three major stress-activated pathways are the mitogen-activated protein kinase (MAPK) pathway (also commonly referred to as extracellular signal-regulated kinases [ERK] pathway), the stress-activated protein kinase (SAPK) pathway (also commonly referred to as Jun N-terminal kinase [JNK] pathway), and the p38 pathway [58–60].

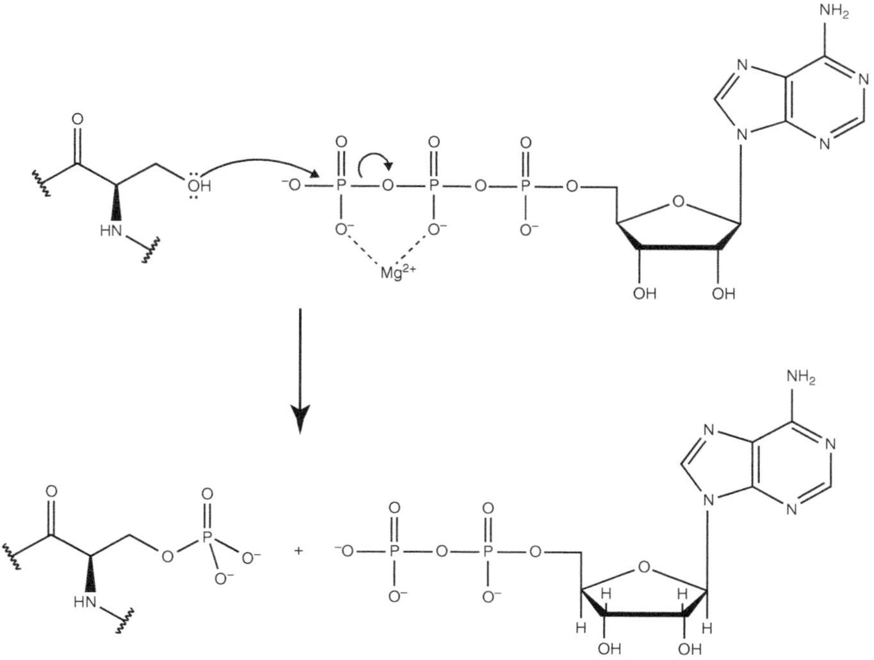

Figure 8.3 Mechanism of phosphorylation. This reaction scheme uses serine as the example residue; however, this mechanism is true for tyrosine and threonine phosphorylation as well.

These stress pathways are endogenously activated, controlling cellular fate via transcriptional upregulation and/or inhibition of genes responsible for either survival/proliferation or cell death [44]. These pathways were traditionally thought of as discrete linear signaling cascades; however, it is now well known that they are merely components of a dynamic and highly interconnected network of pathways that contain many crosstalk, feedback, and feedforward mechanisms to adequately and efficiently respond to a variety of stress-inducing stimuli [11, 61]. Therefore, it is the delicate balance/imbalance of these pathways that decides the cell's ultimate fate [55, 62]. The highly dynamic and interconnected nature of signaling networks has made it increasingly difficult to elucidate and predict network responses to xenobiotic stress. However, advances in computational and network biology; high-throughput experimental techniques such as -omics investigations using mass spectrometry, multiplex bead-based enzyme-linked immunosorbent assay (ELISA) suspension array systems, and microchip arrays; and public databases have greatly improved signaling research. With these advances, new toxicological risk assessment approaches for determining potential toxic outcome of a xenobiotic or mixture using mechanistic information can be attempted, as postulated by the NRC report [40].

8.3.2 Bioenergetic Changes Post-exposure

Kinase signaling is an energy-demanding process, and its reliance on phosphorylation results in the consumption of substantial amounts of available ATP [63]. ATP production governs ATP-consuming processes, such as signal transduction in mammalian cells, and this production is primarily driven by oxidative phosphorylation within mitochondria [64, 65]. Mitochondria are the energy production hub of a cell via production of ATP and, of equal importance, are key mediators in kinase signal transduction, regulating cell survival, proliferation, differentiation, and death [63, 66]. Due to their critical role in many cellular processes, mitochondria are also susceptible to xenobiotic exposure effects. Additionally, irreversible processes leading to cell death primarily rely on two mitochondria-related phenomena: (i) the inability to reverse mitochondrial dysfunction, resulting in ATP depletion, and (ii) the disturbance of membrane function (both mitochondrial membrane and plasma membrane) [67]. Therefore, maintenance of mitochondrial bioenergetics and integrity is critical to cellular fate.

Mitochondria are composed of several key features: the outer mitochondrial membrane, intermembrane space, inner mitochondrial membrane, and matrix. The outer mitochondrial membrane is permeable to small molecules (molecular weight <5000 g/mol) and ions, which readily diffuse through transmembrane porin proteins, such as voltage-gated anion channel (VDAC) [68]. The inner membrane, however, is impermeable to most molecules and ions, including protons and ATP [69] unless actively aided via a transporter. The inner membrane contains the proteins that make up the ETC. Additionally, the inner membrane contains numerous transport proteins that allow metabolites to pass through into the matrix and to export ATP generated by the ETC into the intermembrane space [68] (the space between the two membranes). The intermembrane contains a small heme-containing protein called cytochrome c within the intermembrane side of inner membrane folds (referred to as intracristae space), which acts as an electron carrier for complex III on the ETC and, when released from the outer membrane, can initiate caspase-dependent apoptosis [70, 71]. The protein adenine nucleotide translocase (ANT) is the transporter responsible for shuttling ATP from the matrix to the intermembrane space [68]. Additionally, ANT forms a complex with VDAC (an outer mitochondrial membrane transporter), commonly referred to as the permeability transition pore (PTP) that can compromise the impermeability of the mitochondrial membranes. If the PTP is open, an influx of ions and water can bombard the matrix, causing swelling, loss of membrane potential, and uncoupling of oxidative phosphorylation, eventually leading to cell death [72]. Finally, the matrix, which is in the space contained by the inner membrane, houses all energy-yielding oxidative reactions, such as the citric acid cycle, fatty acid oxidation, and the ETC [68]. The matrix also contains important ions (magnesium, calcium, and potassium), metabolic intermediates, ATP/ADP, and mitochondrial DNA (that is transcribed and translated within the matrix).

The number of mitochondria in each eukaryotic cell can vary depending upon the cell type. Red blood cells (erythrocytes) do not have mitochondria; however, the heart, kidney, and liver are considered mitochondria rich [73]. Mitochondria-rich cell types are especially sensitive to xenobiotics that target the ETC. There are two classes of ETC inhibitors: (i) xenobiotics that block the transport of electrons via binding to ETC enzyme complexes (e.g. deguelin binding to complex I) and (ii) xenobiotics that stimulate/inhibit the flow of electrons at one portion of the ETC but shunt electrons away from their normal route by acting as an electron acceptor. Disruption of normal ETC function and mitochondrial bioenergetics can have deleterious effects (both acute and chronic) on the target tissue as well as whole organism. Examples of pathological conditions where mitochondrial dysfunction plays a critical role are neurodegenerative diseases, neurotoxicity, heart disease (myocardial infarction and atherosclerosis), liver injury (ischemic injury and cholestasis), obesity, and cancer [71, 74]. Thus, mitochondrial bioenergetics should be carefully considered and included in toxicity risk assessment.

8.3.3 Time Scale of Exposure Effects

The toxicity of a xenobiotic or mixture at their intended molecular/tissue target(s) can vary over exposure time, and any change is commonly known as time-dependent toxicity [75]. The initial toxicodynamic response to any xenobiotic or mixture exposure is predominantly coordinated by signal transduction networks, which can initiate response within the first few seconds to minutes of exposure. The time course from initial toxicodynamic response(s) to cell death following exposure can have a vast range [76]. For example, the time delay between xenobiotic exposure and execution of apoptosis can take from several hours to over a day [77, 78]. With this in mind, monitoring early cellular changes that contribute to adaptive stress response (survival and new homeostatic state) or lead to adverse effects (apoptosis, carcinogenicity) may aid toxicological understanding and ultimately the prediction of potential adverse effects from xenobiotic or mixture exposures.

8.4 Experimentally Testing Early Cellular Changes that May Contribute to Exposure Sensing and Response

Traditional risk assessment testing techniques involve screening potential agents using *in vivo* and *in vitro* endpoint experiments, such as neurotoxicity or developmental toxicity, and mode of action analysis, such as cytotoxicity or mutagenicity [79]. However, the recent paradigm shift charged by the NRC report has initiated a new chemical risk assessment approach that utilizes high-throughput *in vitro* screening assays to exploit early cellular changes (such as signaling pathway perturbations and alterations in cellular bioenergetics) to reveal mechanistic information about adverse or adaptive effects after xenobiotic exposure.

8.4.1 Paradigm Shift Toward *In Vitro* Cell Culture

The large number of potential toxic chemical agents that have yet to be fully characterized poses a significant problem for risk assessment testing [40]. Traditional *in vivo* methods cannot be solely relied on for risk assessment testing due to the low throughput, high financial and time cost, and the sheer number of animals necessary to test the thousands of chemicals yet to be evaluated. Further, the utility of mechanistic information from animal studies has been questionable due to disappointing cross-species extrapolation for real-life low-dose human exposure effects [47]. Thus, a paradigm shift from traditional *in vivo* methods to an *in vitro* approach utilizing assays to collect mechanistic information for pathway/network analyses and eventual prediction modeling with computational toxicology has been initiated by the NRC report [40]. The NRC suggested human *in vitro* high-throughput screening assays to measure early cellular effects, such as perturbations of critical signaling pathways related to survival and death (referred to as potential adverse outcome pathways by the NRC), across a wide dosing range and multiple cell lines, both immortalized and primary cell culture [80].

The paradigm shift from traditional *in vivo* animal studies to new *in vitro* high-throughput screening assays using human cell lines raises some new questions. What cell type should be used (immortalized versus primary)? Primary cell culture is culture that is initiated immediately following tissue extraction from the sample organism. Once the cells reach confluency (typically 80% of the culture-containing flask covered by cells, without overlapping each other), cells need to be subcultured (also known as passaging) by dividing the cells into multiple culture flasks for continued growth. Primary cell culture can typically only be subcultured a few times before they can no longer be used. On the other hand, immortalized (continuous) culture is when cells are cultured for a theoretically infinite number of subcultures, which is achieved via transformation (spontaneously transformed via cancerous cell lines or chemically induced immortalization). Immortalized/continuous cell lines offer a low cost *in vitro* experimental setup but may result in disparate responses than those achieved in primary culture or *in vivo* due to their continuous cell cycle progression. Even though the response may be slightly different from immortalized to primary or *in vivo* studies, useful mechanistic information can be collected for toxicity risk assessment. Fortunately, there are several cell lines commercially available that have libraries of data, such as human hepatocellular carcinoma (HepG2) for liver toxicity [81], Michigan Cancer Foundation-7 (MCF-7) (human breast cancer) for estrogen-responsive toxicity studies [82], and human embryonic kidney cells (HEK293) for kidney toxicity [83].

Liver injury and toxicity due to xenobiotic exposure is a major concern for pharmaceutical regulatory and toxicological risk assessment [84]. Additionally, the liver plays a vital role in xenobiotic biotransformation after exposure [85]. Thus, *in vitro* high-throughput approaches using human-derived liver cell lines to probe important early cellular exposure effects and potential pathway perturbations are necessary for toxicological risk assessment. The US EPA ToxCast research program was initiated to screen previously untested environmental chemicals for adverse effects using a large number of high-throughput bioassays. Data collected from ToxCast assays are made available through the ToxCast data library to disseminate chemical exposure profiles to further toxicology research and gain a mechanistic understanding of chemical exposure [86]. The ToxCast library has two liver models for toxicity testing: primary rat hepatocytes and HepG2 cells [84]. Both models offer important information about hepatotoxicity. Due to the limited availability of human primary liver cells, rat primary hepatocytes are often used for hepatotoxicity risk assessments. Previous studies have shown that xenobiotic metabolizing enzymes have significant interspecies variation [87, 88]. An accepted alternative to primary rat hepatocytes for liver toxicity research is the HepG2 cell line. A benefit of using HepG2 versus primary rat hepatocytes is that they are human derived. Most importantly, the HepG2 cell line retains endogenous xenobiotic metabolizing enzymes, whereas primary hepatocyte culture typically loses these vital enzymes [89].

8.4.2 Real-Time *In Vitro* Assays to Measure Early Cellular Changes

As discussed previously, initial cellular responses to xenobiotic exposure are rapid, dynamic, and highly integrated for determining eventual cellular fate. Assays capable of capturing these dynamic processes and relating them to apical outcomes (cell death, survival, plasma membrane degradation) are necessary for xenobiotic risk assessment predictions. Most real-time assays used are related to mitochondrial bioenergetics (reduced form of nicotinamide adenine dinucleotide [NADH] production, cellular oxygen consumption, and estimations of ATP from NADH and oxygen consumption) and cell death (plasma membrane degradation).

8.4.2.1 Using NADH and Oxygen Consumption to Predict ATP Generation

One of the most critical components of bioenergetics, ATP, can be measured to monitor cellular perturbations after xenobiotic exposure. Chapter 6 discussed assays for intracellular ATP that have employed fluorescent tags; however, this can potentially disrupt endogenous intracellular activity (e.g. Förster resonance energy transfer [FRET]) [90] or involve cell lysis (e.g. luciferase assay), making real-time *in vitro* ATP measurements not possible [91]. Monitoring relative ATP generation in response to chemical insult without potentially disrupting sensitive intracellular activity is imperative. As mentioned in Chapter 6, NADH is extremely useful because it strongly absorbs at 340 nm and does not require any fluorescent tag or probe; therefore, real-time kinetic measurements of cellular NADH can be obtained [92, 93]. To this end, the Boyd Lab has developed an extracellular approach to estimate ATP production with data collected from real-time oxygen consumption and NADH production assays. These datasets allow for stoichiometric determinations of ATP production in real time. Theoretically, mitochondrial oxidative phosphorylation is responsible for producing a substantial portion of cellular ATP, and traditionally, NADH and oxygen are related to ATP production, as shown below [69, 94]:

$$1\,\text{NADH} + \frac{1}{2}\text{O}_2 = 3\,\text{ATP} \tag{8.3}$$

Since our assay monitors oxygen consumption, the equation can be doubled:

$$2\,\text{NADH} + \text{O}_2 = 6\,\text{ATP} \tag{8.4}$$

Additionally, cells can produce ATP via glycolysis, where 1 ATP is generated for each available NADH. This may be particularly true for HepG2 cells, since many cancer cells have been shown to have a high reliance on glycolysis due to the Warburg effect [95, 96]. To account for ATP production when there is limited or no oxygen, an "if-then-else loop" was used to calculate theoretical ATP generation:

$$\text{If } 6\times[\text{O}_2]_{\text{sample}} - 2\times[\text{NADH}]_{\text{sample}} > 0;$$

$$\text{Then } \frac{\left[6\times[\text{O}_2]_{\text{sample}}\right] - \left[2\times[\text{NADH}]_{\text{sample}}\right]}{4}$$

$$\text{Else } \frac{[\text{NADH}]_{\text{sample}}}{4}$$

This method has proven successful for two disparate xenobiotics, demonstrating a strong correlation to relative ATP measurements collected via the luciferase assay [93].

8.4.3 Prediction of Posttranslational Phosphorylation Response for Mixtures

As previously mentioned, experimentally determining all possible mixture combinations for a range of doses at different time points is not possible. Therefore, prediction models capable of predicting mixture responses are necessary; however, even relatively simple binary mixture interactions can be difficult to predict based on the current lack of spatiotemporal mechanisms for the individual components. New risk assessment approaches would include a suite of assays that cast a wide net on early cellular changes, such as changes in cellular bioenergetics, and various pathway perturbations, such as alterations in PTMs post-exposure, to fully understand the cellular response, whether that be adaptation after exposure to a new homeostatic state, or cell death [43]. Understanding and eventually predicting chemical mixture effects could not only elucidate potential risk factors associated with exposures but also offer the opportunity to estimate mixtures toxicity at various doses without measuring the actual mixture. With this new risk assessment methodology in mind, Boyd et al. [97] have shown that early pathway perturbation responses, such as phosphorylation PTMs, at time points relevant to critical signaling events, as indicated by an increase in relative oxygen consumption, to chemical mixture exposures (ETC inhibitors as well as broad kinase inhibitors) can be predicted *in vitro* using Bliss independence.

8.4.3.1 Using Bliss Independence (Response Addition) to Predict Relative Phosphorylation During Critical Signaling Events

To illustrate the utility of this method, two mitochondrial ETC inhibitors were used: deguelin (complex I ETC inhibitor) and potassium cyanide (KCN) (complex IV ETC inhibitor). The critical signaling event for deguelin and KCN was estimated to be 400 minutes post-exposure using HepG2 cells and relative oxygen consumption observations [97]. Since alterations in oxygen consumption (cellular respiration) may suggest early perturbations of cellular bioenergetics and potentially adverse effects, Boyd et al. [97] chose a time point where all dosing conditions experienced the highest degree of change. At the identified time points post-exposure, a snapshot of the toxicodynamic response to deguelin and KCN exposure can be determined with a high-throughput approach. Inclusion of valuable proteins that cast a wide net on cell death or survival pathways at this time point is necessary to capture intracellular processes related to adverse or adaptive perturbation responses. After exposure to cellular stressors (e.g. xenobiotics, endogenous molecules), many kinase pathways converge upon mitochondria, which can result in mitochondrial membrane permeabilization (MMP)-mediated death [71, 76]. From this, the list of potential proteins was narrowed to those that are only a few kinase steps removed from key survival or death proteins that are known to alter mitochondrial activity. To this end, a multiplex bead-based ELISA assay was designed and used to simultaneously determine different protein phosphorylation responses for deguelin and KCN (Figure 8.4) at the estimated critical signaling event of 400 minutes post-exposure. Cells were lysed and measured for phosphoprotein response at 400 minutes post-exposure to deguelin and KCN

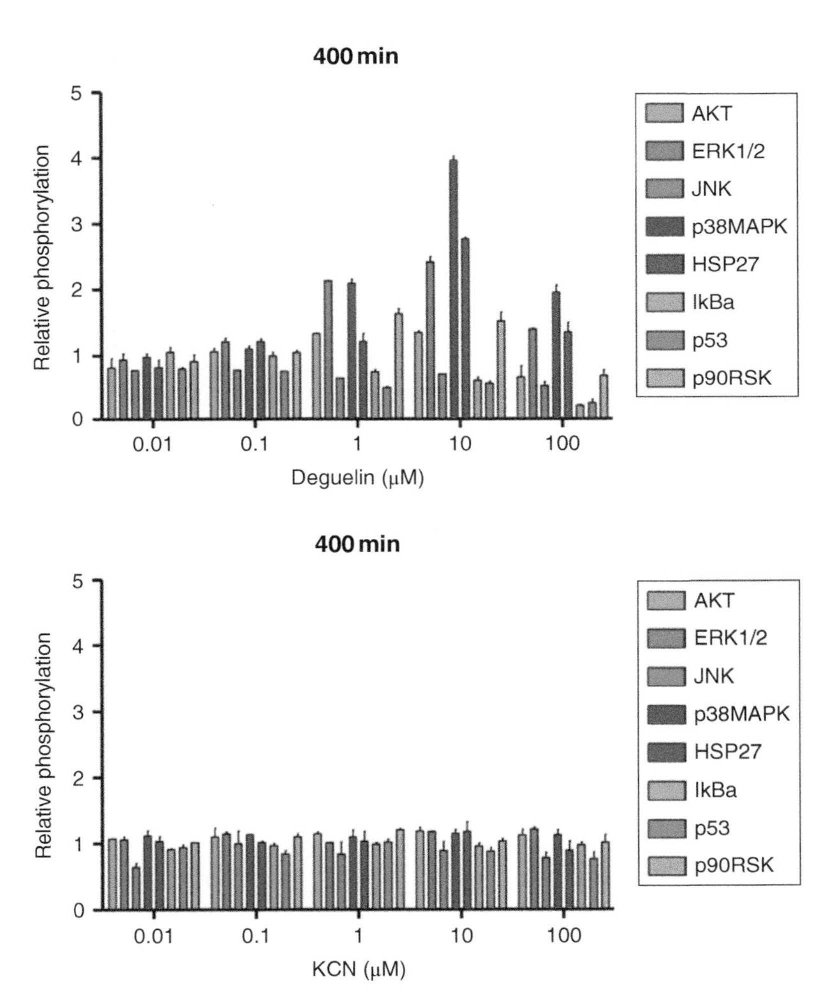

Figure 8.4 Relative phosphorylation responses of HepG2 cells exposed to deguelin or KCN. Relative protein phosphorylation was determined by dosing HepG2 cells with various doses of deguelin (0.01–100 μM) or KCN (0.01–100 μM) and lysing the cell membrane at 400 minutes post-exposure. Relative phosphorylation was calculated by normalizing to vehicle controls. Error bars reflect standard error of the mean (S.E.M.). Assay was performed in duplicate.

alone (Figure 8.4) or as a chemical mixture (Appendix A). Phosphoprotein responses were determined relative to control, which received dosing vehicle (<1% dimethyl sulfoxide [DMSO] for deguelin or water for KCN).

For the same mixture and dosing regimen, "predicted" relative phosphorylation responses at the key time points of interest were determined using Bliss independence (Appendix A) [97]. Experimentally determined phosphoprotein responses of individual chemicals were used to calculate predicted values at the time points of interest. To determine the accuracy of the prediction method, correlation analysis was used for observed and predicted relative phosphorylation responses (Figure 8.5). Prediction of relative phosphoprotein responses to deguelin and KCN mixtures at 400 minutes post-exposure was strongly correlated to experimentally determined mixture phosphoprotein responses ($r = 0.904$, $N = 200$, $P < 0.0001$).

By utilizing downstream convergence of intracellular signal transduction to simplify the understanding of xenobiotics that, when combined, result in toxicity, this method accurately predicted mixture interactions (87.5%). Not only was this method successful for mixture predictions on two chemicals with similar activity (inhibition of ETC activity), but it also proved to be successful for two disparate chemicals, deguelin and staurosporine, which is a broad kinase inhibitor. For deguelin and staurosporine mixtures, this method had

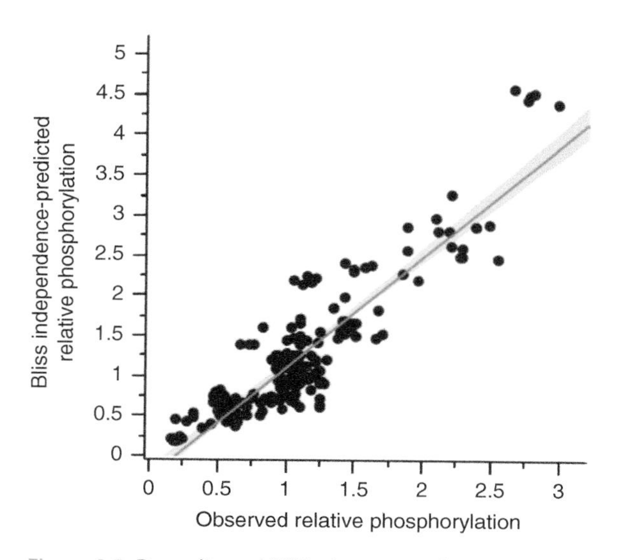

Figure 8.5 Deguelin and KCN mixture correlation. Correlation plot of observed and Bliss independence-predicted phosphoprotein mixture responses at 400 minutes post-exposure. Data points represent the mean of the mixture response. For this mixture dataset, predicted values were strongly correlated to observed values, $r(200) = 0.904$, $P < 0.0001$.

77.5% accuracy [97]. Thus, by monitoring a limited number of pathways, one can screen xenobiotics for potential mixture interactions.

In conclusion, this methodology may offer a new model for risk assessment capable of harnessing early cellular changes post-exposure to predict potential mixture effects related to cytotoxicity in order to predict cellular fate. The approach presented in this chapter could be used to test the tens of thousands of chemicals currently used and predict all combinations at multiple doses that have the potential to be synergistic (i.e. greater than expected from additivity) or antagonistic (i.e. less than expected from additivity). This would then allow current risk assessment to narrow the experimental efforts to only those doses and chemicals of concern for further *in vivo* testing.

References

1 Casarett, L.J. and Klaassen, C.D. (eds.) (2008). *Casarett and Doull's Toxicology: The Basic Science of Poisons*. New York: McGraw-Hill Medical.

2 Pottenger, L.H. and Gollapudi, B.B. (2009). A case for a new paradigm in genetic toxicology testing. *Mutat. Res.* 678 (2): 148–151.

3 Edwards, I.R. and Aronson, J.K. (2000). Adverse drug reactions: definitions, diagnosis, and management. *Lancet* 356 (9237): 1255–1259.

4 Calabrese, E.J. and Baldwin, L.A. (2003). Toxicology rethinks its central belief. *Nature* 421 (6924): 691–692.

5 Calabrese, E.J. and Baldwin, L.A. (2003). The hormetic dose-response model is more common than the threshold model in toxicology. *Toxicol. Sci.* 71 (2): 246–250.

6 Cox, C. (1987). Threshold dose-response models in toxicology. *Biometrics* 43 (3): 511–523.

7 Allen, B.C., Kavlock, R.J., Kimmel, C.A., and Faustman, E.M. (1994). Dose-response assessment for developmental toxicity. II. Comparison of generic benchmark dose estimates with no observed adverse effect levels. *Fundam. Appl. Toxicol.* 23 (4): 487–495.

8 Calabrese, E.J. (2009). Getting the dose-response wrong: why hormesis became marginalized and the threshold model accepted. *Arch. Toxicol.* 83 (3): 227–247.

9 Calabrese, E.J. (2008). Hormesis: why it is important to toxicology and toxicologists. *Environ. Toxicol. Chem.* 27 (7): 1451–1474.

10 Calabrese, E.J. (2001). Overcompensation stimulation: a mechanism for hormetic effects. *Crit. Rev. Toxicol.* 31 (4–5): 425–470.

11 Kholodenko, B.N., Hancock, J.F., and Kolch, W. (2010). Signalling ballet in space and time. *Nat. Rev. Mol. Cell Biol.* 11 (6): 414–426.

12 Cedergreen, N., Christensen, A.M., Kamper, A. et al. (2008). A review of independent action compared to concentration addition as reference models

for mixtures of compounds with different molecular target sites. *Environ. Toxicol. Chem.* 27 (7): 1621–1632.

13 Groten, J.P., Feron, V.J., and Suhnel, J. (2001). Toxicology of simple and complex mixtures. *Trends Pharmacol. Sci.* 22 (6): 316–322.

14 Feron, V.J., Groten, J.P., and van Bladeren, P.J. (1998). Exposure of humans to complex chemical mixtures: hazard identification and risk assessment. In: *Diversification in Toxicology – Man and Environment*, Archives of Toxicology, vol. 20 (ed. J.P. Seiler, J.L. Autrup and H. Autrup), 363–373. Berlin, Heidelberg: Springer.

15 McCarty, L.S. and Borgert, C.J. (2006). Review of the toxicity of chemical mixtures: theory, policy, and regulatory practice. *Regul. Toxicol. Pharmacol.* 45 (2): 119–143.

16 Borgert, C.J., Quill, T.F., McCarty, L.S., and Mason, A.M. (2004). Can mode of action predict mixture toxicity for risk assessment? *Toxicol. Appl. Pharmacol.* 201 (2): 85–96.

17 Loewe, S. and Muischnek, H. (1926). Effect of combinations: mathematical basis of problem. *Arch. Exp. Pathol. Pharmacol.* 114: 313–326.

18 Berenbaum, M.C. (1989). What is synergy? *Pharmacol. Rev.* 41 (2): 93–141.

19 Rajapakse, N., Ong, D., and Kortenkamp, A. (2001). Defining the impact of weakly estrogenic chemicals on the action of steroidal estrogens. *Toxicol. Sci.* 60 (2): 296–304.

20 Bliss, C.I. (1939). The toxicity of poisons applied jointly. *Ann. Appl. Biol.* 26 (3): 585–615.

21 Boyd, J., Saksena, A., Patrone, J.B. et al. (2011). Exploring the boundaries of additivity: mixtures of NADH: quinone oxidoreductase inhibitors. *Chem. Res. Toxicol.* 24 (8): 1242–1250.

22 U.S. Environmental Protection Agency (1986). *Guidelines for the Health Risk Assessment of Chemical Mixtures.* Washington, DC: EPA.

23 U.S. Environmental Protection Agency (1999). *Guidance for Identifying Pesticide Chemicals That Have a Common Mechanism of Toxicity.* Washington, DC: EPA.

24 U.S. Environmental Protection Agency (2000). *Supplementary Guidance for Conducting Health Risk Assessment of Chemical Mixtures.* Washington, DC: EPA.

25 U.S. Environmental Protection Agency (2002). *Guidance on Cumulative Risk Assessment of Pesticide Chemicals That Have a Common Mechanism of Toxicity.* Washington, DC: EPA.

26 Agency for Toxic Substance and Disease Registry (2001). *Guidance for the Preparation of an Interaction Profile.* Atlanta, GA: HHS.

27 Agency for Toxic Substance and Disease Registry (2001). *Guidance Manual for the Assessment of Joint Toxic Action of Chemical Mixtures.* Atlanta, GA: HHS.

28 Aptula, A.O. and Roberts, D.W. (2006). Mechanistic applicability domains for nonanimal-based prediction of toxicological end points: general principles and application to reactive toxicity. *Chem. Res. Toxicol.* 19 (8): 1097–1105.

29 Spurgeon, D.J., Jones, O.A., Dorne, J.L. et al. (2010). Systems toxicology approaches for understanding the joint effects of environmental chemical mixtures. *Sci. Total Environ.* 408 (18): 3725–3734.

30 Butterworth, B.E., Conolly, R.B., and Morgan, K.T. (1995). A strategy for establishing mode of action of chemical carcinogens as a guide for approaches to risk assessments. *Cancer Lett.* 93 (1): 129–146.

31 Schlosser, P.M. and Bogdanffy, M.S. (1999). Determining modes of action for biologically based risk assessments. *Regul. Toxicol. Pharmacol.* 30 (1): 75–79.

32 Dellarco, V.L. and Wiltse, J.A. (1998). US Environmental Protection Agency's revised guidelines for Carcinogen Risk Assessment: incorporating mode of action data. *Mutat. Res.* 405 (2): 273–277.

33 Jonker, M.J., Svendsen, C., Bedaux, J.J. et al. (2005). Significance testing of synergistic/antagonistic, dose level-dependent, or dose ratio-dependent effects in mixture dose-response analysis. *Environ. Toxicol. Chem.* 24 (10): 2701–2713.

34 Meled, M., Thrasyvoulou, A., and Belzunces, L.P. (1998). Seasonal variations in susceptibility of *Apis mellifera* to the synergistic action of prochloraz and deltamethrin. *Environ. Toxicol. Chem.* 17 (12): 2517–2520.

35 Forget, J., Pavillon, J.F., Beliaeff, B., and Bocquene, G. (1999). Joint action of pollutant combinations (pesticides and metals) on survival (LC50 values) and acetylcholinesterase activity of *Tigriopus brevicornis* (Copepoda, Harpacticoida). *Environ. Toxicol. Chem.* 18 (5): 912–918.

36 Posthuma, L., Baerselman, R., Van Veen, R.P., and Dirven-Van Breemen, E.M. (1997). Single and joint toxic effects of copper and zinc on reproduction of Enchytraeus crypticus in relation to sorption of metals in soils. *Ecotoxicol. Environ. Saf.* 38 (2): 108–121.

37 Van Gestel, C.A.M. and Hensbergen, P.J. (1997). Interaction of Cd and Zn toxicity for *Folsomia candida Willem* (Collembola:Isotomidae) in relation to bioavailability in soil. *Environ. Toxicol. Chem.* 16 (6): 1177–1186.

38 Gennings, C., Carter, W., Campain, J. et al. (2002). Statistical analysis of interactive cytotoxicity in human epidermal keratinocytes following exposure to a mixture of four metals. *J. Agric. Biol. Environ. Stat.* 7: 58–73.

39 Jonker, M.J., Piskiewicz, A.M., Ivorra i Castella, N., and Kammenga, J.E. (2004). Toxicity of binary mixtures of cadmium-copper and carbendazim-copper to the nematode *Caenorhabditis elegans. Environ. Toxicol. Chem.* 23 (6): 1529–1537.

40 National Research Council (2007). *Toxicity Testing in the 21st Century: A Vision and A Strategy*. Washington, DC: National Academies Press.

41 Bhattacharya, S., Zhang, Q., Carmichael, P.L. et al. (2011). Toxicity testing in the 21 century: defining new risk assessment approaches based on perturbation of intracellular toxicity pathways. *PLoS One* 6 (6): e20887.

42 Attene-Ramos, M.S., Miller, N., Huang, R. et al. (2013). The Tox21 robotic platform for the assessment of environmental chemicals – from vision to reality. *Drug Discov. Today* 18 (15–16): 716–723.

43 Andersen, M.E. and Krewski, D. (2009). Toxicity testing in the 21st century: bringing the vision to life. *Toxicol. Sci.* 107 (2): 324–330.
44 Boelsterli, U.A. (2007). *Mechanistic Toxicology: The Molecular Basis of How Chemicals Disrupt Biological Targets*, 2e. Boca Raton, FL: CRC Press.
45 Andersen, M.E. (2010). Calling on science: making "alternatives" the new gold standard. *ALTEX* 27: 29–37.
46 Kholodenko, B.N. (2006). Cell-signalling dynamics in time and space. *Nat. Rev. Mol. Cell. Biol.* 7 (3): 165–176.
47 Houck, K.A. and Kavlock, R.J. (2008). Understanding mechanisms of toxicity: insights from drug discovery research. *Toxicol. Appl. Pharmacol.* 227 (2): 163–178.
48 Murphy, L.O., MacKeigan, J.P., and Blenis, J. (2004). A network of immediate early gene products propagates subtle differences in mitogen-activated protein kinase signal amplitude and duration. *Mol. Cell. Biol.* 24 (1): 144–153.
49 von Kriegsheim, A., Baiocchi, D., Birtwistle, M. et al. (2009). Cell fate decisions are specified by the dynamic ERK interactome. *Nat. Cell. Biol.* 11 (12): 1458–1464.
50 Wold, F. (1981). In vivo chemical modification of proteins (post-translational modification). *Annu. Rev. Biochem.* 50: 783–814.
51 Aye-Han, N.N., Ni, Q., and Zhang, J. (2009). Fluorescent biosensors for real-time tracking of post-translational modification dynamics. *Curr. Opin. Chem. Biol.* 13 (4): 392–397.
52 Wang, Y.C., Peterson, S.E., and Loring, J.F. (2014). Protein post-translational modifications and regulation of pluripotency in human stem cells. *Cell Res.* 24 (2): 143–160.
53 Kumar, D., Srikanth, R., Ahlfors, H. et al. (2007). Capturing cell-fate decisions from the molecular signatures of a receptor-dependent signaling response. *Mol. Syst. Biol.* 3: 150.
54 Schilling, M., Maiwald, T., Hengl, S. et al. (2009). Theoretical and experimental analysis links isoform-specific ERK signalling to cell fate decisions. *Mol. Syst. Biol.* 5: 334.
55 Bononi, A., Agnoletto, C., De Marchi, E. et al. (2011). Protein kinases and phosphatases in the control of cell fate. *Enzyme Res.* 2011: 329098.
56 Endicott, J.A., Noble, M.E., and Johnson, L.N. (2012). The structural basis for control of eukaryotic protein kinases. *Annu. Rev. Biochem.* 81: 587–613.
57 Wang, Y., Eddy, J.A., and Price, N.D. (2012). Reconstruction of genome-scale metabolic models for 126 human tissues using mCADRE. *BMC Syst. Biol.* 6: 153.
58 Paul, A., Wilson, S., Belham, C.M. et al. (1997). Stress-activated protein kinases: activation, regulation and function. *Cell Signal.* 9 (6): 403–410.
59 Tibbles, L.A. and Woodgett, J.R. (1999). The stress-activated protein kinase pathways. *Cell. Mol. Life Sci.* 55 (10): 1230–1254.
60 Pouyssegur, J., Volmat, V., and Lenormand, P. (2002). Fidelity and spatio-temporal control in MAP kinase (ERKs) signalling. *Biochem. Pharmacol.* 64 (5–6): 755–763.

61 Junttila, M.R., Li, S.P., and Westermarck, J. (2008). Phosphatase-mediated crosstalk between MAPK signaling pathways in the regulation of cell survival. *FASEB J* 22 (4): 954–965.
62 Currie, H.N., Vrana, J.A., Han, A.A. et al. (2014). An approach to investigate intracellular protein network responses. *Chem. Res. Toxicol.* 27 (1): 17–26.
63 Hammerman, P.S., Fox, C.J., and Thompson, C.B. (2004). Beginnings of a signal-transduction pathway for bioenergetic control of cell survival. *Trends Biochem. Sci.* 29 (11): 586–592.
64 Buttgereit, F. and Brand, M.D. (1995). A hierarchy of ATP-consuming processes in mammalian cells. *Biochem. J.* 312 (Pt 1): 163–167.
65 Ainscow, E.K. and Brand, M.D. (1999). Top-down control analysis of ATP turnover, glycolysis and oxidative phosphorylation in rat hepatocytes. *Eur. J. Biochem.* 263 (3): 671–685.
66 Mohamed, S.M., Veeranarayanan, S., Minegishi, H. et al. (2014). Cytological and subcellular response of cells exposed to the type-1 RIP curcin and its hemocompatibility analysis. *Sci. Rep.* 4: 5747.
67 Law, M. and Elmore, S. (2008). Mechanisms of cell death. In: *Molecular and Biochemical Toxicology*, 5e (ed. R.C. Smart and E. Hodgson), 327–369. Hoboken: Wiley.
68 Ninomiya-Tsuji, J. (2008). Mitochondrial dysfunction. In: *Molecular and Biochemical Toxicology*, 5e (ed. R.C. Smart and E. Hodgson), 371–389. Hoboken: Wiley.
69 Lehninger, A., Cox, M.M., and Nelson, D.L. (eds.) (2008). *Lehninger Principles of Biochemistry, W.* New York: H. Freeman.
70 Desagher, S. and Martinou, J.C. (2000). Mitochondria as the central control point of apoptosis. *Trends Cell. Biol.* 10 (9): 369–377.
71 Kroemer, G., Galluzzi, L., and Brenner, C. (2007). Mitochondrial membrane permeabilization in cell death. *Physiol. Rev.* 87 (1): 99–163.
72 Fosslien, E. (2001). Mitochondrial medicine—molecular pathology of defective oxidative phosphorylation. *Ann. Clin. Lab. Sci.* 31 (1): 25–67.
73 Veltri, K.L., Espiritu, M., and Singh, G. (1990). Distinct genomic copy number in mitochondria of different mammalian organs. *J. Cell. Physiol.* 143 (1): 160–164.
74 Nunnari, J. and Suomalainen, A. (2012). Mitochondria: in sickness and in health. *Cell* 148 (6): 1145–1159.
75 Dawson, D.A., Allen, E.M., Allen, J.L. et al. (2014). Time-dependence in mixture toxicity prediction. *Toxicology* 326: 153–163.
76 Rehm, M., Huber, H.J., Hellwig, C.T. et al. (2009). Dynamics of outer mitochondrial membrane permeabilization during apoptosis. *Cell Death Differ.* 16 (4): 613–623.
77 Lemasters, J.J., Nieminen, A.L., Qian, T. et al. (1998). The mitochondrial permeability transition in cell death: a common mechanism in necrosis, apoptosis and autophagy. *Biochim. Biophys. Acta.* 1366 (1–2): 177–196.

78 Messam, C.A. and Pittman, R.N. (1998). Asynchrony and commitment to die during apoptosis. *Exp. Cell. Res.* 238 (2): 389–398.
79 Dix, D.J., Houck, K.A., Martin, M.T. et al. (2007). The ToxCast program for prioritizing toxicity testing of environmental chemicals. *Toxicol. Sci.* 95 (1): 5–12.
80 Andersen, M.E., Al-Zoughool, M., Croteau, M. et al. (2010). The future of toxicity testing. *J. Toxicol. Environ. Health B. Crit. Rev.* 13 (2-4): 163–196.
81 O'Brien, P.J., Irwin, W., Diaz, D. et al. (2006). High concordance of drug-induced human hepatotoxicity with in vitro cytotoxicity measured in a novel cell-based model using high content screening. *Arch. Toxicol.* 80 (9): 580–604.
82 Holliday, D.L. and Speirs, V. (2011). Choosing the right cell line for breast cancer research. *Breast Cancer Res.* 13 (4): 215.
83 Sasaki, A., Oshima, Y., and Fujimura, A. (2007). An approach to elucidate potential mechanism of renal toxicity of arsenic trioxide. *Exp. Hematol.* 35 (2): 252–262.
84 Kavlock, R., Chandler, K., Houck, K. et al. (2012). Update on EPA's ToxCast program: providing high throughput decision support tools for chemical risk management. *Chem. Res. Toxicol.* 25 (7): 1287–1302.
85 Mersch-Sundermann, V., Knasmuller, S., Wu, X.J. et al. (2004). Use of a human-derived liver cell line for the detection of cytoprotective, antigenotoxic and cogenotoxic agents. *Toxicology* 198 (1–3): 329–340.
86 Sipes, N.S., Martin, M.T., Kothiya, P. et al. (2013). Profiling 976 ToxCast chemicals across 331 enzymatic and receptor signaling assays. *Chem. Res. Toxicol.* 26 (6): 878–895.
87 Selkirk, J.K. (1977). Divergence of metabolic activation systems for short-term mutagenesis assays. *Nature* 270: 604–607.
88 Maslansky, C.J. and Williams, G.M. (1982). Primary cultures and the levels of cytochrome P450 in hepatocytes from mouse, rat, hamster, and rabbit liver. *In Vitro* 18 (8): 683–693.
89 Knasmuller, S., Parzefall, W., Sanyal, R. et al. (1998). Use of metabolically competent human hepatoma cells for the detection of mutagens and antimutagens. *Mutat. Res.* 402 (1–2): 185–202.
90 Berg, J., Hung, Y.P., and Yellen, G. (2009). A genetically encoded fluorescent reporter of ATP:ADP ratio. *Nat. Methods* 6 (2): 161–166.
91 Imamura, H., Nhat, K.P., Togawa, H. et al. (2009). Visualization of ATP levels inside single living cells with fluorescence resonance energy transfer-based genetically encoded indicators. *Proc. Natl. Acad. Sci. U.S.A.* 106 (37): 15651–15656.
92 McComb, R.B., Bond, L.W., Burnett, R.W. et al. (1976). Determination of the molar absorptivity of NADH. *Clin. Chem.* 22 (2): 141–150.
93 Vrana, J.A., Currie, H.N., Han, A.A., and Boyd, J. (2014). Forecasting cell death dose-response from early signal transduction responses in vitro. *Toxicol. Sci.* 140 (2): 338–351.

94 Kadenbach, B. (1986). Regulation of respiration and ATP synthesis in higher organisms: hypothesis. *J. Bioenerg. Biomembr.* 18 (1): 39–54.
95 Vander Heiden, M.G., Cantley, L.C., and Thompson, C.B. (2009). Understanding the Warburg effect: the metabolic requirements of cell proliferation. *Science* 324 (5930): 1029–1033.
96 Warburg, O. (1956). On the origin of cancer cells. *Science* 123 (3191): 309–314.
97 Boyd, J., Vrana, J.A., and Williams, H.N. (2013). In vitro approach to predict post-translational phosphorylation response to mixtures. *Toxicology* 313 (2–3): 113–121.

Appendix A

Observed relative phosphorylation of deguelin and KCN mixtures (400 minutes post-exposure).

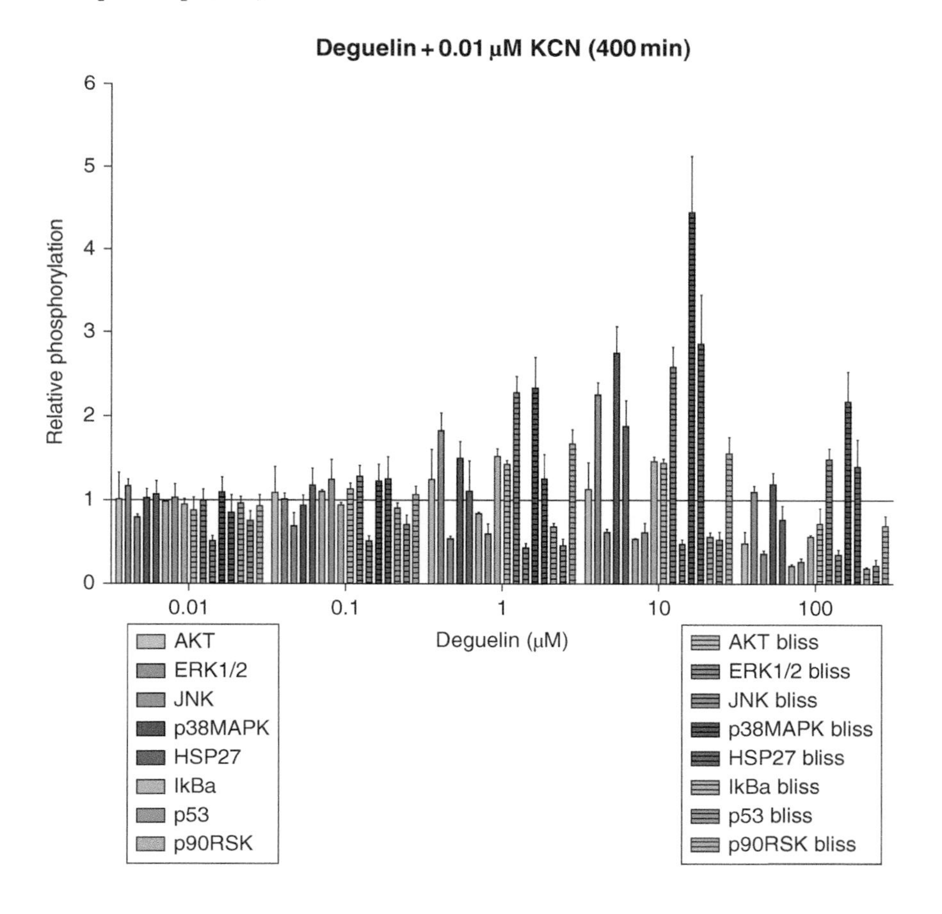

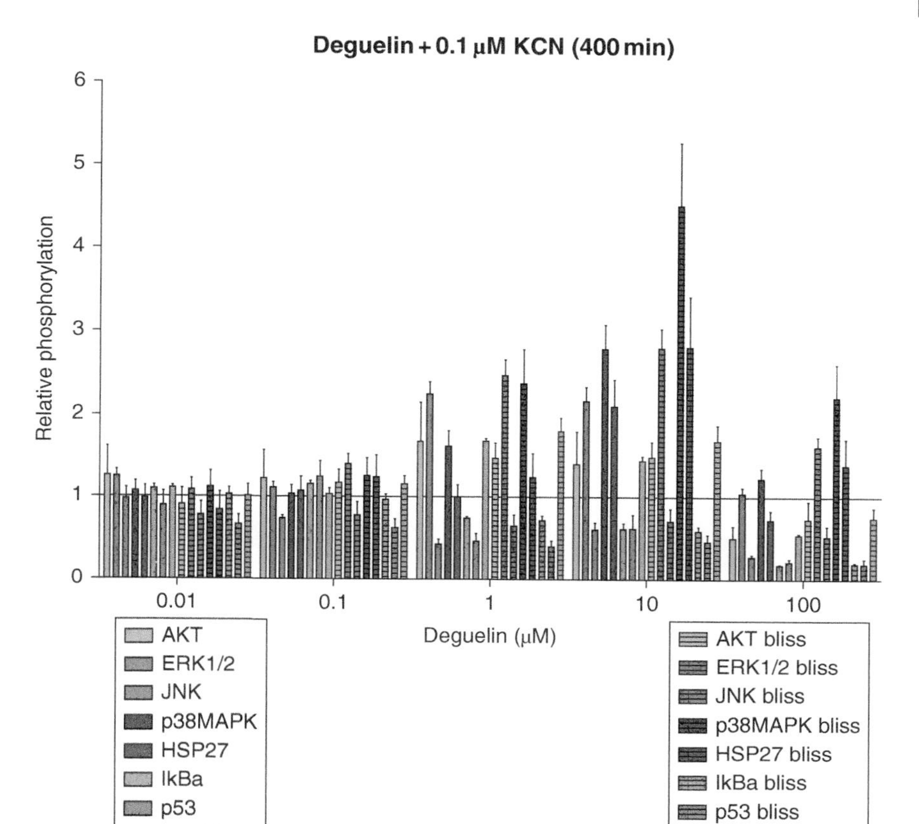
Deguelin + 0.1 μM KCN (400 min)
Relative phosphorylation
6
5
4
3
2
1
0
0.01
0.1
1
10
100
Deguelin (μM)
AKT
ERK1/2
JNK
p38MAPK
HSP27
IkBa
p53
p90RSK
AKT bliss
ERK1/2 bliss
JNK bliss
p38MAPK bliss
HSP27 bliss
IkBa bliss
p53 bliss
p90RSK bliss

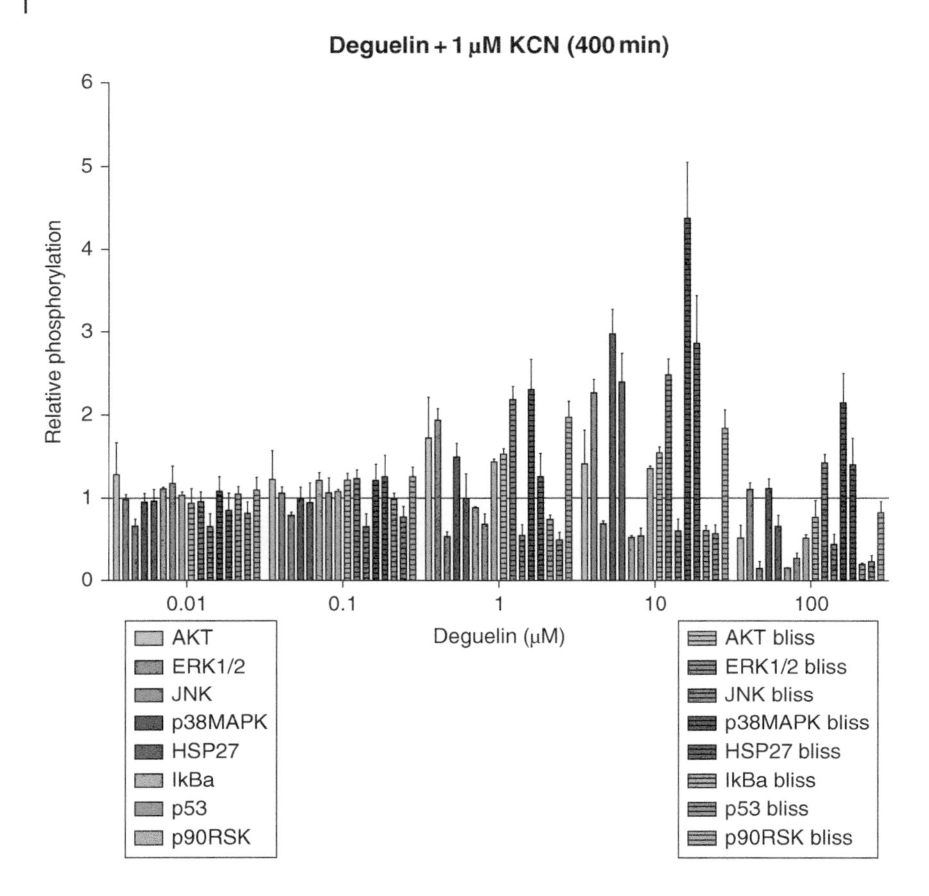
Deguelin + 1 μM KCN (400 min)
6
5
4
3
2
1
0
Relative phosphorylation
0.01
0.1
1
10
100
Deguelin (μM)
AKT
ERK1/2
JNK
p38MAPK
HSP27
IkBa
p53
p90RSK
AKT bliss
ERK1/2 bliss
JNK bliss
p38MAPK bliss
HSP27 bliss
IkBa bliss
p53 bliss
p90RSK bliss

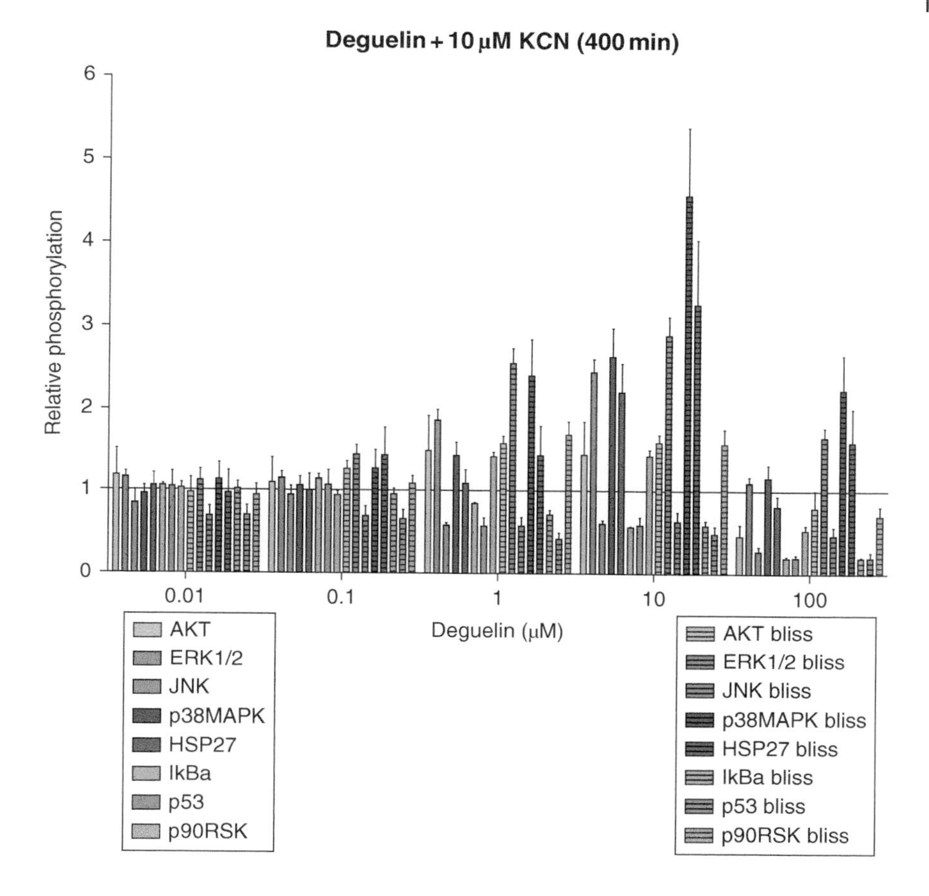
Deguelin + 10 μM KCN (400 min)
Relative phosphorylation
6
5
4
3
2
1
0
0.01
0.1
1
10
100
Deguelin (μM)
AKT
ERK1/2
JNK
p38MAPK
HSP27
IkBa
p53
p90RSK
AKT bliss
ERK1/2 bliss
JNK bliss
p38MAPK bliss
HSP27 bliss
IkBa bliss
p53 bliss
p90RSK bliss

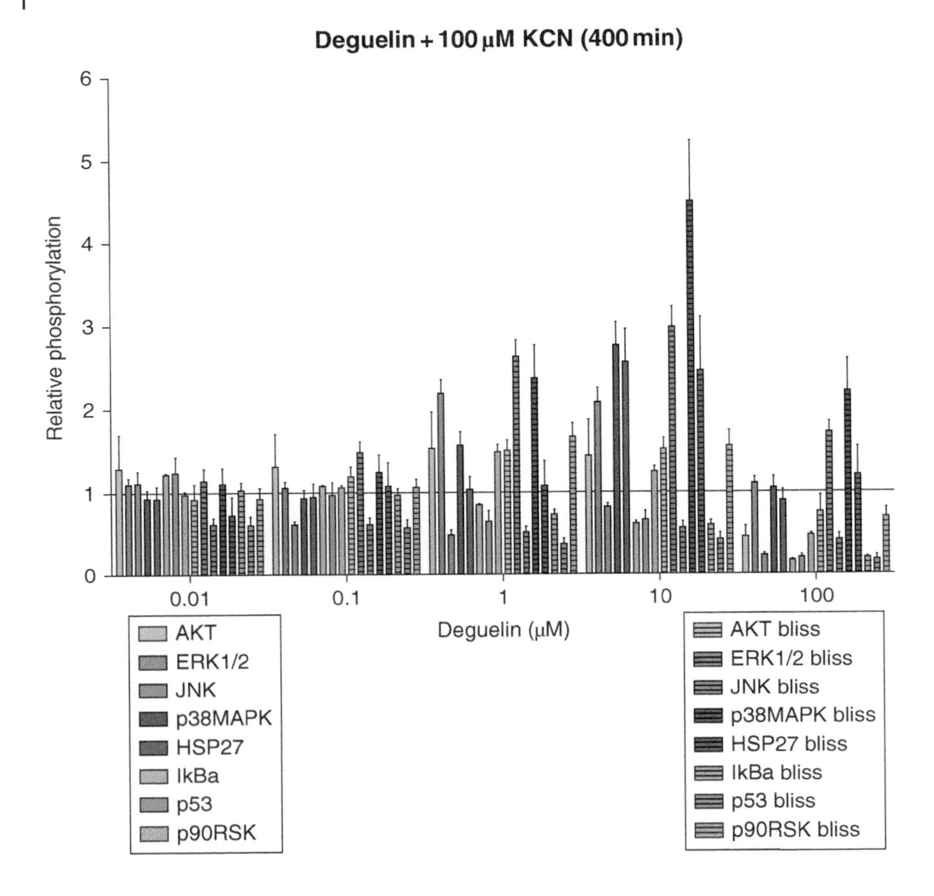
Deguelin + 100 μM KCN (400 min)
Relative phosphorylation
0
1
2
3
4
5
6
0.01
0.1
1
10
100
Deguelin (μM)
AKT
ERK1/2
JNK
p38MAPK
HSP27
IkBa
p53
p90RSK
AKT bliss
ERK1/2 bliss
JNK bliss
p38MAPK bliss
HSP27 bliss
IkBa bliss
p53 bliss
p90RSK bliss

9

Future Research in Signaling

Jonathan W. Boyd[1], *Nicole Prince*[2], *and Marc Birringer*[3]

[1]*Department of Orthopaedics and Department of Physiology and Pharmacology, West Virginia University School of Medicine, Morgantown, WV, USA*
[2]*Department of Orthopaedics, West Virginia University School of Medicine, Morgantown, WV, USA*
[3]*Department of Oecotrophologie, Fulda University of Applied Sciences, Fulda, Hesse, Germany*

The aim of this book is to inform the reader about the intimate relationship between signal transduction and cellular survival/death, but the overarching goal of all mechanistic toxicology and pharmacology is to better understand both the causes of disease and potential treatments. With this goal in mind, it is important to identify areas of future research that could improve our understanding of cell signaling and serve as a bridge that better connects cellular response with disease states. There are some interesting recent studies that provide unique perspectives on signal transduction, which we hope will provide both insight and inspiration to further enhance our understanding of the integration of cellular signaling networks into tissues, organs, and whole organisms.

9.1 Translational Research and a Spatiotemporal Understanding of Signal Transduction

For cellular signal transduction research to reach its potential in toxicology and pharmacology, it is necessary to bridge the gap between early responses and end disease states. Individual cells are preprogrammed with a certain amount of robustness that allows them to respond to many different types of exposures and conditions, but in a eukaryotic organism, they also possess the ability to sense and respond to a multicellular environment. Signaling responses to changes in the localized environment can lead to critical cellular bifurcations, which may result in altered processes or functions for the individual cell. The modified functions of this single cell may impact other local cells, which

Cellular Signal Transduction in Toxicology and Pharmacology: Data Collection, Analysis, and Interpretation, First Edition. Edited by Jonathan W. Boyd and Richard R. Neubig.

changes *their* environment and their subsequent responses and functions. The ultimate result of these functional adaptations could be the presentation of clinical symptoms associated with disease states, which might be treatable given an understanding of the alterations that initiated the process. However, prior to any clinical application, the translation of mechanistic information into practical use requires confirmation that (i) laboratory studies are relevant to clinical/field trials and (ii) cellular-based results are relevant and actionable in the whole organism.

From a clinical translation perspective, many drugs based on alteration of signal transduction cascades have received Food and Drug Administration (FDA) approval following successful clinical trials. Kinase inhibitors make up the largest set of these compounds, and the transformation of our perceived understanding of selectivity and efficacy offers an interesting example of the potential pitfalls of translating mechanistic information into practice. Originally, selectivity was the central theme in identifying new drug candidates from screening of kinase inhibitors. However, along the way, we have discovered that promiscuous inhibitors may hold the greatest therapeutic potential. A hallmark of many disease states is the overexpression of protein kinases, so designing therapeutics with a degree of decreased selectivity can be an attribute that allows kinase inhibitors with multiple targets to overcome the innate cellular resistance to inhibition. The kinase inhibitor imatinib was designed to specifically target one type of cancer, but during its implementation, it was found to inhibit the activity of several additional kinases and proteins, which turned out to be beneficial in reaching its goal [1]. While it may seem counterintuitive to introduce this type of pan-kinase inhibitor, it has shown great promise for tumor treatment and is advantageous when resistance is an issue. Another example of multikinase inhibition as a strategy for treatment is sorafenib, an anticancer drug, which targets Raf kinases, vascular endothelial growth factor receptors (VEGFR), and platelet-derived growth factor (PDGF) receptors that regulate the survival and proliferation signaling of cells. Sorafenib has successfully treated tumors by inhibiting the antiapoptotic signaling that allows both angiogenesis and tumorigenesis [2].

Kinase inhibitor strategies led to investigation of other immune-targeted therapies. Immune signaling proteins, such as cytokines, function by transmitting signals through kinases, and this is thought to be one mechanism behind autoimmune diseases. Inhibiting kinase activity could achieve immunosuppression without causing nonimmunological side effects, and it has recently become a target research area for autoimmune disease, transplant rejection, and anticancer drug development. Further, immune signaling is frequently altered along the route to carcinogenesis through a blockade of immune checkpoints that control the duration and amplitude of physiological immune response. Early immune signaling can drive the balance of stimulatory and inhibitory responses downstream, but tumor cells antagonize these inhibitory pathways to induce tolerance and promote overproliferation [3]. Knowledge of early signal transduction events can allow clinicians to develop cancer therapies that take advantage of the patient's own immune response capability.

Immunotherapy includes any means to increase the endogenous activity of the immune system, including vaccines or cytokine, antibody or T cell proliferation; the desired effect is the enhanced, but "natural," stimulation of the antitumor response or the alterations in tumors that cause heightened susceptibility to the native immune system. For example, combining the durable response of an immunotherapeutic (in some cases, found to reduce recurrence rates by ~10% [4]) with the potential globally high response rates of kinase inhibitors (often greater than 50% for short periods of time before cellular adaptations induce resistance) may produce an efficacious response for some of the most terminal cases of cancer like melanomas [5]. An interesting example is sunitinib, which has approval to treat renal cell cancer (RCC). Sunitinib is a multikinase inhibitor with promiscuous activity against PDGF, c-Kit, VEGFR2, RET-associated kinases, and fms-like tyrosine kinase 3 (FLt3), which also reverses type I immune suppression via an increase in interferon (IFN)-γ producing cells and a decrease in interleukin (IL)-4 producing T cells [6]; it is currently hypothesized that its activity at VEGFR2 and c-Kit may be the mechanistic link to its immunotherapeutic benefits via a decrease in myeloid-derived suppressor cells present in tumors, which are thought to confer immunosuppression [7]. Thus, it is easy to comprehend the critical importance of an understanding of the intricate signaling events that occur with disparate, but locally present cells, in order to apply this knowledge for enhanced clinical outcomes.

From the lessons learned in the drug development, it appears that there is much more research to be done before we can fully capitalize on cellular signal transduction research and translate it into clinical practice. In fact, more basic research associated with validating *in vitro* results in an *in vivo* system could provide a greater understanding of both the mechanisms that are being studied and the organism itself. While this is an enormous undertaking, one that is currently being pursued by researchers in many areas of pharmacology and toxicology, one particular focus area (within this translational arena) that requires additional research is the spatial understanding of signaling at scales large and small.

To fully understand early signaling responses to stressors, one must monitor the signal transduction as it exists in whole tissues. A spatial understanding of signaling in response to a physical stressor provides a facile means to explore the integrated response. For example, immune response after traumatic injury can be probed down a spatial gradient to understand the organism's response as a whole. *In vivo* studies of the levels of phosphorylated proteins in response to traumatic injury provide a bridge to spatial understanding of immune signaling. Phosphorylation is a key to initiating cellular signaling, and researchers can capitalize on this knowledge by testing for phosphorylated protein levels in tissues to probe initial response and repair mechanisms. Han et al. [8] cross-referenced spatial phosphorylation data with earlier data on cytokine response to map a correlation between cellular response and area of injury. Spatial studies in conjunction with temporal data provide a picture of the progression from initial signal through functional response, and this knowledge may be applied to adapt and refine current treatment procedures.

Small-scale differences in spatially relevant signaling have recently been explored and highlight the potential influence of location on the heterogeneity of potential cascades. Several studies have recently sought to understand how spatial patterns can modulate signal transduction, and they have found that the signaling state within a cascade is not only determined by engagement with a target but is also influenced by spatial factors. Studies of T cells and their targets investigated immunological synapses to show how signaling reactions use spatial organization on micrometer scales to change signaling outcomes [9]. In another example, the tumor suppressor Merlin serves a critical function in spatial organization for Hippo signaling, as this pathway is highly dependent on the recruitment of factors to the plasma membrane [10]. Finally, isoform-specific signal outputs have been observed as a consequence of the spatial organization of lipids in Ras signaling [11]. Zhou and colleagues [11] found that K-Ras and H-Ras isoforms produce a specific signal output based on the spatial organization, whereby the signaling profiles result from the compositionally distinct nanocluster assembly of lipids.

9.2 Integrating Second Messengers into Signal Transduction

Beginning with a triggering event (e.g. binding of an exogenous compound with a receptor or protein), second messengers can be thought of as the great mediators of cellular response. Capable of amplifying or dampening a signal in fractions of a second, these small and varied "foot soldiers" are capable of altering the fate of a cell based on their available concentrations. The rising and falling concentrations of a variety of second messengers form what is clearly a chemical code to be interpreted by peptides and proteins, but we unfortunately know very little about this language.

To crack this code, one must first be able to measure and visualize second messengers and their signaling partners, spatially and temporally in real time. There are a few key examples (listed below) that demonstrate the potential that is unlocked once we determine the exact mechanism from initial exposure to functional response, but more techniques are needed. If we hope to better elucidate disease progression or discover potential therapeutics, it is essential that we continue to study the relationships between second messengers and their respective signaling partners that contribute to cellular survival.

H_2O_2 is a central hub in redox signaling and oxidative stress, especially in insulin signaling and growth factor-induced signaling cascades. H_2O_2 diffuses through tissues to initiate immediate cellular effects, such as cell shape changes and recruitment of immune signals, and researchers can monitor this in real time with a variety of probes [12, 13]. The cellular response measured in terms of the production of H_2O_2 can be mapped spatially and temporally to understand tissue healing and cellular proliferation mechanisms in tadpoles as a model for mammalian responses [14]. Following amputation of the tail, there is a sustained production of peroxides during regeneration; further, lowering peroxide levels (either pharmacologically or genetically) will reduce cell generation impair regeneration via the impacts of this potent, localized second messenger on Wnt/β-catenin signaling [15].

Another example is cyclic adenosine monophosphate (cAMP), which is a key second messenger that can affect the β-adrenergic receptor and contribute to fight-or-flight nervous system responses. Temporal studies of cAMP-dependent pathways validate its study for activity of G-protein coupled receptors (GPCRs). By studying the time course of cAMP response, researchers have been able to further understand the variety of factors that influence nervous system signaling [16]. Briefly, a Förster resonance energy transfer (FRET)-based sensor for cAMP was utilized to understand dynamics associated with stimulation and desensitization of a β_2-adrenergic receptor (β_2AR) in human embryonic kidney cells (HEK-293); G-protein-coupled receptor kinase (GRK6) was also explored for its ability to phosphorylate β-arrestin and render it inactive following stimulation. It was found that while cAMP degradation is capable of desensitization, its primary role is the initiation of signals downstream because its activity was much slower than GRK6. This firmly places the inactivation of a receptor into the hands of a closely associated kinase and demonstrates the utility of dynamic measurements to construct signal transduction mechanisms.

Zinc has also been identified as a novel second messenger, and it is thought to operate through two distinct mechanisms. Late zinc signaling is dependent on a change in the transcription of zinc transporters, as zinc is a structural constituent in a great number of proteins, including enzymes of cellular signaling pathways and transcription factors. In contrast, the "zinc wave" is thought to be an early zinc response that is directly induced by an extracellular stimulus with implications for cytokine production [17]. Of critical importance, researchers have been able to visualize this process through time course studies of the proximity and amplitude of response to zinc release, which have shown that zinc is essential for the downstream phosphorylation of kinases that control cell survival and proliferation [18]. Future research endeavors should seek to understand the spatial and temporal dependence of second messengers, as examples like these demonstrate the power that real-time, visual insights can provide to the field.

9.3 Understanding Crosstalk in Signal Transduction

Beyond improving our ability to interpret second messengers, the need to understand signaling crosstalk is a necessity for disentangling the observational complexity associated with response to exposures that could lead to, or treat, disease. Crosstalk is a form of signal integration, where seemingly independent signaling cascades are able to alter each other in response to stimulus. The underlying rationale for crosstalk is unknown, but it has been postulated that interacting species may play several different roles within cellular response: amplification via convergence toward a functional endpoint and directing spatiotemporal activity via divergence that may simply be driven by saturated primary response pathways could also be the result of multifunctional signaling proteins responding with coordinated efforts that create dove-tailed, context-specific signaling.

In order to improve our ability to appreciate exposures from a toxicological perspective or modify cellular response from a pharmacological perspective, we must fully understand all of the disparate variables that impact crosstalk. To this end, there has been some fantastic work to investigate crosstalk in a variety of cell signaling pathways. For example, feedback loop interdependence has shown that crosstalk is context dependent for mitogenic Ras/mitogen-activated protein kinase (MAPK) and survival PI3K/Akt pathways [19]; it was demonstrated that both pathways can activate or inhibit one another through multiple signaling routes and external cues control the response, as this crosstalk mechanism is highly dependent on growth factors and availability of receptors. A separate study of extracellular signal-related kinase (ERK), c-Jun N-terminal kinase (JNK), and p38MAPK showed that crosstalk may be able to actually describe the sometimes conflicting data regarding the roles of specific signal transduction proteins; in this case, it was demonstrated that p38MAPK is pro-survival and its increase in concentration following chemical insults represents a failed attempt at survival, rather than the incorrect, but often cited, interpretation that p38MAPK is proapoptotic [20] (Figure 9.1).

In regards to disease, many researchers have investigated crosstalk interactions between steroid receptors and growth factor receptors in efforts to understand breast cancer progression. Among the wealth of findings, crosstalk between the estrogen receptor (ER) and the epidermal growth factor receptor (EGFR/HER1)/HER2 signaling contributes to the development of resistance to classical endocrine therapies (e.g. tamoxifen) [21]. Importantly, this crosstalk induces both *de novo* (up front) and acquired resistance of breast cancer cells through different mechanisms. For the *de novo* mechanism, short-term

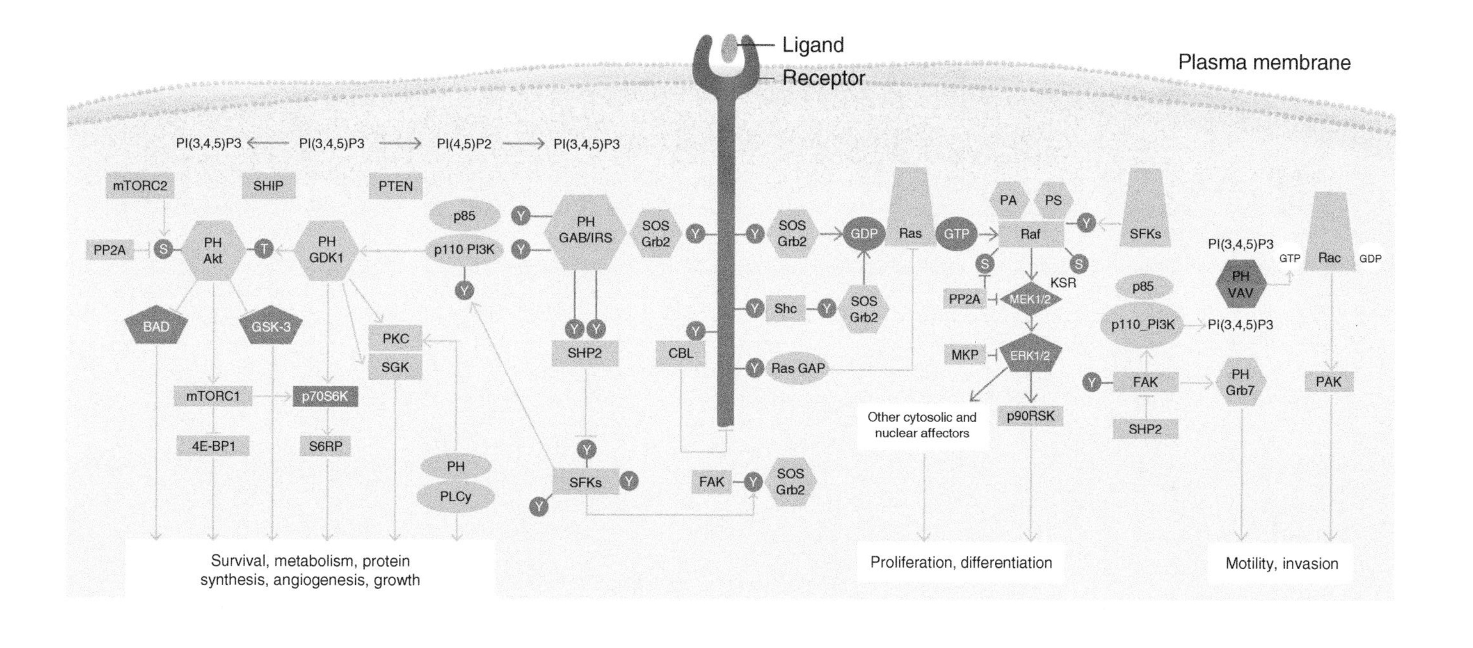

Figure 9.1 Example of crosstalk associated with multiple cell signaling pathways.

exposure to tamoxifen (which is metabolized to 4-hyroxytamoxifen and *N*-desmethyl-4-hydroxytamoxifen, the active metabolites with higher affinity for the ER and actively compete with estradiol) can induce rapid, non-genomic activation of EGFR/HER2 downstream pathways, including the activation of p42/44 MAPK and AKT; these proteins, once activated, then phosphorylate the ER and the coactivator AIB1, which results in increased expression of estradiol [22]. For long-term acquired resistance to tamoxifen, it appears that IGFR activation coupled with EGFR/HER2 activation (and its downstream responses of p42/44 MAPK, AKT, etc.) leads to the upregulation of both ER and estradiol, thereby conferring resistance to breast cancer cells [23, 24]. These findings are all in agreement with clinical studies that have noted that tumors that co-express HER2 and AIB1 have poor outcomes when treated with tamoxifen alone [25, 26]. Overall, it is clear that multiple signal transduction networks modulate ER activity, and while there is much more work to be done, they are critical for future drug design associated with treating breast cancer. Ultimately, while investigating these crosstalk interactions is complex, it is vital to understand the variety of factors involved in toxicological and pharmacological responses.

9.4 Posttranslational Modifications (PTMs) and Target Identification in Signal Transduction

In order to improve our understanding of crosstalk, we must identify potential targets, including all of the potential sites that are available for posttranslational modifications (PTMs). PTMs are firmly integrated within signaling networks, but our ability to determine site availability in a context-specific manner is lacking. Given the dynamic nature of protein/peptide conformation, this is a monumental task, albeit a necessary effort if one intends to mechanistically deconstruct disease states or modify signaling cascades as potential therapeutics.

We are just now beginning to understand that large functional consequences involving PTMs that present as disease are typically the result of many slight modifications associated with clusters of proteins; in many cases, an initial alteration in one protein conformation leads to over- or undercompensation of other proteins within a network, which in turn results in further modifications across a span of proteins that are involved in a particular function. Therefore, in the realm of therapeutics, altering functional outcomes becomes much more complicated than simple inhibition of a single protein, which can lead to either lack of efficacy or unintended side effects. Rather, many new therapeutic regimens are attempting to specifically target particular sites on multiple proteins in order to have a desired effect.

Historically, many of our most effective therapeutics have demonstrated activity at multiple sites, but our knowledge of this was limited. For example, acetylsalicylic acid (aspirin) extracted from willow bark, which was described as a therapeutic more than 2000 years ago, acts as an anti-inflammatory drug not only through activity that includes irreversible inhibition of cyclooxygenase (COX-1) but also through modifying the enzymatic activity of COX-2, resulting in its transition from a cyclooxygenase to a lipoxygenase. This initiates the formation of several mediators (e.g. lipoxins, resolvins, and maresins) thought to contribute to its anti-inflammatory activity [27]. Finally, additional protein targets are thought to contribute aspirin's overall anti-inflammatory functional response when administered to patients for pain or fever, including activity as a mitochondrial electron transport chain uncoupler, a modulator of NF-κB signaling, and stimulus for the increased formation of NO radicals [28–30].

An additional complexity that should be considered when attempting to disentangle the multifunctional roles of PTMs is their ability to not only alter downstream signaling networks but also direct the availability of a protein to be involved in an interaction via protein trafficking. An example is Ras modulation of signaling pathways, which is regulated through PTMs. It was once believed that farnesylation of the CAAX motif alone was responsible for RAS trafficking to the cellular membrane via an irreversible change in conformation that increases lipophilicity. However, it is now known that there are many modifications that alter trafficking, including phosphorylation, nitrosylation, ubiquitination, and peptidyl-prolyl isomerization; further, trafficking is not limited to the plasma membrane, but additional subcellular organelles including the Golgi apparatus where RAS can directly affect GTP–GDP exchange functionality and initiate signaling cascades [31].

Detection of PTMs has been an ongoing effort, with much of the focus relying on mass spectrometry (MS) as a platform to simultaneously measure several different modifications in a sample. Proteomics has expanded the scope of signaling via PTMs to deal with tens of thousands of potentially functional sites, allowing for the discovery of new biologically relevant signaling pathways. For example, lysine acetylation, which regulates chromatin architecture and p53 function, has been thoroughly studied using this technique, and MS has allowed for a deeper understanding of the functional networks associated with this process in an unbiased manner [32]. Recently, this field has expanded to investigate the wide variety of PTMs known to occur physiologically, including phosphorylation, SUMOylation, glycosylation, and beyond. Not only is it possible to quantify these processes, but also through computational methods, researchers can understand the spatial and temporal regulation of PTMs. For example, studies of the EGF receptor have shown ubiquitination directs degradation through lysosome targeting of receptors, all through proteomics techniques [33]. Likewise, monitoring phosphorylation events after EGF stimulation has given insight into the temporal cellular response and consequence of PTM availability [34].

While many methods exist to compute the availability of these sites, the field currently lacks follow-up due to the breadth of information. Certain types of PTMs have been thoroughly studied, such as phosphorylation mapping to identify specific regions of activity, but more classes of PTMs need to be explored to encompass the vast range of possible binding sites and the variability in function that this leads to in eukaryotic organisms [35]. PTM availability is critical for understanding crosstalk, and proteomics has helped set a firm foundation. However, there is much more to be done on this front before a true relationship can be drawn to elucidate toxicological endpoints and therapeutic targets. Eventually, targeted MS strategies may provide a full identification and quantification of all PTMs, which when coupled with network mapping could lead to new therapies, including the potential for customized regiments at the individual level [36].

9.5 Epigenetic Endpoints in Signal Transduction

A potentially transformative new group of signaling targets involving epigenetic regulatory endpoints represents an exciting avenue of research; epigenetic regulation involves a set of processes that affect gene expression without any change to the underlying DNA sequence. While signaling cascades have long been related to initiation of gene expression, research highlighting the ability to impact heritable alterations in expression via modifications to DNA, chromatin, and histones provides a new perspective to cellular signal transduction. The exploration of the relationship between signaling and epigenetics is relatively recent, but there have been, and will continue to be, valuable discoveries that lead to an enhanced understanding of cellular response. For example, it is well known that activated signaling networks can modify the expression of both DNA-methylating enzymes and histone chaperones, which then impact epigenetics, and kinases can directly modify histone tail residues that modulate transcription in a transient epigenetic-like manner [37].

Cells must be able to pass along important short-lived but heritable information regarding their optimization with a certain local environment. This includes not only differentiation and developmental alterations but also information regarding previous responses to localized signaling cues that lead to altered anabolism and catabolism for a particular cell at a distinct time. From a metabolic view, nutrient availability is one of the primary drivers of a cell's response to its local environment, and cells utilize epigenetics to efficiently turn on and off genes for sensing and transport using as little energy as possible. For example, HIF1α and HIF1β bind to histone acetyltransferase p300, resulting in increased local nucleosomal acetylation and increased expression of hypoxia-responsive genes. Further, during times of starvation, ketone bodies (e.g. β-hydroxybutyrate) increase and inhibit histone deacetylase activity (HDAC) [38].

Probing epigenetic alterations promotes an understanding of crosstalk between the genome and epigenome, which also allows researchers to further investigate cellular paths to disease. Studies have shown that while the epigenetic path to cancer is driven by chromatin structure, it is interweaved with genetic alterations that can control the epigenetic response [39]. Transcription factors act with chromatin regulators to produce epigenetic changes in response to stimuli, and these changes can persist long after the signal is gone. For example, AKT phosphorylates several different chromatin-modifying enzymes (EZH2 [40], BMI1 [41], p300 [42]) to reveal or hide DNA sequences, and this process is mediated by growth factors, cytokines, and chemokines, resulting in activity that has been linked to some forms of carcinogenesis [37]. Further, epigenetic enzymes show promise for future cancer therapeutics, as they are prime targets of AKT signaling, ultimately leading to future drugs that alter chromatin modifications to regulate growth and differentiation. Additionally, cellular responses to growth hormones via surface receptors have demonstrated a strong linkage with epigenetic modifications (chromatin modifications, methylation, and noncoding RNA) associated with several downstream signal transduction targets, including Janus-family tyrosine kinase-2 (JAK2), signal transducers and activators of transcription (STAT-1, -3, and -5), MAPKs, insulin receptor substrate-1 (IRS1), phosphoinosotidyinositol-3-phosphate kinase (PI3K), diacylglycerol (DAG), protein kinase-C (PKC), insulin-like growth factor-1 (IGF-1), and even intracellular Ca^{2+} concentrations [43]. Inappropriate activation and inhibition of cell signaling through these epigenetic modifications can lead to disruption of the cell cycle that may eventually manifest as disease. Therefore, understanding precisely how epigenetic modifiers alter gene expression can enhance therapeutic strategies. Overall, signal transduction targets offer a plethora of therapeutic avenues to better design both epigenetic drugs and adjuvants, as well as strategies to enhance treatment regimens that lie at the nexus of several disease states.

Epigenetic alterations are also valuable immunomodulators, especially when applied to areas such as cellular stress response. Acetylation studies of genes that control histone modification in macrophages showed that epigenetic modifications control the selective silencing of inflammatory genes at the chromatin level, illustrating that inflammatory response can be a direct result of epigenetic modifications rather than signaling alone [44]. In this case, a single receptor (TLR4) elicited differential downstream genes (pro-inflammatory vs. antimicrobial) based on chromatin acetylations associated with repeated exposures to lipopolysaccharide that essentially allow for tolerance to inflammation while still inducing antimicrobial activity. An alternative example is *Leishmania*, which has been proposed to induce epigenetic modifications (i.e. DNA methylation) at sites that prevent activation of host defense mechanisms, but the exact mechanism and ramifications on signal transduction remains elusive [45].

In summary, to enhance the field of signal transduction, it is critical to consider epigenetic alterations by investigating the direct communication between signaling molecules and chromatin-modifying enzymes and indirect mechanisms via methylation and even microRNA [46]. One critical concern is defining the time scale of some of these epigenetic modifications that result from signal transduction cascades. Are they transient (i.e. not heritable), or are they longer lasting and capable of being passed down to future generations (of both cells and progeny)? Determining which modifications are heritable will lead to an enhanced understanding of the modifications that can both lead to disease and those which may lead to a cure. Ultimately, this understanding of the intimate relationship between signal transduction, epigenetics, and disease mechanisms holds the promise of leading to more targeted therapies with a greater efficacy (Figure 9.2).

9.6 The Integration of Nutrition and Signal Transduction

In the last few decades, nutrition research has undergone a dramatic transformation. It has evolved from a more or less phenotypical description of nutrient-associated diseases into an "omics"-based research discipline covering all aspects of modern pharmacological approaches. Terms such as "nutrigenomics," "nutraceuticals," or "personalized nutrition" reflect these changes in recent literature. Influences of macro- and micronutrients to signal transduction responses on a cellular level are well documented, but the interplay between nutrients, gut microbiota, and enteroendocrine cells, which senses nutrients and fosters feedback circuits controlling gut motility/secretion, has not been well studied. Further, linking these nutritional systems, with signaling to the brain that can alter eating behavior, demands systems biology approaches. While recent research on model organisms such as nematodes (*Caenorhabditis elegans*), fruit flies (*Drosophila melanogaster*), mice, and rats has filled some of the knowledge gaps, the goal of modern nutrition research will be the translation of basic research into behavioral changes of whole populations.

Understanding energy metabolism in living organisms (including humans) is of fundamental interest since an unbalanced nutrient intake leads to undernutrition or overfeeding. In its recent report, the World Health Organization (WHO) stated that in 2014, more than 1.9 billion adults were overweight or obese, while 462 million were underweight via chronic malnutrition. This dichotomy is often referred to as the double burden of malnutrition, which is characterized by the coexistence of conditions that lead to populations experiencing under- and overnutrition. Directly related to this burden, diet-related noncommunicable diseases (NCD) have increased, headed by metabolic disorders such as type 2 diabetes or cardiovascular diseases [47].

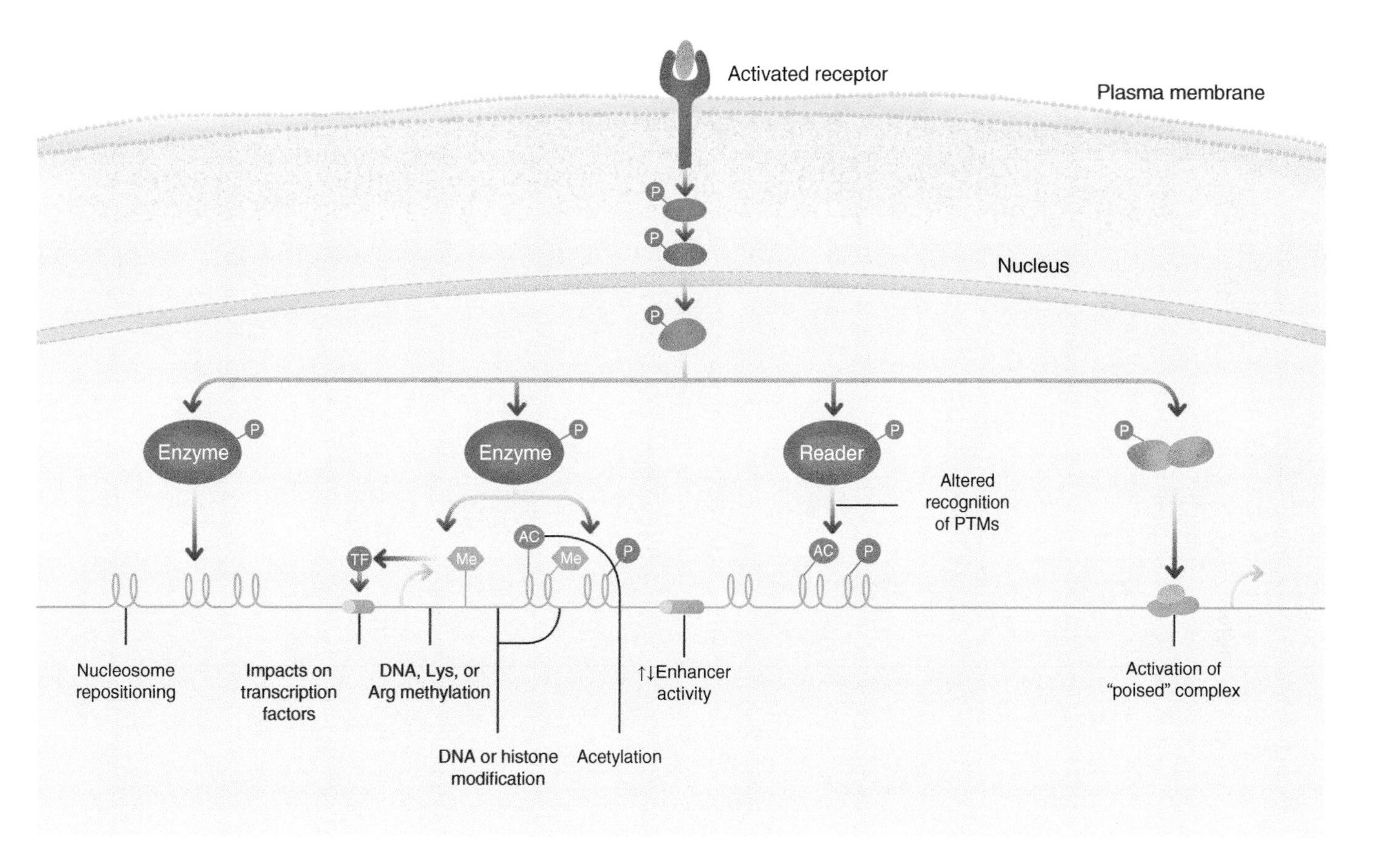

Figure 9.2 Epigenetic routes from signal transduction to chromatin modifications.

Ingestion of food is a complex process that starts with nutrient perception of the five basic taste qualities (sour, sweet, umami, bitter, and salty) in the lingual epithelium (Figure 9.3, for review see Chapter 1). As a brief review, many GPCRs have been assigned to smell and taste signaling in the last decades [48]. The taste receptors of the gustatory buds of the tongue are involved in sweet, umami, and bitter sensation and belong to receptor types T1R and T2R, where a combination of heterodimers of T1R1 and T1R3 are responsible for umami sensation, while T1R2 and T1R3 are responsible for the detection of sweet. Approximately 25–30 different T2R receptors are able to sense more than 700 bitter tastes. All the taste receptors have a large extracellular domain with a Venus flytrap ligand domain (VFT) to scavenge small molecules. Both the T1R2/T1R3 and T1R1/T1R3 heterodimers are coupled to the G-protein gustducin to transmit intracellular signals mediated by second messengers such as cAMP or IP_3-mediated Ca^{2+} release [49] and subsequent opening of transient receptor potential channel (TRMP5) channels, resulting in a depolarization of the taste cell membrane [50]. Interestingly, T1R1, T1R2, and T1R3 are also expressed in enteroendocrine cells (Figure 9.3).

In the last three decades, much of the emphasis in research was driven by investigations into the regulation of hunger and satiety. In general, nutrient sensing is the ability of a cell to react to the presence or absence of a critical nutrient in order to maintain cellular homeostasis. In addition to intracellular signaling, cells of the intestine are able to secrete endocrine, paracrine, and neurocrine signals to communicate within the intestine or between the gut and other organs such as the liver or the brain stem, including the hypothalamic network. Endocrine or neurocrine signaling of the gut is mediated by expression of several cell-specific peptide hormones that are responsible for the regulation of the orexic network [51]. The most investigated peptides include glucose-dependent insulinotropic polypeptide (GIP), cholecystokinin (CCK), glucagon-like peptide-1 (GLP-1), and secretin (see Figure 9.3). More recently, fat tissue has been validated as an essential endocrine organ itself, secreting neuropeptides such as appetite-regulating leptin or adiponectin that strongly correlate with insulin sensitivity of cells [51]. Further, several key cellular sensors are able to regulate intracellular downstream processes such as AMP-activated kinase (AMPK) that acts as a master energy sensor or the target of rapamycin (TOR), which regulates cell growth by monitoring amino acids concentration [52]; both of these factors are highly conserved across several species [53–55]. In addition, often underestimated is the role of oxygen as a nutrient. Under hypoxic conditions, hypoxia-inducible factor α (HIFα) induces the expression of glycolytic genes, allowing cells to use anaerobic glycolysis instead of oxygen-dependent oxidative phosphorylation [56].

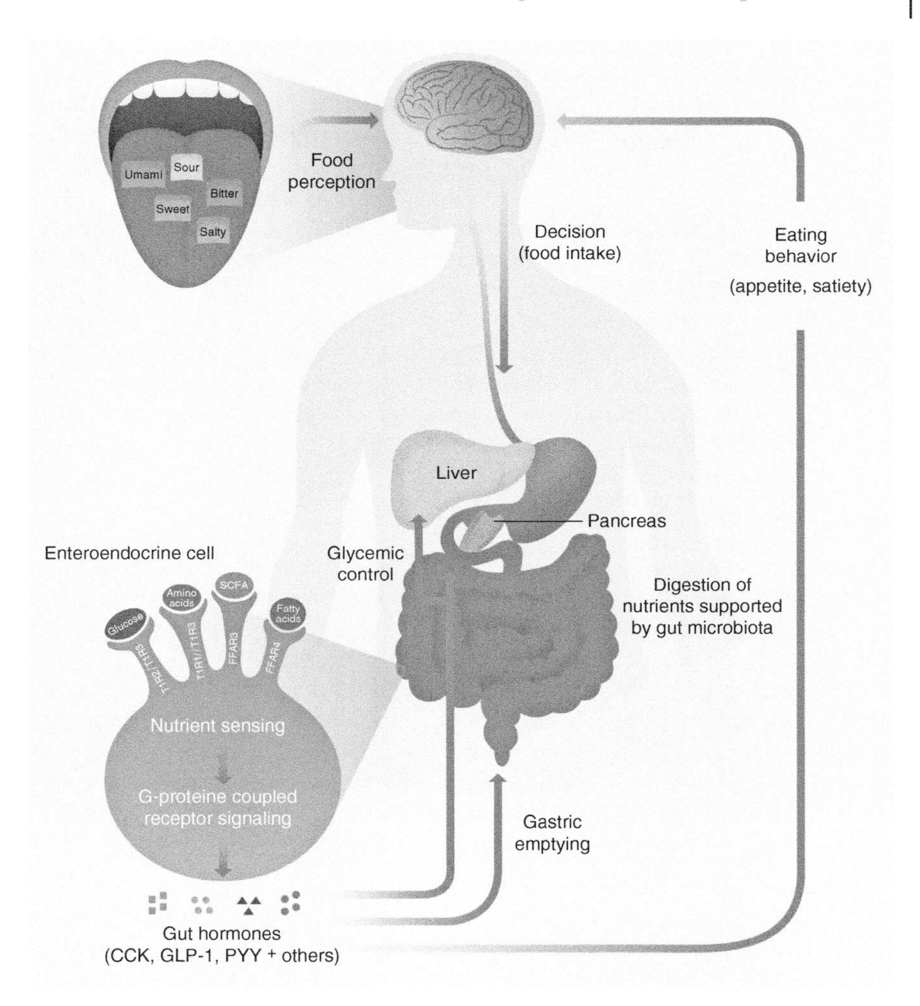

Figure 9.3 The effects of food ingestion on hormonal signaling in enteroendocrine cells.

9.6.1 Cellular AMPK Signaling

The most important "energy currency" of the living cell is ATP. Hydrolysis of ATP to ADP and phosphate delivers chemical energy that is used by countless biological processes in all living organisms. The ratio of ATP to ADP (or to AMP) is tightly monitored by AMPK to prevent an energy deficit of the cell. As a result of a misbalance by increased energy demands, AMPK activates energy delivering processes and inhibits energy-consuming anabolic pathways. For example, activation of AMPK has been shown to increase glucose uptake by upregulating the glucose transporter (GLUT-4) gene expression; inhibiting the production of malonyl-CoA, an inhibitor of fatty acid oxidation; and activating peroxisome proliferator-activated receptor-γ coactivator 1α (PGC1α), a key transcriptional factor for mitochondrial biogenesis. As a consequence, the cell oxidative capacity is elevated and ultimately leads to a higher glucose utilization and ATP production. At the same time, AMPK activation induces the inhibition of energy-consuming processes in the cell. As such, activated AMPK phosphorylates and therefore inhibits acetyl-CoA-carboxylase 1 (ACC1), the key regulatory enzyme in fatty acid synthesis. In addition, 3-hydroxy-3-methyl-glutaryl-coenzyme A (HMGCoA)-reductase (a key enzyme of cholesterol synthesis) is inhibited as well as protein synthesis, gluconeogenesis, and glycogen synthesis. Finally, AMPK can also influence glucose homeostasis by inhibiting mammalian TOR complex 1 (mTORC1) [52, 55].

9.6.2 Cellular TOR Signaling

The TOR is conserved in all eukaryotes and modulates cell growth depending on amino acid availability [53, 57]. TOR acts as a protein kinase to modulate signaling pathways. It belongs to two larger protein complexes, mTORC1 and mTORC2, which are distinguished by adaptor proteins, Raptor and Rictor, respectively. Only mTORC1 senses amino acids and is activated under nutrient-sufficient conditions to enable anabolic processes, such as protein synthesis. At the same time, catabolic processes such as autophagy, lysosome synthesis, or proteasome assembly are inhibited [54]. Environmental stressors, including low ATP levels, hypoxia, or reactive oxygen species (ROS), reduce mTORC1 activity, partly by Raptor phosphorylation through activated AMPK.

9.6.3 Gut Microbiota

To understand the gut–brain signaling, one must understand the role of the gut flora since there is an indissociable link between the ingested food, the composition of intestinal bacteria species, food fermentation, and supply of nutrients that is sensed by endocrine cells. The gut flora or gut microbiota consists of up to 10^{12} cells/g of gastrointestinal mass and is composed of up to 1000 distinct bacterial species [58, 59]. The majority of the gut microbiota is composed of the phyla *Firmicutes* and *Bacteroidetes* with less from *Proteobacteria* and *Actinobacteria*. The microbiota lives in symbiosis with the host and provides macro- and micronutrients by fermentation of food and supports the host during inflammatory and autoimmune conditions [60]. Several strains are associated with distinct eating behaviors. A diet high in carbohydrates, protein, and saturated fats correlates with the amount of *Bacteroides* species, whereas *Prevotella* species are associated with long-term carbohydrate intake. Anaerobic fermentation of dietary fibers and complex carbohydrates by *Bifidobacteria* in the large intestine yields short-chain fatty acids (SCFAs) such as acetate, propionate, and butyrate that are available as nutrient source for the host. SCFAs can easily cross the epithelial barrier and are recognized by intestinal cells, adipocytes, and hepatocytes where they are used for energy supply. Acetate accounts for 60% of all SCFAs in microbial fermentation and is used by hepatocytes for cholesterol and triglyceride synthesis [61]. In addition, SCFAs have a strong impact on gut–brain/organ signaling (see below). Emerging evidence shows that dietary intake and eating behavior (i.e. vegan, vegetarian, or carnivore) alters the phyla composition of the gut microbiota. Nondigestible carbohydrates and fibers influence the host's energy balance through multiple mechanisms, including alterations in energy supply, gut transit, energy intake, and energy expenditure.

9.6.4 The Integration of Endocrine Gut Signaling

Nutrient sensing in the gut is observed by specialized enteroendocrine cells, such as K and L cells. The number of endocrine cells comprises only 1% of the intestinal tract cells; however, as a functional unit, this number of cells sums up to the biggest endocrine organ in the body. More than 20 hormones have been associated with this functional system, including those associated with endocrine, paracrine, and neurocrine activity. As mentioned above, enteroendocrine cells are equipped with nutrient sensors, such as GPCR-mediated signaling cascades or electrogenic channels and transporters that ultimately release neurotransmitters and/or gut hormones [50]. Most interestingly, sweet receptors T1R2/T1R3 were found in intestinal K and L cells [48, 62]; these receptors stimulate the release of gut peptides via coupling to membrane-bound heterotrimeric guanine nucleotide-binding G-protein gustducin. Among these, glucose-dependent insulinotropic peptide (GIP) and GLPs 1 and 2 (GLP-1 and GLP-2), both known as incretins, enhance insulin secretion after a carbohydrate-rich meal to buffer blood glucose levels. The ability of the gut to regulate the insulin secretion is known as the incretin effect and has been a primary objective in many studies and pharmacological approaches for type 2 diabetes research; for example, exenatide, a GLP-1 receptor agonist, was approved in 2005 as an antidiabetic drug.

The T1R1/T1R3 heterodimer receptor pair senses amino acids and protein hydrolysates in the gut and induces the expression of CCK [63]. Medium- and long-chain fatty acids bind to the GPCR120/free fatty acid receptor 4 (FFAR4) to induce CCK expression [64]. As a consequence, CCK mediates gastric emptying and signals satiation via the CCK receptor in neuronal cells. Of note, CCK is also secreted in response to oral triglycerides and carbohydrates.

SCFAs activate specific GPCRs at the surface of multiple cell types including mucosa cells, adipocytes, hepatocytes, and neuronal cells [65]. In enteroendocrine cells, GPR41/FFAR3 senses SCFAs and induces the secretion of the neuropeptide hormone peptide tyrosine tyrosine (PYY), whereas GPR41 on adipocytes increases adipogenesis and induces leptin, a key endocrine hormone of fat cells that induces satiety in the brain stem and hypothalamus, respectively [66]. Interestingly, PYY is able to slow the transit time of the intestinal content and thereby increases the energy recovery from food, which can contribute to obesity. In addition, PYY activates several neuropeptide Y-family receptors such as neuropeptide Y receptor type 2 (NPY2R), which is expressed in regions of the hypothalamus that are responsible for hunger regulation (satiety). However, recent data fail to support the hypothesis that PYY inhibits eating in humans or animals [51, 67, 68].

In summary, the gut is now readily accepted as an endocrine organ that senses nutrients and responds with the secretion of hormones. The interplay of the intestinal cells with the gut microbiota is currently an exciting research objective and demands multi-omics approaches. The contribution of the gut microbiome in combination with food composition not only on the double burden of malnutrition but also on inflammation and immunity will undoubtedly be part of future research. Examples include newer animal knockout models to provide an in-depth understanding of basic signaling pathways and supported cell-based hypotheses. In addition, gnotobiotic animals will unravel the relationship between gut microbiota and host and possibly the mechanism by which gut microbiota is associated with the onset of insulin resistance and inflammation, a phenomenon called "metabolic endotoxemia" [69]. Further, investigating the use and safety of fecal transplantation from host-to-host will possibly open a new therapeutic discipline within medicine. As an example, *Clostridium difficile* infections were successfully treated with fecal microbiota from healthy volunteers [70]. In conclusion, interdisciplinary approaches combining microbiology, gastroenterology, and epidemiology promise exciting progress in the coming decades that will open the door for an enhanced understanding of the connectivity and cooperativity of all organisms via signal transduction networks.

References

1 Ghoreschi, K., Laurence, A., and O'Shea, J.J. (2009). Selectivity and therapeutic inhibition of kinases: to be or not to be? *Nat. Immunol.* 10 (4): 356–360.

2 Wilhelm, S., Carter, C., Lynch, M. et al. (2006). Discovery and development of sorafenib: a multikinase inhibitor for treating cancer. *Nat. Rev. Drug Discov.* 5 (10): 835–844.

3 Pardoll, D.M. (2012). The blockade of immune checkpoints in cancer immunotherapy. *Nat. Rev. Cancer* 12 (4): 252–264.

4 Tarhini, A.A. and Agarwala, S.S. (2006). Cutaneous melanoma: available therapy for metastatic disease. *Dermatol. Ther.* 19 (1): 19–25.
5 Shada, A.L., Molhoek, K.R., and Slingluff, C.L. Jr. (2010). Interface of signal transduction inhibition and immunotherapy in melanoma. *Cancer J.* 16 (4): 360–366.
6 Finke, J.H., Rini, B., Ireland, J. et al. (2008). Sunitinib reverses type-1 immune suppression and decreases T-regulatory cells in renal cell carcinoma patients. *Clin. Cancer Res.* 14 (20): 6674–6682.
7 Ko, J.S., Zea, A.H., Rini, B.I. et al. (2009). Sunitinib mediates reversal of myeloid-derived suppressor cell accumulation in renal cell carcinoma patients. *Clin. Cancer Res.* 15 (6): 2148–2157.
8 Han, A.A., Currie, H.N., Loos, M.S. et al. (2016). Spatiotemporal phosphoprotein distribution and associated cytokine response of a traumatic injury. *Cytokine* 79: 12–22.
9 Manz, B.N. and Groves, J.T. (2010). Spatial organization and signal transduction at intercellular junctions. *Nat. Rev. Mol. Cell Biol.* 11 (5): 342–352.
10 Yin, F., Yu, J., Zheng, Y. et al. (2013). Spatial organization of Hippo signaling at the plasma membrane mediated by the tumor suppressor Merlin/NF2. *Cell* 154 (6): 1342–1355.
11 Zhou, Y., Liang, H., Rodkey, T. et al. (2014). Signal integration by lipid-mediated spatial cross talk between Ras nanoclusters. *Mol. Cell. Biol.* 34 (5): 862–876.
12 Lukyanov, K.A. and Belousov, V.V. (2014). Genetically encoded fluorescent redox sensors. *Biochim. Biophys. Acta* 1840 (2): 745–756.
13 Ezerina, D., Morgan, B., and Dick, T.P. (2014). Imaging dynamic redox processes with genetically encoded probes. *J. Mol. Cell. Cardiol.* 73: 43–49.
14 Sies, H. (2014). Role of metabolic H2O2 generation: redox signaling and oxidative stress. *J. Biol. Chem.* 289 (13): 8735–8741.
15 Love, N.R., Chen, Y., Ishibashi, S. et al. (2013). Amputation-induced reactive oxygen species (ROS) are required for successful *Xenopus* tadpole tail regeneration. *Nat. Cell Biol.* 15 (2): 222–228.
16 Violin, J.D., DiPilato, L.M., Yildirim, N. et al. (2008). B2-adrenergic receptor signaling and desensitization elucidated by quantitative modeling of real time cAMP dynamics. *J. Biol. Chem.* 283 (5): 2949–2961.
17 Yamasaki, S., Sakata-Sogawa, K., Hasegawa, A. et al. (2007). Zinc is a novel intracellular second messenger. *J. Cell Biol.* 177 (4): 637–645.
18 Taylor, K.M., Hiscox, S., Nicholson, R.I. et al. (2012). Protein kinase CK2 triggers cytosolic zinc signaling pathways by phosphorylation of zinc channel ZIP7. *Sci. Signal.* 5 (210): ra11.
19 Aksamitiene, E., Kiyatkin, A., and Kholodenko, B.N. (2012). Cross-talk between mitogenic Ras/MAPK and survival PI3K/Akt pathways: a fine balance. *Biochem. Soc. Trans.* 40 (1): 139–146.

20 Currie, H.N., Vrana, J.A., Han, A.A. et al. (2014). An approach to investigate intracellular protein network responses. *Chem. Res. Toxicol.* 27 (1): 17–26.

21 Arpino, G., Wiechmann, L., Osborne, C.K., and Schiff, R. (2008). Crosstalk between the estrogen receptor and the HER tyrosine kinase receptor family: molecular mechanism and clinical implications for endocrine therapy resistance. *Endocr. Rev.* 29 (2): 217–233.

22 Shou, J., Massarweh, S., Osborne, C.K. et al. (2004). Mechanisms of tamoxifen resistance: increased estrogen receptor-HER2/neu cross-talk in ER/HER2-positive breast cancer. *J. Natl. Cancer Inst.* 96 (12): 926–935.

23 Knowlden, J.M., Hutcheson, I.R., Barrow, D. et al. (2005). Insulin-like growth factor-I receptor signaling in tamoxifen-resistant breast cancer: a supporting role to the epidermal growth factor receptor. *Endocrinology* 146 (11): 4609–4618.

24 Jordan, N.J., Gee, J.M., Barrow, D. et al. (2004). Increased constitutive activity of PKB/Akt in tamoxifen resistant breast cancer MCF-7 cells. *Breast Cancer Res. Treat.* 87 (2): 167–180.

25 Osborne, C.K., Bardou, V., Hopp, T.A. et al. (2003). Role of the estrogen receptor coactivator AIB1 (SRC-3) and HER-2/neu in tamoxifen resistance in breast cancer. *J. Natl. Cancer Inst.* 95 (5): 353–361.

26 Kirkegaard, T., McGlynn, L.M., Campbell, F.M. et al. (2007). Amplified in breast cancer 1 in human epidermal growth factor receptor – positive tumors of tamoxifen-treated breast cancer patients. *Clin. Cancer Res.* 13 (5): 1405–1411.

27 Koeberle, A. and Werz, O. (2014). Multi-target approach for natural products in inflammation. *Drug Discov. Today* 19 (12): 1871–1882.

28 Somasundaram, S., Sigthorsson, G., Simpson, R.J. et al. (2000). Uncoupling of intestinal mitochondrial oxidative phosphorylation and inhibition of cyclooxygenase are required for the development of NSAID-enteropathy in the rat. *Aliment. Pharmacol. Ther.* 14 (5): 639–650.

29 Paul-Clark, M.J., Van Cao, T., Moradi-Bidhendi, N. et al. (2004). 15-epi-lipoxin A4-mediated induction of nitric oxide explains how aspirin inhibits acute inflammation. *J. Exp. Med.* 200 (1): 69–78.

30 McCarty, M.F. and Block, K.I. (2006). Preadministration of high-dose salicylates, suppressors of NF-kappaB activation, may increase the chemosensitivity of many cancers: an example of proapoptotic signal modulation therapy. *Integr. Cancer Ther.* 5 (3): 252–268.

31 Ahearn, I.M., Haigis, K., Bar-Sagi, D., and Philips, M.R. (2014). Regulating the regulator: post-translational modification of Ras. *Nat. Rev. Mol. Cell Biol.* 13 (1): 39–51.

32 Choudhary, C., Kumar, C., Gnad, F. et al. (2009). Lysine acetylation targets protein complexes and co-regulates major cellular functions. *Science* 325 (5942): 834–840.

33 Huang, F., Kirkpatrick, D., Jiang, X. et al. (2006). Differential regulation of EGF receptor internalization and degradation by multiubiquitination within the kinase domain. *Mol. Cell* 21 (6): 737–748.
34 Dengjel, J., Akimov, V., Olsen, J.V. et al. (2007). Quantitative proteomic assessment of very early cellular signaling events. *Nat. Biotechnol.* 25 (5): 566–568.
35 Vandermarliere, E. and Martens, L. (2013). Protein structure as a means to triage proposed PTM sites. *Proteomics* 13 (6): 1028–1035.
36 Olsen, J.V. and Mann, M. (2013). Status of large-scale analysis of post-translational modification by mass spectrometry. *Mol. Cell. Proteomics* 12 (12): 3444–3452.
37 Badeaux, A.I. and Shi, Y. (2013). Emerging roles for chromatin as a signal integration and storage platform. *Nat. Rev. Mol. Cell Biol.* 14 (4): 211–224.
38 Shimazu, T., Hirschey, M.D., Newman, J. et al. (2013). Suppression of oxidative stress by β-hydroxybutyrate, an endogenous histone deacetylase inhibitor. *Science* 339 (6116): 211–214.
39 You, J.S. and Jones, P.A. (2012). Cancer genetics and epigenetics: two sides of the same coin? *Cancer Cell* 22 (1): 9–20.
40 Cha, T.L., Zhou, B.P., Xia, W. et al. (2005). Akt-mediated phosphorylation of EZH2 suppresses methylation of lysine 27 in histone H3. *Science* 310 (5746): 306–310.
41 Nacerddine, K., Beaudry, J.B., Ginjala, V. et al. (2012). Akt-mediated phosphorylation of Bmi1 modulates its oncogenic potential, E3 ligase activity, and DNA damage repair activity in mouse prostate cancer. *J. Clin. Invest.* 122 (5): 1920–1932.
42 Huang, W.C. and Chen, C.C. (2005). Akt phosphorylation of p300 at Ser-1834 is essential for its histone acetyltransferase and transcriptional activity. *Mol. Cell. Biol.* 25 (15): 6592–6602.
43 Alvarez-Nava, F. and Lanes, R. (2017). GH/IGF-1 signaling and current knowledge of epigenetics; a review and considerations on possible therapeutic options. *Int. J. Mol. Sci.* 18 (10): 1624.
44 Foster, S.L., Hargreaves, D.C., and Medzhitov, R. (2007). Gene-specific control of inflammation by TLR-induced chromatin modifications. *Nature* 447 (7147): 972–978.
45 Robert, M.W., Morrison, C.J., and Kobor, M.S. (2016). Epigenetics: a new model for intracellular parasite–host cell regulation. *Trends Parasitol.* 32 (7): 515–521.
46 Sato, F., Tsuchiya, S., Meltzer, S.J., and Shimizu, K. (2011). MicroRNAs and epigenetics. *FEBS J.* 278 (10): 1598–1609.
47 Global Nutrition Report (2017). Nourishing the SDGs. http://globalnutritionreport.org/the-report/.
48 Chandrashekar, J., Hoon, M.A., Ryba, N.J., and Zuker, C.S. (2006). The receptors and cells for mammalian taste. *Nature* 444 (7117): 288–294.

49 Lindemann, B. (1996). Taste reception. *Physiol. Rev.* 76 (3): 719–766.

50 Reimann, F., Tolhurst, G., and Gribble, F.M. (2012). G-protein-coupled receptors in intestinal chemosensation. *Cell Metab.* 15 (4): 421–431.

51 Steinert, R.E., Feinle-Bisset, C., Asarian, L. et al. (2017). Ghrelin, CCK, GLP-1, and PYY(3-36): secretory controls and physiological roles in eating and glycemia in health, obesity, and after RYGB. *Physiol. Rev.* 97 (1): 411–463.

52 Yuan, H.X., Xiong, Y., and Guan, K.L. (2013). Nutrient sensing, metabolism, and cell growth control. *Mol. Cell* 49 (3): 379–387.

53 Saxton, R.A. and Sabatini, D.M. (2017). mTOR signaling in growth, metabolism, and disease. *Cell* 169 (2): 361–371.

54 Rabanal-Ruiz, Y., Otten, E.G., and Korolchuk, V.I. (2017). mTORC1 as the main gateway to autophagy. *Essays Biochem.* 61 (6): 565–584.

55 Garcia, D. and Shaw, R.J. (2017). AMPK. mechanisms of cellular energy sensing and restoration of metabolic balance. *Mol. Cell* 66 (6): 789–800.

56 Gordan, J.D., Thompson, C.B., and Simon, M.C. (2007). HIF and c-Myc: Sibling rivals for control of cancer cell metabolism and proliferation. *Cancer Cell* 12 (2): 108–113.

57 Düvel, K., Yecies, J.L., Menon, S. et al. (2010). Activation of a metabolic gene regulatory network downstream of mTOR complex 1. *Mol. Cell* 39 (2): 171–183.

58 Flint, H.J., Duncan, S.H., Scott, K.P., and Louis, P. (2015). Links between diet, gut microbiota composition and gut metabolism. *Proc. Nutr. Soc.* 74 (1): 13–22.

59 Flint, H.J., Scott, K.P., Louis, P., and Duncan, S.H. (2012). The role of the gut microbiota in nutrition and health. *Nat. Rev. Gastroenterol. Hepatol.* 9 (10): 577–589.

60 van de Wouw, M., Schellekens, H., Dinan, T.G., and Cryan, J.F. (2017). Microbiota-gut-brain axis: modulator of host metabolism and appetite. *J. Nutr.* 147 (5): 727–745.

61 Wong, J.M., de Souza, R., Kendall, C.W. et al. (2006). Colonic health: fermentation and short chain fatty acids. *J. Clin. Gastroenterol.* 40 (3): 235–243.

62 Shirazi-Beechey, S.P., Daly, K., Al-Rammahi, M. et al. (2014). Role of nutrient-sensing taste 1 receptor (T1R) family members in gastrointestinal chemosensing. *Br. J. Nutr.* 111 (Suppl 1): S8–S15.

63 Daly, K., Al-Rammahi, M., Moran, A. et al. (2013). Sensing of amino acids by the gut-expressed taste receptor T1R1-T1R3 stimulates CCK secretion. *Am. J. Physiol. Gastrointest. Liver Physiol.* 304 (3): 271–282.

64 Gribble, F.M., Diakogiannaki, E., and Reimann, F. (2017). Gut hormone regulation and secretion via FFA1 and FFA4. *Handb. Exp. Pharmacol.* 236: 181–203.

65 Mohajeri, M.H., Brummer, R.J.M., Rastall, R.A. et al. (2018). The role of the microbiome for human health: from basic science to clinical applications. *Eur. J. Nutr.* S1–S14.

66 Ichimura, A., Hirasawa, A., Hara, T., and Tsujimoto, G. (2009). Free fatty acid receptors act as nutrient sensors to regulate energy homeostasis. *Prostaglandins Other Lipid Mediat.* 89 (3–4): 82–88.

67 Cox, L.M. and Blaser, M.J. (2013). Pathways in microbe-induced obesity. *Cell Metab.* 17 (6): 883–894.

68 Boggiano, M.M., Chandler, P.C., Oswald, K.D. et al. (2005). PYY3-36 as an anti-obesity drug target. *Obes. Rev.* 6 (4): 307–322.

69 Plovier, H. and Cani, P.D. (2017). Microbial impact on host metabolism: opportunities for novel treatments of nutritional disorders? *Microbiol. Spectr.* 5 (3): BAD-0002-2016.

70 Baktash, A., Terveer, E.M., Zwittink, R.D. et al. (2018). Mechanistic insights in the success of fecal microbiota transplants for the treatment of *Clostridium difficile* infections. *Front. Microbiol.* 9: 1242.

Index

Cellular Signal Transduction in Toxicology and Pharmacology: Data Collection, Analysis, and Interpretation, First Edition. Edited by Jonathan W. Boyd and Richard R. Neubig.

d

e

t